Maintenance and Troubleshooting

In Industrial Automation

Revision 1.0

ISBN 978-0-578-38751-2

By

Frank Lamb

Information contained in this work has been obtained from sources believed to be reliable, however, the author shall not be responsible for any errors, omissions or damages arising out of use of this information. This work is published with the understanding that Automation Consulting, LLC and the author are supplying information but are not rendering engineering or other professional services. If such services are required, the assistance of an appropriate professional should be sought.

AUTOMATION CONSULTING, LLC

www.automationllc.com

This Page Intentionally Left Blank

Contents

Introduction

This book was written partially in response to a request for a troubleshooting course from Automation NTH, who I now work for as "Lead Trainer". I have spent the past eight or so years teaching PLC (Programmable Logic Controller) and HMI (Human-Machine Interface) programming classes across North America, and most of the people in those classes weren't there to learn programming at all, they were maintenance people there to learn how to diagnose and take care of equipment.

Jeff Buck of Automation NTH had approached me with the idea of creating a course specifically for troubleshooting. I had held classes for this customer before, and students had ranged from engineers straight out of college, to maintenance technicians, to project managers. My first book, *Industrial Automation: Hands On (McGraw-Hill, 2013)* was recommended reading for these classes, but it is a hardcover book and pretty expensive.

Troubleshooting in general is a difficult subject to address in a book. There are many different types of machines and systems, and students have different levels of knowledge in many different disciplines. I decided to cover a little bit about many of the subjects that technicians need to know about without going into too much detail.

This book provides fundamental information about machine and system maintenance. It also serves as an overview of the philosophy behind troubleshooting industrial automation systems and machinery. Basic industrial automation topics are covered such as mechanical systems, pneumatics, hydraulics, electricity and control systems. Much of this general knowledge is necessary before determining what problems might occur in industrial machinery.

Since I had already published a book on PLC programming, *Advanced PLC Hardware and Programming (Automation Consulting, LLC, 2019)*, I was able to re-use some of the content from the general PLC section.

The book also introduces the reader to practical techniques that can be used in troubleshooting industrial and technical equipment.

If maintenance is thought of in terms of a battle or war; there is a **strategy** that is used to approach how this battle and future battles may be fought, and there are the **tactics** that are used to accomplish the strategy. There are also common "weapons" or **tools** that are used in achieving the goal. This book is separated into three sections; Strategy, Tactics and Tools.

What is: NTH University?

Automation NTH is a system integration, engineering and machine building company located near Nashville, Tennessee. They have been in business since 1999 and have grown rapidly.

Good, experienced controls engineers are hard to find. Both manufacturers and system integrators struggle to hire enough controls engineers to keep up with demand. Hiring engineers straight out of school presents its own challenges, too. It typically takes years of on-the-job training for a controls engineer to be able to work independently on an automation project.

Frequently training will just come in the form of "learning by osmosis" through shadowing experienced engineers on projects. This approach results in highly varied skill development because the learning is greatly shaped by the specific project they're working on and the senior engineer they are paired with.

A number of years ago, they decided that they needed something better than the "learning by osmosis" approach. They wanted their engineers to develop more quickly and ensure a consistent baseline of controls engineering skills among their staff. For this reason, they developed a program called "NTH University" to accelerate and standardize controls engineering training.

Many of their full-time engineers started through the NTH University internship program, where interns rotate through hands-on training including PLC and HMI programming, controls design, and panel fabrication. The internship culminates with a project where the intern designs and builds a training box from scratch and then programs it to run a sorting application on a small conveyor.

NTH University coursework also includes topics such as machine safety, pneumatics, system startup, ethernet/IP networking, vision systems, and robotics. New hires straight from school start their full-time work by completing more self-guided courses, and engineers who finish this program earn what they consider to be a "Masters in Automation." NTH University courses also help experienced hires fill in knowledge gaps.

Automation NTH estimates that engineers who complete their internship program and continue with more advanced NTH University courses develop at 2x-3x the rate at which they would with the traditional "learning by osmosis" approach. This enables them to give their new hires substantial responsibilities on big, complex projects within their first year of full-time work.

This book helps in supporting this program.

Strategy

Maintenance strategy involves learning the prerequisites that are required to successfully maintain and solve problems in the equipment that you will be working on.

There are many different fields of industrial automation and technology, and each has its own rules and type of equipment. It is unlikely that someone who is an expert in the petroleum industry could step into a high-tech semiconductor factory and immediately understand how the equipment operates, and vice-versa. At the same time, there are elements such as basic electrical and mechanical knowledge that are common to both. This section includes some of the basic elements that are required to successfully solve problems with industrial equipment but is by no means a full course on any topic.

This section includes information that engineers sometimes use to design machines and control systems, but as stated in the front of this book *"the author shall not be responsible for any errors, omissions or damages arising out of use of this information. This work is published with the understanding that Automation Consulting, LLC and the author are supplying information but are not rendering engineering or other professional services. If such services are required, the assistance of an appropriate professional should be sought."*

Use this section as an introduction to the field of industrial automation and investigate the subjects further as required.

Maintenance

Machine and system maintenance is sometimes referred to by the acronym **MRO**, an abbreviation for **M**aintenance, **R**epair and **O**verhaul. Functions within this area include replacing, adjusting, and repairing elements of a machine or system so that it can properly perform its required functions.

Some of the common types of maintenance that fall under MRO include Preventive Maintenance, also known as PM, Corrective Maintenance, and Predictive Maintenance.

Preventive Maintenance

PM is a routine for periodically inspecting and servicing equipment. The goal is to prevent failure of components by replacing worn items, adjusting, and lubricating parts of the equipment. Technicians also have the opportunity to examine equipment in detail to identify fatigued areas and notice problems before they cause breakdown. This helps to prevent downtime and allow equipment to make it from one planned service to the next.

Other terms related to this are **Planned** or **Scheduled** maintenance, where a scheduled service may take place based on time, or on the availability of a specialist. This can also be based on equipment's running hours or the distance traveled.

Condition-based maintenance occurs when a technician or machine operator notices a problem, either during scheduled service or any other time before the equipment actually breaks down. This may require an unscheduled stoppage of machinery depending on the severity of the problem, but ideally this can be scheduled near-term during non-production time.

Corrective Maintenance

When equipment breaks down unexpectedly, it must be repaired or replaced before it can resume operation. This can be the most expensive type of maintenance, due to the production revenue lost and because equipment failure can cause damage to other parts of machinery. This can be differentiated from condition-based maintenance by the machine breaking before the problem is noticed.

If a repair part or redundant equipment is not available, additional downtime can result. For this reason, corrective maintenance as a planned event is usually undesirable. Having duplicate or spare equipment on standby can be a form of mitigation for this condition.

The form of corrective maintenance is usually a step-by-step procedure where the failure of the equipment triggers the event. Documentation for the equipment is used to either repair it or identify spare part numbers.

Predictive Maintenance

Data collection and analysis can help determine the condition of in-service equipment and estimate when maintenance should be performed. At a minimum, predictive maintenance can highlight spare parts that should be ready if a failure occurs before maintenance can be scheduled and performed. This can save money over other types of maintenance since tasks are only performed when warranted.

Equipment condition can be monitored by periodic (offline) or continuous (online) sensing of components. Vibration, temperature, power consumption or even visual condition via cameras can be included. While this can increase the reliability of equipment, it requires an investment in hardware and software.

Computerized Maintenance Management Systems (CMMS)

CMMS software maintains a database of information that can be used to schedule maintenance for preventive and predictive maintenance, and also locate spare parts.

Some companies also require verification of regulatory compliance. This means that maintenance activities need to be recorded and parts tracked, automating these functions can save money and time.

CMMS can also help management make informed decisions, for instance calculating cost differences between corrective maintenance and preventive maintenance. They can produce status reports and keep records with details of maintenance activities and part usage. This software may be installed on a company's computer servers or be cloud based with a subscription model.

Exercise 1

1. What are the three major types of maintenance?

    _____
    _____
    _____

2. Of the three categories listed above, which is the most expensive?

    _____

3. What software can be used to schedule preventive or scheduled maintenance?

    _____

Machine/System Theory

One of the first and most important elements of maintaining a machine or system is understanding how it is _supposed_ to work. This knowledge can be gained in various ways, from reading the documentation, to observing it in action, to researching it in books or online.

Observation

Most machinery has certain things in common. It is usually used to move, process, or change a product. There is usually a place or places where material enters the machine, and a place where product exits.

Watching a machine in operation is one of the best ways to learn how it works. How does the product enter and exit the machine, and what is done to it as it is processed? How fast does the machine or system run in parts or volume of material per minute?

Does a machine operator have to load or unload parts or material, and does their speed affect the operation of the machinery? Do they have to fill hoppers, part feeders or dispensing equipment?

Does the machinery stop operating because of machine faults? If so, what has to be done to get the machinery operating again? Is there an HMI or display to help diagnose problems?

Consider **taking notes** as you observe the machinery in operation. It is best to learn about the machinery and process while it is in operation, before something goes wrong and you have to fix it! The time to learn your machinery is _before_ you have a problem!

Use Your Senses

Visual signs of problems may be the easiest to find. Leaking fluids, metal shavings or black sooty powder are signs that machinery needs attention. Burnt and blackened components on circuit boards, warped or melted plastic and wear points on tooling and components can all be indicators. Look for anything that out of the "ordinary" or normal condition.

Machines can be video recorded to analyze their operation, motion can be slowed to examine functionality in detail. Cameras can also be strategically placed around and even inside a machine for full time monitoring.

Thermal imaging can also be used to identify hot spots in an electrical cabinet, such as determining if connections or terminations are tight. Follow proper safety procedures when working with exposed electrical devices!

In addition to watching the machinery with your eyes, your other senses can be useful to diagnose problems.

Listen to the sounds a machine makes as it operates. Sounds will change as problems appear, grinding or high-pitched squealing can be a sign that something is wearing or misaligned. Knowing the sound of your machinery when it is operating correctly can help identify problems when the sound changes. You could even record the sound so that you have a reference!

The sense of **touch** can be helpful in identifying warmth or vibration in equipment. Of course, this can be dangerous. If you suspect something is hot, approach it with your hand slowly, you can often sense heat without actually touching the object. Use the **back of your hand** if you must touch something.

Many components vibrate even when they are operating correctly, knowing how much vibration is normal can be helpful for comparison. **Sensors** can also be used to monitor vibration or temperature; **trends** are useful to identify wear.

Smells can also indicate problems. Burning or melting rubber has a distinctive smell, as does hot lubricants or metal. Again, knowing the smells of machinery when it is operating correctly can be helpful for comparison.

Products and Types of Machinery

Knowing about the **product** being processed is also a good place to start. Is it a machined part or assembly? Is it a package or box? A liquid or powder? All of these products have characteristics that can change the way they move in machinery.

There are also **classifications** of machinery that it can be useful to know about. There are many different types of packaging machines, robots, pumps, conveyors and material handling machines that have similar concepts behind them. Vibratory bowls, escapements, dial tables and indexers are other general classifications of equipment. Learning about these general categories as they apply to your equipment can be helpful.

Exercise 2

1.   What kind of products do you manufacture or process?

_____
_____
_____
_____

2.   Are your products consistently the same, or can they vary depending on the process?

_____
_____

_____
_____

3. What kind of machines and standard subassemblies do you have in your plant? (Classifications)

_____
_____
_____
_____

4. Does your machinery require a person to operate it?

_____
_____
_____
_____

5. Is there an Operator Interface or HMI on the machine to operate the machinery with and help diagnose faults?

_____
_____
_____
_____

6. List some of the characteristics of your machinery that can be sensed visually, by sound, touch, or smell.

a. _____

b. _____

c. _____

d. _____

e. _____

Documentation

Your equipment ideally will have documentation accompanying it. Usually, it will be provided by the manufacturer and may even have a troubleshooting guide.

Recommended maintenance instructions will often help identify areas where problems can occur in the machinery. Lubrication, belt tensioning and replacement of worn tooling are examples of maintenance procedures that are supposed to be performed on a periodic basis.

Where recommended maintenance is to be performed can also help to identify problem spots on a machine. Since these are items that move or wear out, they are also areas where things can go wrong.

Replacement parts lists or a bill of materials also often list components such as sensors or other electrical components that may need to be replaced. Again, these are potential problem areas.

Exploded parts views can show the internal workings of elements of machinery. Sometimes these are included as part of the maintenance or component replacement instructions. These can be very educational in the examination of machine operation and also help to identify potential failure points.

Electrical schematics show the detailed wiring of the machine or system, and also often have a bill of material attached. The first few pages can be scanned to identify the major sections and components of the machine. Power distribution identifies larger subassemblies by name; variable frequency drives, robots, controllers, servos and auxiliary equipment can be seen as an overview.

Piping and Instrumentation Diagrams (P&IDs) may also be a part of your documentation, especially if you are in process control.

The **PLC program** can be examined as a guide to the system layout also. Looking at the **hardware configuration** can show where communication-based remote I/O nodes are located, and programs are often written with subroutines that correspond to different operational parts of the machine or system.

Programs are also often separated into input and output routines, these sections list the I/O points and their addresses, often also containing descriptions that tell the function and location of devices.

Knowing where all of these documents and software are located can save valuable time when there is a problem. The time to learn where your documentation is located is *before* it breaks down!

A word of caution: older PLCs can lose their program when the battery is low and power is lost. It is *highly recommended* that backups are made regularly for all programs!

Exercise 3

1. What are some of the documents and programs you have that may help you with maintaining your equipment? Where is it kept?

_____
_____
_____
_____

Research

There is a nearly unlimited amount of information about different types of machinery and components **online**. Manufacturer and vendor websites, interest groups on social media and blogs can all be sources of information. While not all of the information is accurate, there are millions of articles, documents and videos to choose from.

Books have also been written about many of the subjects related to maintenance and troubleshooting of equipment. While some of these are in expensive hardback textbook form, e-books and less expensive options exist also.

Catalogs can be a great source for component information. Some also have tutorials about the technical aspects of the product, or application examples.

Exercise 4

1. What are some of the product brands you use in your plant? Consider things like PLCs, robots and sensors.

_____
_____
_____

2. What websites have you heard of that are good resources for product or subject matter information? Ask others what they have used.

_____
_____
_____
_____

Things You Need to Know

In order to maintain and troubleshoot industrial equipment, there are several disciplines or fields of study that you need to know. While it is not necessary to have an engineering level of understanding, there is a level of mechanical and electrical knowledge that is required.

This book covers some of the basic elements of these disciplines, but the reader should study these subjects in depth for a more complete understanding.

Mechanical

There are many different types of mechanical assemblies and elements that you will encounter in industrial automation. These are just a few to consider learning more about.

Power Transmission

Bearings and bushings allow shafts to rotate or slide smoothly. Bearings often have some type of rollers or balls that require lubrication, there will be a fitting on the outside of the bearing housing if so. Grease or other lubricants are applied periodically using a grease gun, or it may be done automatically.

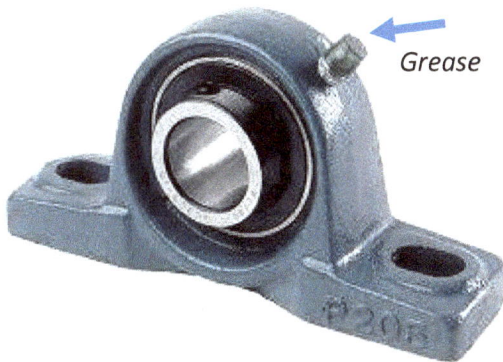

Grease

Bushings may also contain balls or rollers, but typically do not have grease fittings. They are often just softer metal collars that a shaft can rotate or slide smoothly in.

As bearings or bushings wear, they can often make noise or create heat. This is an example of using your senses to detect problems with these devices.

Figure 1 - Bearing

Gears and pulleys, like bearings, tend to be wear points. Gears can either mesh with other gears or engage with **chains**. These also need to be lubricated periodically, either as part of a maintenance routine or as an automated system.

Pulleys are used with **belts**; the belt is usually the weakest link that must be replaced periodically, and proper tension maintained. As with other moving parts, noise and heat are signs of possible problems, but belts may also create an odor of melting or burning rubber as they heat up!

Motors also require periodic maintenance. In addition to bearings, motors may also have brushes and commutators that make electrical contact as the motor turns.

The windings of a motor are coated with an insulation that can break down, causing the motor to become less efficient and create more heat. This is often detected with a thermal overload that uses current to trip a bimetallic element in the motor starter.

Vibration is also sometimes monitored in order to predict motor failure before an event. Failure of a large or critical motor can cost a lot of money in down time or even damage other equipment when it fails. For this reason, motors are often monitored electronically in several different ways. For large critical systems this can involve expensive and complex monitoring solutions.

Couplings are used to connect the shafts of motors to other devices such as gearboxes and encoders. They are meant to compensate for vibration and slight misalignments, but because they move and can be less sturdy that the shafts themselves, they create another failure point, Spider couplings are made of elastomeric materials, and helical couplings can have very delicate structures. They are also usually located inside a housing, so they can be hard to view and take time to replace.

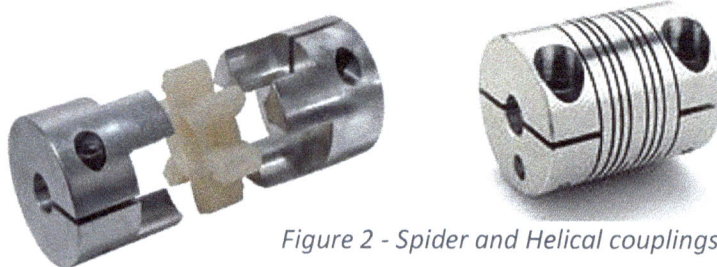

Figure 2 - Spider and Helical couplings

Loose couplings also allow for positional errors in servo systems. If a motor position starts to drift, the coupling should be checked for proper tightness. Keyed couplings are often used to eliminate shaft slippage.

Seals are built into many bearings or other places where shafts extend from equipment. Heat and wear can damage these, allowing dirt or moisture to enter delicate operating areas of components. Replacing seals on a periodic basis is part of a good maintenance program.

A **clutch** is used to engage and disengage power transmission from a driving shaft to a driven shaft, allowing the motor to run at a continuous speed. **Brakes** are used to stop rotating shafts. Both of these operate by using friction against pads or plates, which creates heat and wear. These are often combined in a single mechanism called a **clutch-brake**.

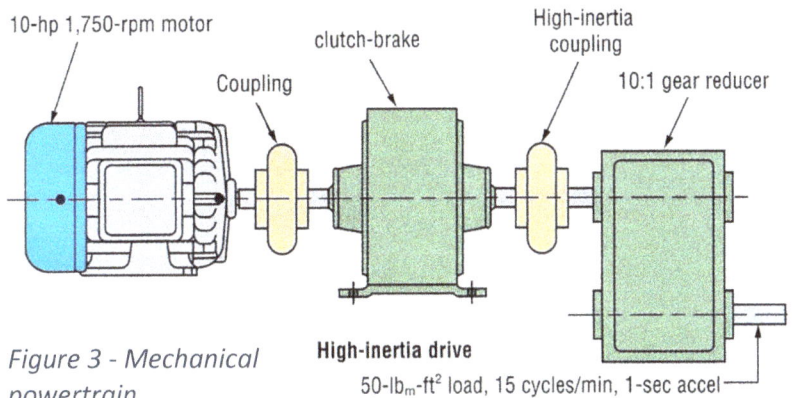

Figure 3 - Mechanical powertrain

Figure 3 shows many of the elements listed in combination. This powertrain changes the output speed and torque from the driving motor and allows the load to be connected and disconnected, bringing it to a quick stop.

Mechanical Adjustments and Maintenance

Motors need maintenance and inspection regularly to prevent problems. Following are some of the things to check on a regular basis:

1. **Ventilation** – Ensure fan area is free of dust, dirt, or foreign objects.

2. **Brushes/Commutators** – Most common in DC motors, these need to be inspected regularly. Springs are used to press the brush against the commutator rings, weaker springs can decrease brush life. Typical brushes last about 7500 hours, minimum life might be 2000-5000 hours, while 10,000 is about the maximum.

3. **Bearings** – some motors have lubrication ports for greasing the bearings. Follow manufacturer's instructions for lubrication; too much and the grease will end up squeezing out of the seals, too little and the bearings will have reduced life.

   When bearings are completely sealed and don't have grease fittings, they should still be inspected, and replaced if worn out. Excessive loading, hot motors and harsh environments can all contribute to bearing wear.

4. **Vibration** – Too much vibration can damage motor windings or create metal fatigue. This can be difficult to detect by hand, but there are vibration sensors available for preventive maintenance.

5. **Windings** may also need to be checked periodically. This can be done visually, or by measuring the resistance with a meter.

6. **Electrical Connections** – Over time, motor connections can loosen leading to higher current and eventual failure. Visible signs can include discoloration, but motor connections should be checked periodically for tightness.

Belts and chains should be monitored regularly for wear and proper **tensioning**. When they are too loose, they can slip, and when too tight they can contribute to reduced life in power train elements. Belts are usually tensioned to where they are taut but don't exert excessive side loading to bearings, chains with gears may have a bit of slack.

Metal chains need to be lubricated regularly; this is often done automatically. Check sprockets for signs of wear and replace links if needed. Belts should be replaced as needed or when recommended by the manufacturer.

Cleaning of machine parts should be done to reduce corrosion and to prevent contamination of product. This can include painting and application of light oil such as WD-40 to fasteners.

A maintenance checklist or preventive maintenance software package can help track activity and keep records.

Exercise 5

1.   List some of the possible failure points in the blower picture shown in Figure 4:

_____
_____
_____
_____
_____

Figure 4 – Blower Assembly

2.   Figure 5 shows a pop-up diverter on a chain pallet conveyor. The pop-up is raised with an air cylinder and its conveyor is a ribbed rubber belt. List some possible maintenance and failure points:

Figure 5 - Pop-up conveyor

_____
_____
_____
_____

3.   What senses can you use to detect problems with components of a mechanical power transmission system?

_____
_____
_____
_____

Pneumatics

Fluid power is a term that is used to describe both pneumatic and hydraulic systems. As with power transmission components, there are shafts, seals, bushings, and bearings that are built into pneumatic or hydraulic actuators.

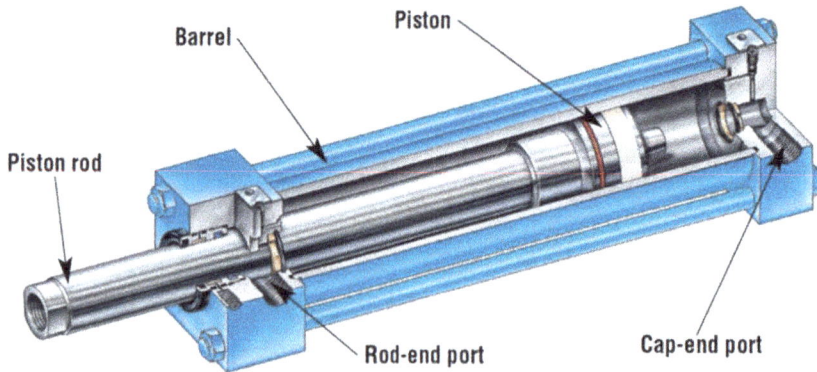

Figure 6 - Pneumatic cylinder

Actuators are used to move or hold products, or to exert a force against it, such as in a press. Pneumatic and hydraulic actuators are typically cylindrical in shape, and so are called **cylinders**.

Cylinders operate by means of the fluid (air or hydraulic) exerting force against a **piston** attached to a **rod**. Figure 6 shows a typical cylinder with two ports; when air enters the cap-end port, the cylinder extends, and when it enters the other port and exhausts from the back or cap-end port it retracts. Actuators such as parallel grippers (Figure 7) may also have internal gearing or other mechanisms built in.

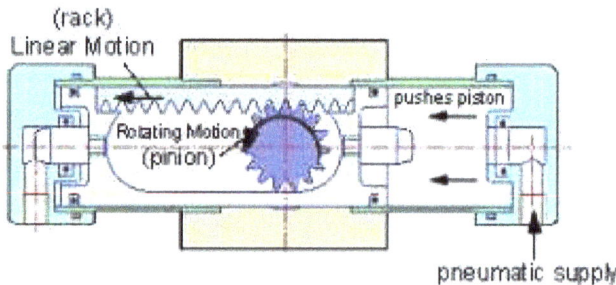

Figure 7 - Parallel Gripper mechanism

The speed and pressure at which the air enters and exits the ports controls the way the cylinder operates. The cross-sectional area and volume of the cylinder also helps determine the force and speed of the rod. The formula for the force exerted by a piston is **F = P x A**, where **P** is the air or fluid pressure exposed to the rod, and **A** is the area of the cylinder face. There are many possible factors that affect this force calculation, consult a good pneumatics reference if exact calculations are needed.

Here is an example of a force calculation for a 1.5" bore air cylinder at 60 lbs/in^2:

To find the area, first calculate the radius, 1.5"/2 = 0.75".

Area = πr^2 = 3.14159 (0.75 x 0.75) = 1.767

60 psi x 1.767 = **106.0287 lbf** (pounds-force)

This can also be converted to Newtons (1 lbf = 4.44822N) or kilograms force (kgf) (1 lbf = 0.453592), so 106.0287 lbf = 471.639N = 48.0938 kgf

Flow controls restrict the air into or out of the ports, this changes the speed of the rod in either the extending or returning direction. Flow controls act differently depending on whether air is being **metered in** (on the cap end port to extend the cylinder) or **metered out** (restricting the flow on the rod end while extending the cylinder). It is generally considered a better control technique to meter out, since this is the best way to slow an unloaded cylinder. Flow controls are usually placed on both ports. Just like tightening a screw, turning the flow control clockwise restricts flow more, slowing the actuator.

Valves are used to turn on and off the airflow to the actuator. There are many different valve configurations available, some have two solenoids to control air to each of the A and B ports and exhaust the other. With a single solenoid valve, the A port is actuated when it is energized, and the B port exhausts. When the solenoid is de-activated, a spring returns the valve position pressurizing the B side and exhausting the A side. Some single acting valves are also configured to exhaust both sides when turned off, using a spring on the actuator itself to return the cylinder.

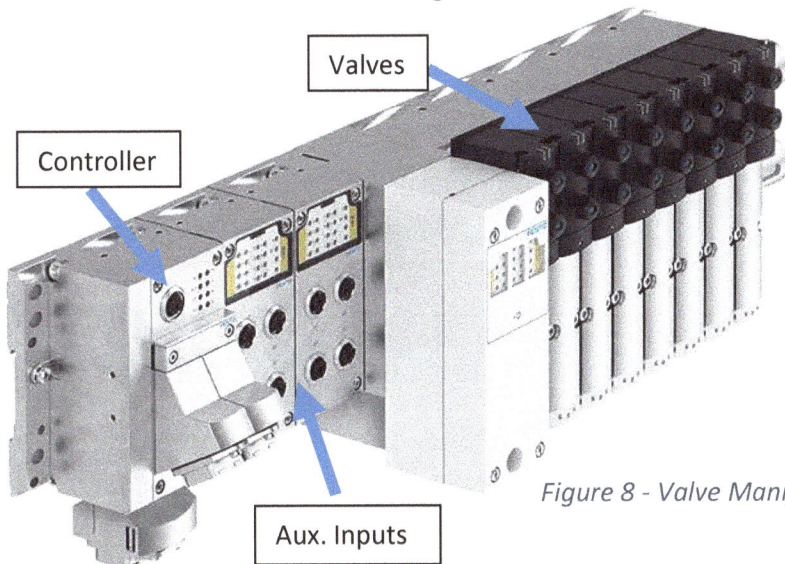

Figure 8 - Valve Manifold

Valves are often arranged in groups on a **manifold** as shown in Figure 8. This allows multiple devices to be controlled with one centralized air supply and control connection. The control may use a multiconductor cable to carry signals to (and from) the unit or use a communication protocol to a control unit.

Many valves require a minimum pressure (usually at least 15-20 psi) to shift the position of the valve. Low pressure applications may require more care in valve selection. **Piloted** valves use a separate higher pressure to shift the valve if the pneumatic circuit's air is not sufficient.

Fittings and plastic air **tubing** are used to connect the valves to the actuators. Fittings often thread onto the ports on the cylinder and manifold and are sometimes built into the valve. Fitting threads are usually wrapped with Teflon tape for a proper seal.

Fittings and hose are available in both standard and metric sizes, and with different threads. Hose diameters such as 6mm and ¼" are very close in size and are easily confused. This can make connections looser or tighter than they would be if using the proper size. **Adapters** are sometimes used to interface from one size of hose or fitting to another. It is important to identify the sizes and threads used in a pneumatic system and not try and force something to fit where it is not designed to.

Fitting threads, hose connections and the hoses themselves can leak air, even when sized correctly. This can usually best be detected by listening for a "hissing" sound. Since all ports are not pressurized at all times, you will have to activate valves to check all of the connections.

Mufflers or **Silencers** reduce noise and prevent foreign matter from entering the exhaust ports. They come in various sizes and threads. They are usually placed on the exhaust ports of valves and are assumed to be present even if not shown in the following diagrams.

Figure 9 - Muffler and Symbol

There are other standardized symbols that are used to represent pneumatic circuits. The following pages are meant to be a short tutorial on the subject and are by no means complete, but there are a few diagrams that will help you get a feel for how to read these symbols and diagrams.

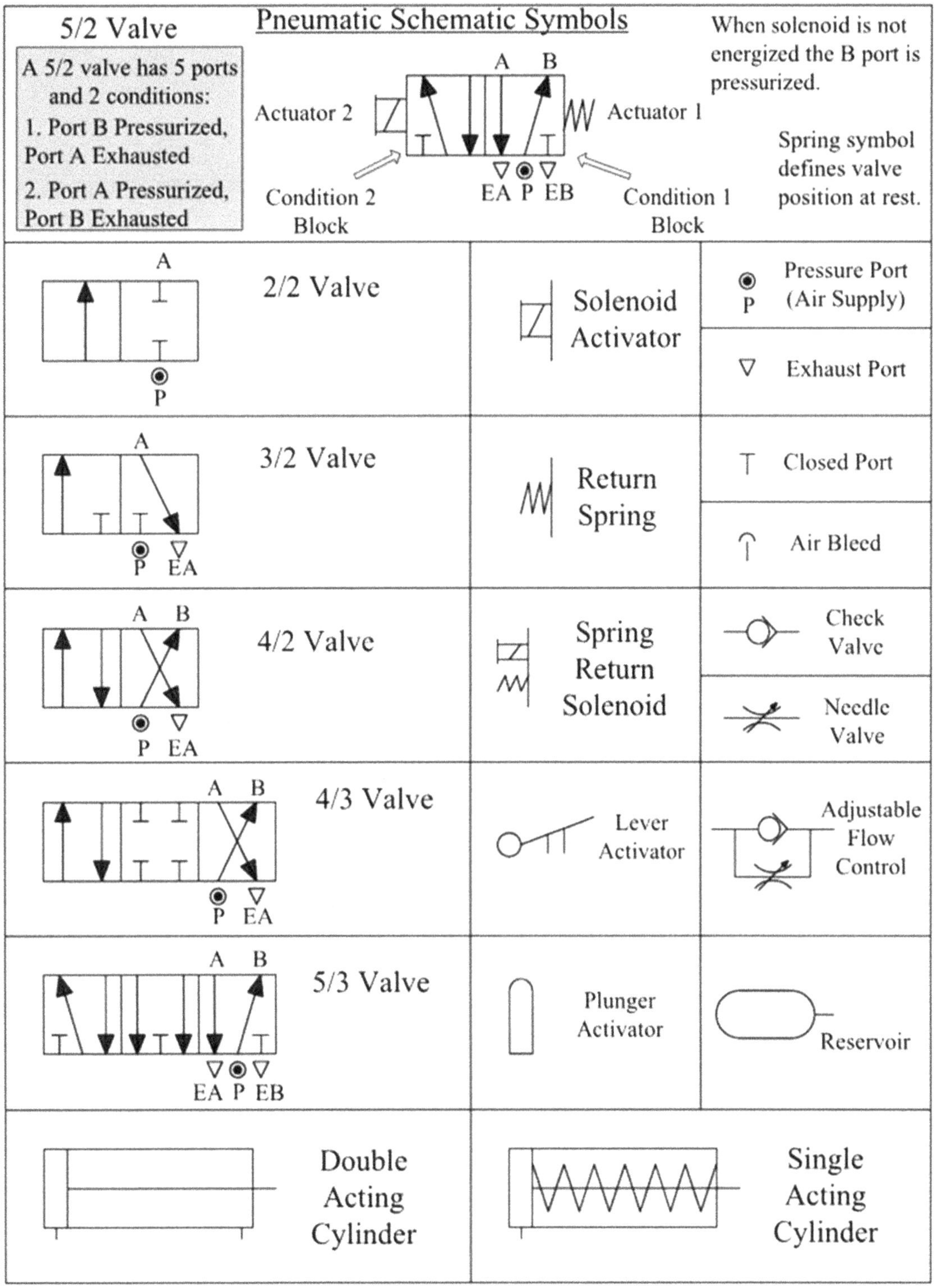

Figure 10 - Pneumatic symbols (Industrial Automation: Hands-On, F. Lamb, McGraw-Hill 2013)

Figure 10 shows some of the symbols used in pneumatic diagrams. The following diagrams show the distribution of air through a system and are sometimes included in machine documentation.

Note that not all the valves shown in Figure 10 are externally actuated. **Needle** valves restrict flow in both directions and **Check** valves allow flow in one direction only. The valves themselves may also have diagrams printed on them showing what type of valve it is. Learning to read these symbols is an important part of pneumatic troubleshooting.

The numbers expressed in the designation are the number of ports and the number of positions for the valve. The blocks show the condition of the air flow when each solenoid is energized and when it is in the off position.

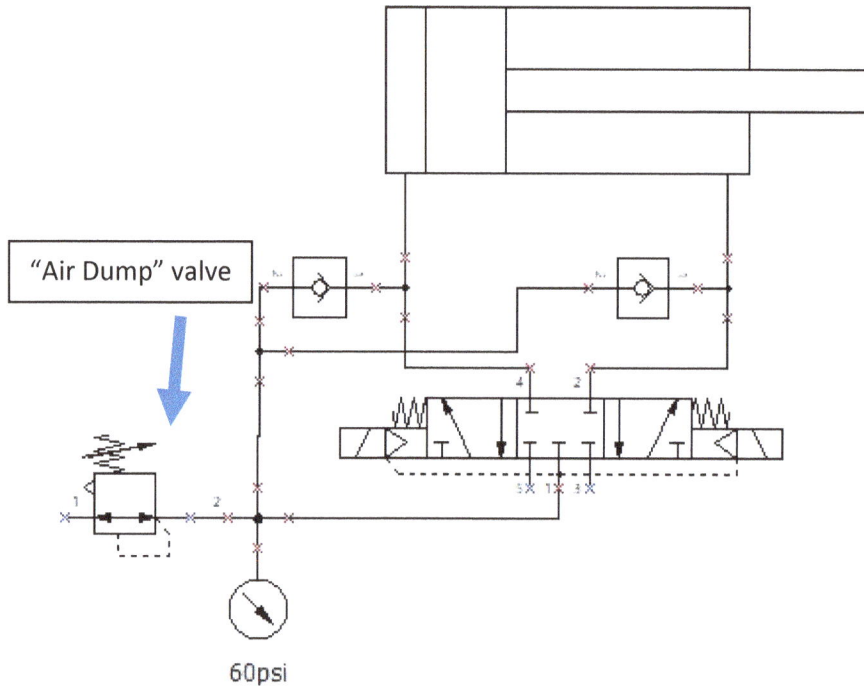

"Air Dump" valve

60psi

Figure 11 - 5/3 valve and cylinder

Figure 11 shows a diagram of a cylinder controlled by a 5/3 center blocked valve that has an additional exhaust solenoid on it, sometimes called an "air dump". The purpose of this circuit is to exhaust residual stored pressure when the rest of the system loses air.

The middle block shows the condition of the flow when the two solenoids are off. The pressure port is labeled 1, and when turned off, the two sides keep the pressure on each side of the piston holding it in position. If the air dump is turned on, the air will escape through the check valves, freeing the cylinder. This is a requirement in some cases for safety, to release the stored energy in the circuit with an emergency stop.

The numbering and port designation can differ depending on the brand of valve being used. Consult the manufacturer's documentation for your product.

Figure 12 shows a pneumatic circuit with a **Soft Start** valve. Soft starts are 3/2 valves that gradually increase air to downstream components when energized. They are commonly used in industrial applications. This exhausts the air when turned off, also often used with emergency stops. They also prevent systems from "banging" when pressure is applied.

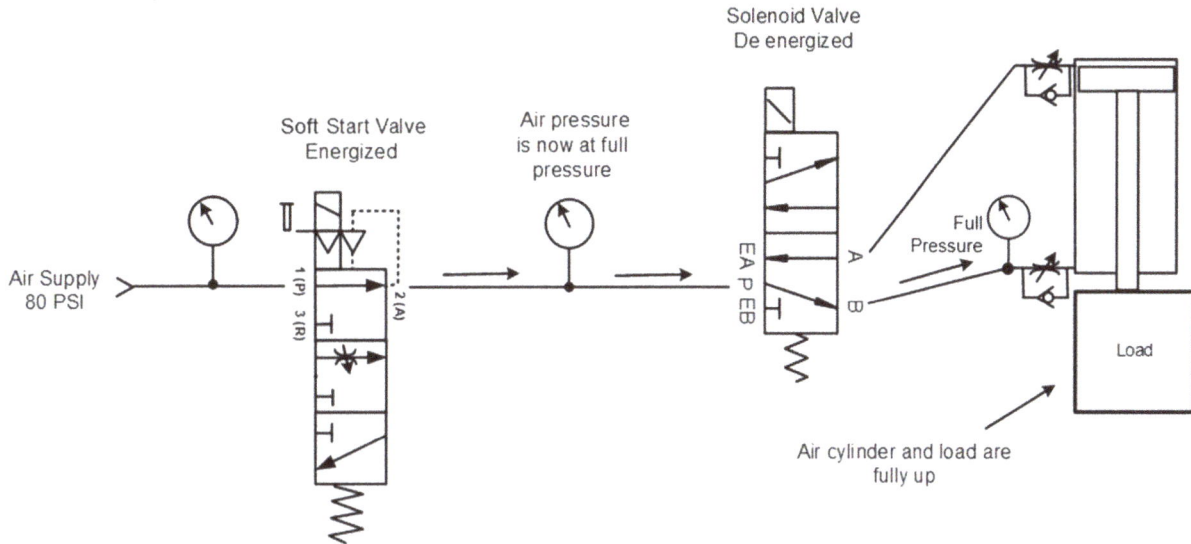

Figure 12 - Soft Start

Air Preparation is a term that describes the components needed to prevent contamination from solids, water, and oil. Plant air from a compressor flows through multiple devices, pipes and fittings that can add particulates, oil, and moisture. Even though air dryers, filters, water separators and regulators are often present at the compressor, the air is should also be treated at the machine to prevent damage.

Figure 13 shows an air preparation assembly made up of several components.

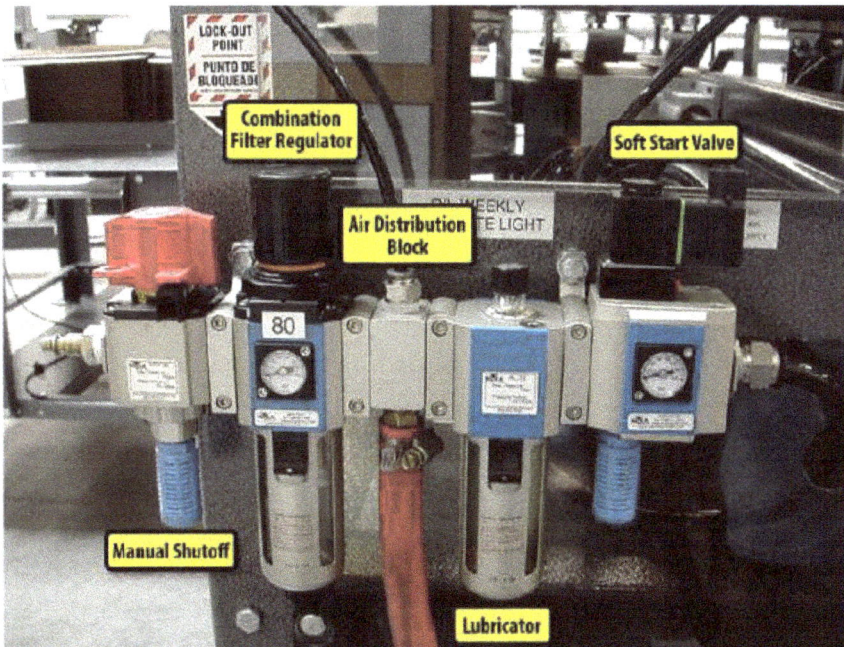

Figure 13 - Air Preparation (courtesy of Automation Direct)

Manual Shutoff valves can be used to remove incoming air from the system. They can also be locked for safety purposes.

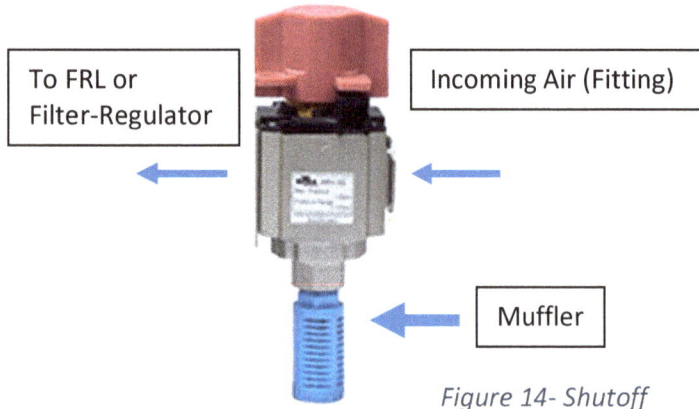

To FRL or Filter-Regulator

Incoming Air (Fitting)

Muffler

Figure 14- Shutoff

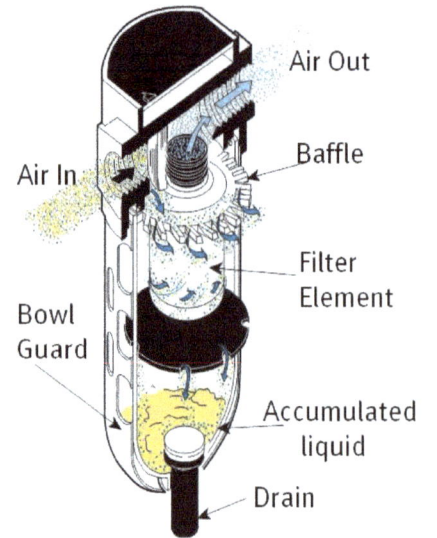

Air Out

Air In

Baffle

Bowl Guard

Filter Element

Accumulated liquid

Drain

Figure 15 - Filter

Filters are used to remove particulates and water from the incoming supply. They have a clear glass or plastic bowl so that the liquid can be seen. **Mist Separators** are used to remove aerosol state oil mist from the exhaust of process air. They can look similar to the filter shown in Figure 15 but may also be large self-contained systems.

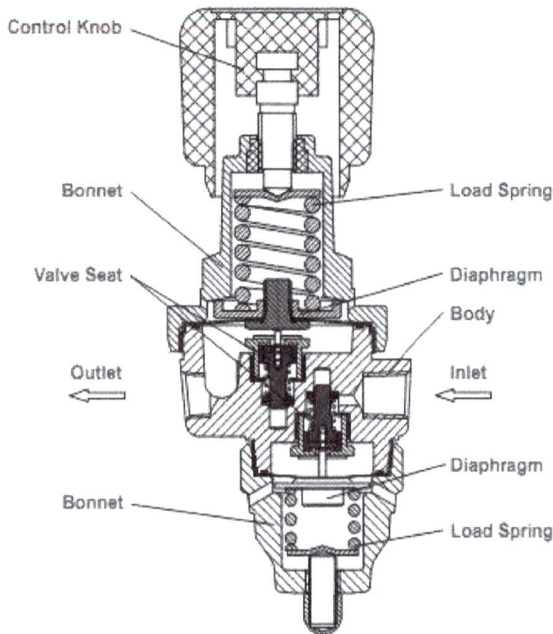

Control Knob

Bonnet

Valve Seat

Outlet

Bonnet

Load Spring

Diaphragm

Body

Inlet

Diaphragm

Load Spring

Regulators reduce and control the incoming pressure for machinery. They use a spring with a diaphragm to allow excess pressure to escape and usually have a gauge attached.

Figure 16 - Pressure Regulator

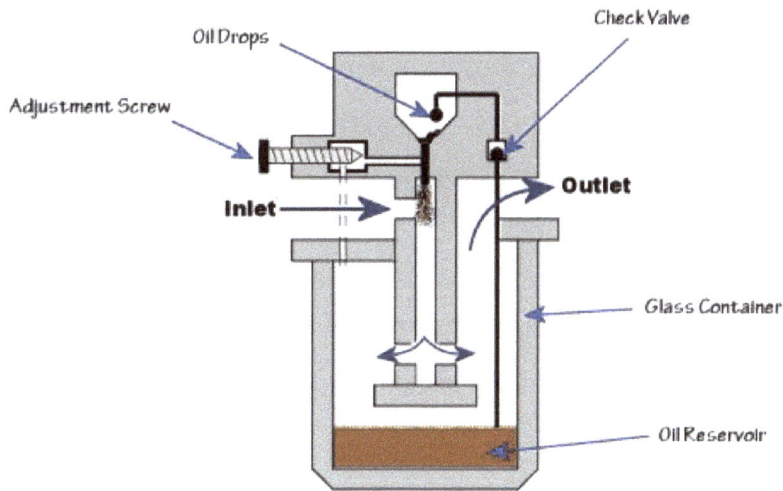

Figure 16 - Lubricator diagram

Lubricators were important at one time to lubricate rubber parts like seals inside pneumatic components. Seal materials have evolved over time however and lubricators are usually only required for pneumatic tools and some air motors or air clutches. *Most modern machines do not use lubricators.*

FRLs/FLRs, or Filter-Regulator-Lubricators, are unitary self-contained units that accomplish filtering, pressure regulation and lubrication.

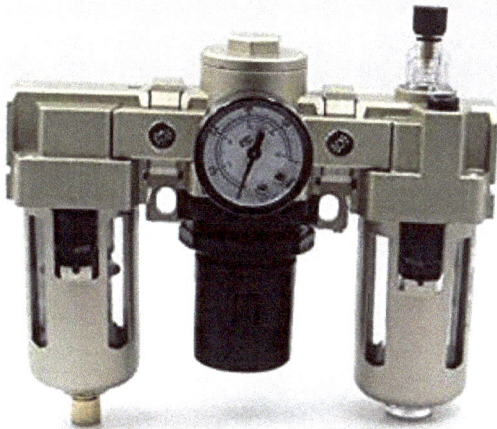

Figure 17 - FRL Assembly

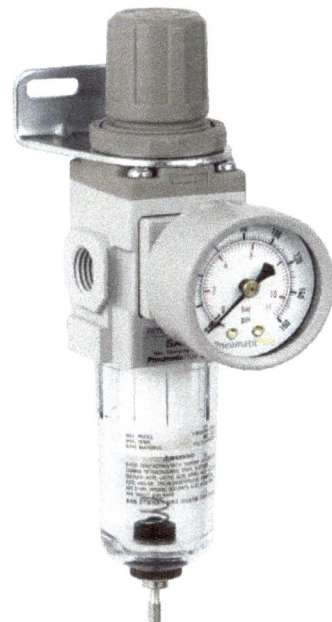

Filter Regulators include filtering and pressure regulation without lubrication. They can be combined into a single unit as shown in Figure 18.

Figure 18 - Filter Regulator

Pneumatic Adjustments and Maintenance

Following are some of the items that may require attention in a pneumatic system:

1. **Supply** - Pneumatic systems require clean dry air to operate properly. Ensuring that the plant air is as uncontaminated and dry as possible is an important part of ensuring machines have as long a life expectancy as possible.

2. **Air Preparation** - Set pressure regulators at the manufacturer's recommended value to ensure proper machine operation. Check filter bowls and drain/clean as necessary.

   If lubricators are required ensure that oil reservoirs are filled with the proper lubricant.

3. **Actuators and Control** - Periodically adjust **flow controls** to ensure proper movement of cylinders and actuators. Faster is not always better, ensure that actuators and products move in a smooth controlled way. There may be flow controls on both ports of air cylinders, remember that metering out (the port used to exhaust the cylinder) is the best way to prevent the jerking movement of a cylinder with no load.

4. **Connections** - Check fittings and hose connections for leaking air. Rewrap fitting threads with Teflon tape and replace hoses or fittings as necessary. Listen for air leaks with solenoids activated in each direction. Solenoids usually have manual activators or buttons to energize them with. Use the manual functions on operator interfaces to test the valves also.

5. **Sensing** - Set sensors on cylinders and tooling **with air applied**, centering the sensor over the internal cylinder magnet or tooling. More will be covered on sensors later.

Exercise 6

1. What is the purpose of the circuit shown in Figure 20?

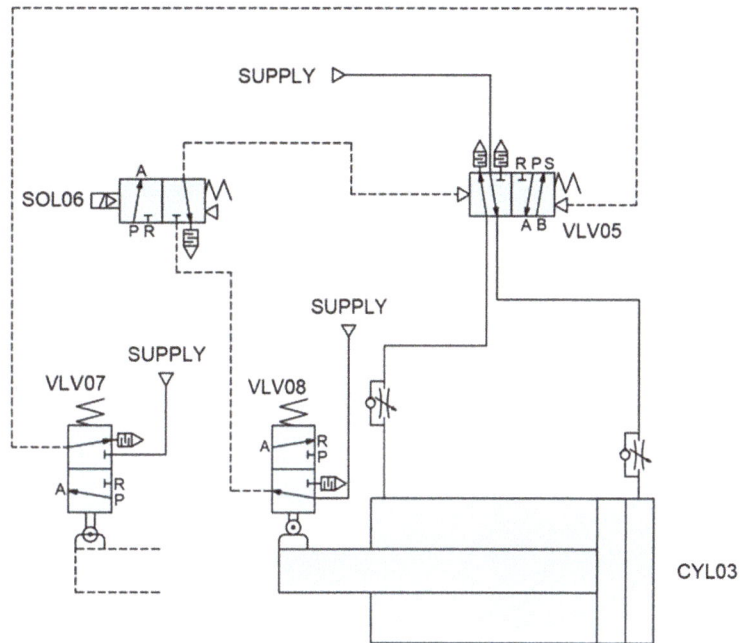

_____

_____

_____

_____

_____

Figure 20 – What does this circuit do?

2. To prevent a cylinder from jerking when air is applied a flow control should be placed on the _____ (Pressure/Exhaust) port of the actuator. This is also called meter _____ (In/Out).

3. If neither solenoid is actuated on the valve shown in Figure 21, will the connected cylinder be able to be moved by hand? _____ Another name for this valve is center _____ (open/blocked). What is the numerical designation for this valve type? _____

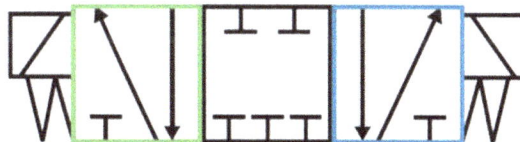

Figure 21 - Double spring return valve

4. Do you need an FRL on most modern machines? _____

5. What type of valve might be used on this pusher cylinder? _____

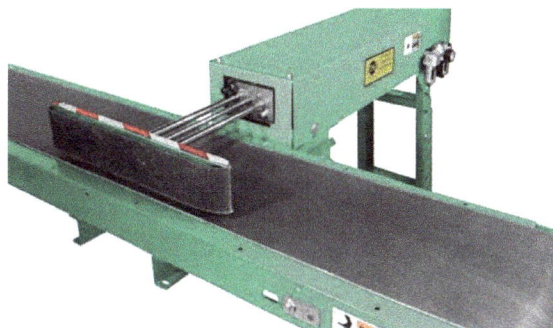

_____

_____

_____

Figure 22 - Pusher

6. Figure 22 does not show any control or safety devices.

    a. What might be used as a signal for the pusher to activate?

    _____

    _____

    b. What problems do you see with this configuration, and what might you use to correct them?

    _____

    _____

    _____

    _____

7. What is the purpose of VLV01 and VLV02 in the pneumatic circuit shown in Figure 23?

_____

_____

_____

_____

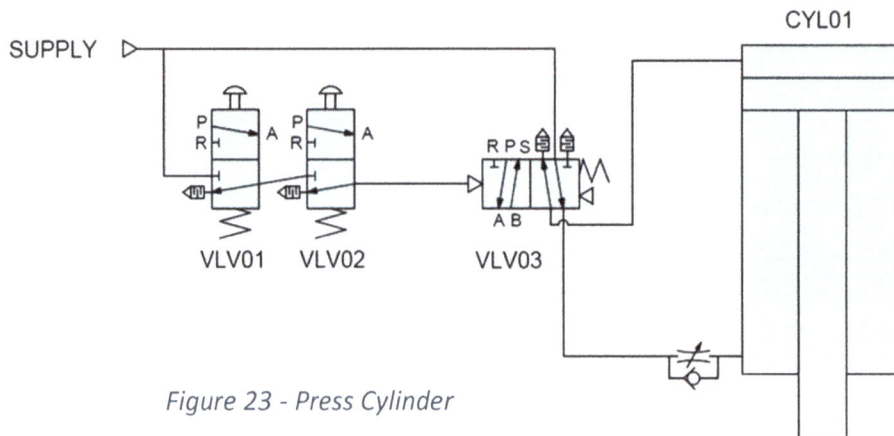

Figure 23 - Press Cylinder

8. The diameter of an air cylinder is 2 inches, and the pressure is 75 psi. What is the applied force at the end of the rod? _____

    a. 185.5 lbf
    b. 212.9 lbf
    c. 235.6 lbf
    d. 271.9 lbf

Hydraulics

Hydraulic power is another form of fluid power. Unlike pneumatics however, hydraulic systems require a contained fluid throughout the system. Whereas a pneumatic system can be connected to a plantwide system of compressed air, hydraulic systems require their own pumps and fluid supply.

Because pneumatic systems use air, which is compressible, there is a delay in actuator movement. With hydraulic systems there is no delay in movement, and the available force is much higher for a similar sized actuator. Where most pneumatic systems operate in the 60-100 psi range, hydraulic systems can provide 1,000-5,000 psi or even more.

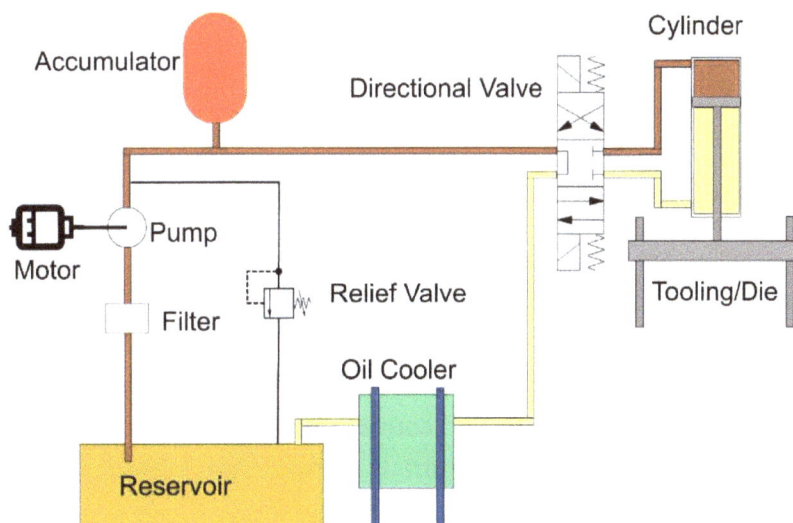

Figure 24 - Hydraulic System

Many of the components of a hydraulic system are similar to those of a pneumatic system, but there are a few differences.

A **reservoir** of hydraulic fluid (oil) is required. Compressed air can be pulled from the atmosphere, whereas fluid has to be supplied.

The **filter** is made of different materials than typical air filters. They remove both water and solid contaminants from the system. More hydraulic failures are a result of contamination than any other cause.

A **pump** is required to provide pressure for the system. This is usually driven by an electric motor, but there are other types, even engines are used for some large systems!

A **relief valve** is used to ensure pressure does not exceed the requirement. Fluid is returned to the reservoir.

An **accumulator** stores energy to maintain pressure, dampen vibrations and pulsing, and improves the efficiency of the system. They are pre-charged with an inert gas, typically nitrogen. A moveable barrier, usually a piston or rubber bladder separates the oil and gas. The gas is usually pressurized to 80-90% of the working pressure of the system.

Directional valves are often used for moving the actuator. These are usually 4/2 or 4/3 configuration. Because of the force required to shift the spool inside the valve body, these may be pilot-actuated, using a pressurized fluid (often about 50 psig) to move the spool.

Proportional valves are sometimes used to control speeds of actuators. These can be controlled from 0-100% by an analog signal from a PLC or controller.

Pressurizing fluids creates heat, which can break down the hydraulic oil and reduce its life. A **hydraulic oil cooler** is a heat exchanger that removes heat to the outside air. These may be air cooled, or water cooled.

Hydraulic Adjustments and Maintenance

Many of the maintenance tasks for pneumatic systems apply to hydraulics also. Because the system runs on hydraulic fluid however there are various areas with additional concerns.

1. **Prevent the system from overheating** – Hydraulic fluid gets hot as it is pushed through pumps, tubing, and relief valves. If the temperature is too low, water can condense in the reservoir, if it is not properly removed by the filter it can cause pump cavitation. If the temperature is too high, oxidation causes varnish and sludge deposits that can clog the filters. Typical plant hydraulic system fluids run in the 110-150 degree Fahrenheit range.

   Perform regular checks of the oil cooler and monitor the temperature in the reservoir.

2. **Keep the system clean** – Keep the reservoir covered and ensure the area around drain lines and breather fill openings are kept free of debris. Dirt, water and metal dust or shavings can make their way through these openings. Clean the cap before replacing it on the reservoir. Store hydraulic fluid in a clean environment.

3. **Keep fluid clean and test for contaminants** – check and change filters on a regular basis. Filter oil added to the system through portable filters. Add hydraulic fluid of the same brand and viscosity grade as needed. Sample the fluid for color, visible signs of contamination and odor.

4. **Visual and other checks:**

   a. Inspect hydraulic hoses, tubing and fittings for leaks and frays.

   b. Inspect the inside of the hydraulic reservoir for signs of aeration. Use a flashlight and into the fill hose for signs of foaming or small whirlpools, these may be a sign of a leak in the suction line or faulty shaft seals.

   c. Check return, pressure and hydraulic filter indicators and pressure gauges against manufacturer's documentation.

   d. Check system temperature independently using an infrared thermometer, also look for "hot spots" on the motor or proportional valves. High temperatures on the valves can be caused by the valve sticking.

   e. Listen to the pump for unusual noise. Cavitation the formation of bubbles or "cavities" of air in the pump. It is caused by areas of lower pressure around an impeller. It will damage the pump, decrease flow and cause vibration if not treated.

   f. Check breather caps, fill screens and all filters regularly.

5. **Hydraulic Safety:** Tubes and hoses can develop pinholes that could cause a high-pressure stream of fluid. As mentioned previously, the fluid could also be hot. You should _never_ run your hands over a hydraulic line to check for leaks. Use a sheet of paper or cloth with gloves.

Exercise 7

1. What is the biggest cause of hydraulic system failure?

_____

2. What is a typical hydraulic line pressure (psi)? _____

3. What is the typical temperature for hydraulic fluid in the reservoir (F)? _____

4. What provides pressure in a hydraulic system? _____

Electrical

In order to maintain and troubleshoot industrial equipment, it is important to be able to read and understand the diagrams and symbols that represent electrical and control circuits. It is also necessary to review some basic electrical theory and principles.

Electrical and Controls Devices and Symbols

Figure 25 - Resistor

Resistors/Resistance/Impedance:

Resistors are a basic component used in many electrical circuits. They are classified by their resistive value, measured in **Ohms**, and their wattage, which is a measure of how much current they can handle at a specific voltage (power). Resistors are marked with colored stripes that provide the resistance value and the tolerance (accuracy, +/- %) of the resistor. A resistor color code chart is shown in Appendix C at the back of this book.

One of the most important functions of resistors is to reduce the current through a device, and/or to establish a specific voltage to power it. In manufacturing it is not as important to consider individual resistors on a circuit board, but instead to think of the equivalent resistance of devices in an electrical circuit. When devices fail in an open state, their resistance becomes infinite and no current flows through them. This can be detected by using a multimeter set to resistance or Ohms. Use of multimeters will be covered in a later chapter.

Impedance is also an important concept. In DC (direct current) circuits, resistance often remains the same for components. In AC (alternating current) circuits, the resistance of a device can vary depending on the frequency of the applied voltage. This is known as the **impedance** of a circuit. This is also known as **reactance**.

Figure 26 - Inductor (left) and capacitor (right)

Inductors and Capacitors:

An inductor is just a coil of wire. One of the properties of electricity is that it creates a magnetic field by flowing through a conductor. When wire is coiled up, the property is intensified, and when wrapped around a magnet, it intensifies further. This can have useful effects un circuits, but also create undesired effects when wires are coiled up in control panels.

Inductance is how inductors or "coils" are measured. The unit of measure is the **Henry**, and typical values are measured in milli or microhenries. The important feature to consider in inductors is that as the frequency increases, the impedance also increases. Inductors also oppose a change in current, which makes the current lag or "fall behind" the voltage.

A capacitor is simply two "plates" of conductive material separated by an insulator called a **dielectric**. The unit of measure for a capacitor is the **Farad**, and typical values are micro or picofarads. This measurement is essentially how much of a charge can be held. A property of a capacitor is that if a voltage is applied to the plates and then disconnected, the plates will hold the charge for a short time, similar to a battery.

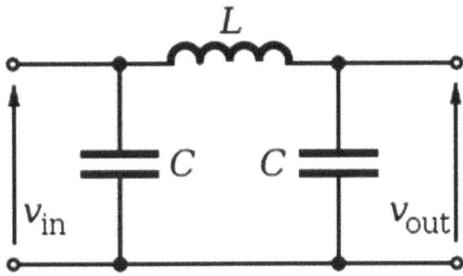

Figure 27 - Lowpass filter

In a DC circuit, a capacitor is effectively an open conductor, but if AC is applied the impedance of the component decreases as the frequency increases. Capacitors have the opposite effect of inductors; the voltage lags the current in a circuit. Because of the properties of inductors and capacitors, they are important elements in **filtering** signals.

Power Sources:

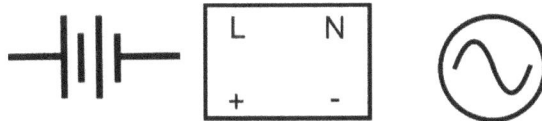

Figure 28 - Battery, DC Power Supply, AC source

The symbols for a battery and an AC power source are often shown in component level circuit diagrams, but most schematic diagrams will show that DC power usually is provided from a power supply. DC power supplies convert AC power into DC, usually at the 24-volt level.

AC power is usually applied to a system or machine at a single point with a disconnect and fusing or circuit breaker. If different voltages are needed within the system, AC voltages can be changed by use of a **transformer**.

Figure 29 - Transformer

A transformer is just two inductors or coils that transfer AC power. By using a different number of windings in each coil, voltage can either be stepped up or stepped down. Converting 480 volts AC to 120 volts for use by internal control components is very common in industrial machinery.

Three phase transformers may also be used to convert 480 to 240 on all three phases if needed, but usually a single phase of 480vac is chosen to convert to 120vac.

Transformers may also be used to isolate ac circuits for safety reasons. They isolate any DC component on the incoming power side from the receiving equipment and can also suppress electrical noise. They are predictably called **isolation transformers**. The output voltage is the same as the incoming voltage.

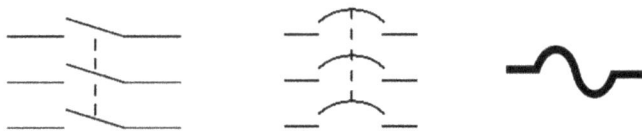

Figure 30 - 3 phase disconnect (left), 3 phase circuit breaker (center), and fuse (right).

Disconnects, Breakers and Fuses:

Disconnects are used to remove power from a circuit or system, usually manually. For safety reasons they usually allow the handle to be controlled from outside the control enclosure. Disconnects may also have fuses built in; the diagram would then show fuses where the diagonal switch elements are shown in Figure 30.

Circuit breakers, like disconnects, may have a handle that allows it to remove power manually. They also will "trip" if too much current passes through a thermal element on any of the phases. Figure 30 shows a circuit breaker that will disconnect all 3 phases either if an operator turns the handle, or if the breaker trips. Disconnects and circuit breakers also often have an auxiliary contact as part of the mechanism that will let the control system know of the status of the device.

Fuses are thermal elements that are destroyed when too much current passes through them. If it is desired to know the condition of the circuit past the fuse, a connection is often made to the circuit after the fuse, which can drive a relay. Fuse holders may also have a light indicator that illuminates if the fuse has blown. Fuses are usually rated both by the maximum current that can flow through them and a voltage class.

Figure 31 - Thermal overload

Thermal Overloads, like fuses, react to the current that flows through them; unlike fuses they are not destroyed when this happens. They are made of a bimetallic strip that serves as a closed contact while current flows through the circuit. If the current exceeds the rating of the overload, the metal bends and the contact opens or "trips". Circuit breakers and motor starter overloads both use this element. The bimetal strip is usually heated indirectly.

Thermal overloads may also be electronic, using temperature sensors or current transformers to sense the amount of current.

Figure 32 - Coil

Coils and Contacts: Coils, like inductors, are just coils of wire wrapped around a piece of iron. When current flows through the wire, the coil acts as an electromagnet and pulls the contact into its opposite state, either open or closed, either completing or disconnecting a circuit. Coils can be operated by applying AC or DC voltage.

Contacts are configured in two different ways, Normally Open (NO) and Normally Closed (NC). This means that without energizing the coil, they are in the open or closed position. When the coil is activated, the contacts will change states.

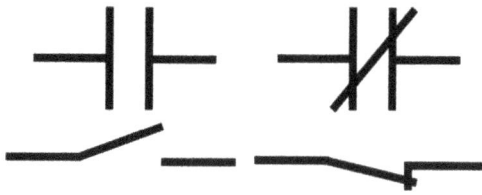

When coils and contacts are placed together, they are called **relays** or **contactors**. They may have multiple sets of contacts, some NO and some NC depending on the requirement.

Figure 33 - Normally Open and Normally Closed contacts

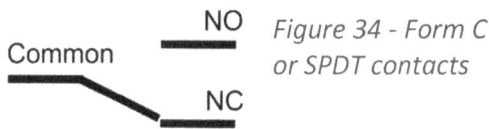

Figure 34 - Form C or SPDT contacts

Contacts may also be configured so that when deenergized a NC contact is made, and when energized a NO contact is made, using the same common terminal. These are often referred to as "Form C" contacts. These are also referred to as "Single Pole Double Throw" (SPDT).

Contactors are usually larger multi-contact relays with multiple "Single Pole Single Throw" (SPST) contacts, used to power motors. These are used along with overloads on all 3 phases to control AC motors, collectively this is called a **motor starter**. More on this will be covered later in this book.

Solid-State Devices: Solid-state components are made of silicon-based material combined with different impurities or chemicals. They can be used as relays to switch signals, amplify or change signals, or even emit and detect light.

Figure 35 - Diode

Diodes allow current to pass in one direction but not the other. If you think of the shape as an arrow, current flows from positive to negative in the direction the arrow points. Diodes are used to rectify signals, i.e., flip the negative half of an AC signal positive, and also as protection for relay coils. When power is removed from a coil, it opposes a change in current, which can cause arcing. Placing a diode in the reverse direction across a coil can create a "flywheel effect" that protects both the coil and the energizing output.

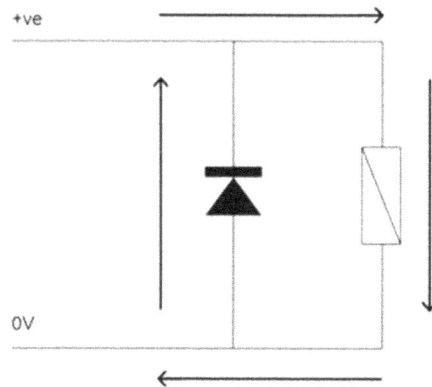

Figure 36 - "Flywheel" diode

Diodes can also be used to emit or produce light (a **Light Emitting Diode** or **LED**). If they react to light by changing conductivity, they are called **photodiodes**. Photo transistors and photoresistors do the same thing but use different symbols.

Figure 37 - LED

Figure 38 - Photodiode

Solid state devices such as diodes are made of materials that have an excess of electrons (N type material) or an excess of "holes" (a space for an electron, P type material). A single piece of silicon or Germanium is "doped" with a chemical impurity such as Boron or Gallium (p-type), or Arsenic or Phosphorus (n-type) on each end to create a PN junction, or diode. These materials are called semiconductors.

If two such junctions are created in a piece of semiconductor material, the device is called a **transistor**. Transistors have three areas: the Emitter, the Base and the Collector. Between each of these areas is a PN or NP junction. The arrangement of these junctions determines what type of transistor it is, either NPN or PNP. A good way to remember which symbol is which is that the arrow points toward the N type material. The arrow is also on the emitter part of the transistor.

Figure 39 - NPN Transistor

Figure 40 - PNP Transistor

The symbols shown in figures 39 and 40 are called junction transistors. There are other types such as MOSFETs, JFETs, BJTs, IGBTs with different symbols, but the use and concepts are the same.

Depending on how the transistor is wired with other components into a circuit, it can act as a relay or remotely operated switch, or as an amplifier to change the amplitude of a signal. They are used in sensors, variable frequency drives, and logic circuits.

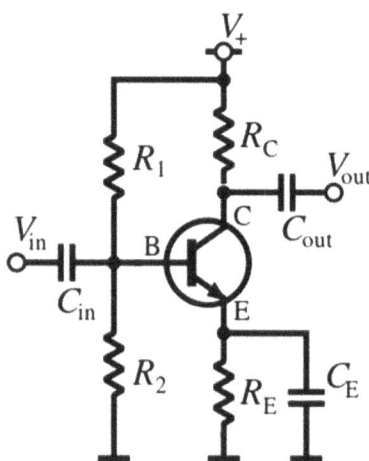

Electrical Circuits: All of the previously listed components can be combined in different ways to accomplish different tasks.

Amplifiers take a small change in voltage in the input side of the circuit and change it to a larger change on the output side. **Switching circuits** can change the output signal from 0 volts to the V+ (usually 24vdc in industrial signals) by applying a voltage to the input. They can also be combined in groups in integrated circuit "chips" and microprocessors to perform logic functions.

Figure 41 - Common Emitter amplifier circuit (NPN)

An important concept with switching circuits is that they can be **sinking** (NPN) or **sourcing** (PNP) in terms of the signal they provide at the output. This will be covered more completely in the sensors part of this book.

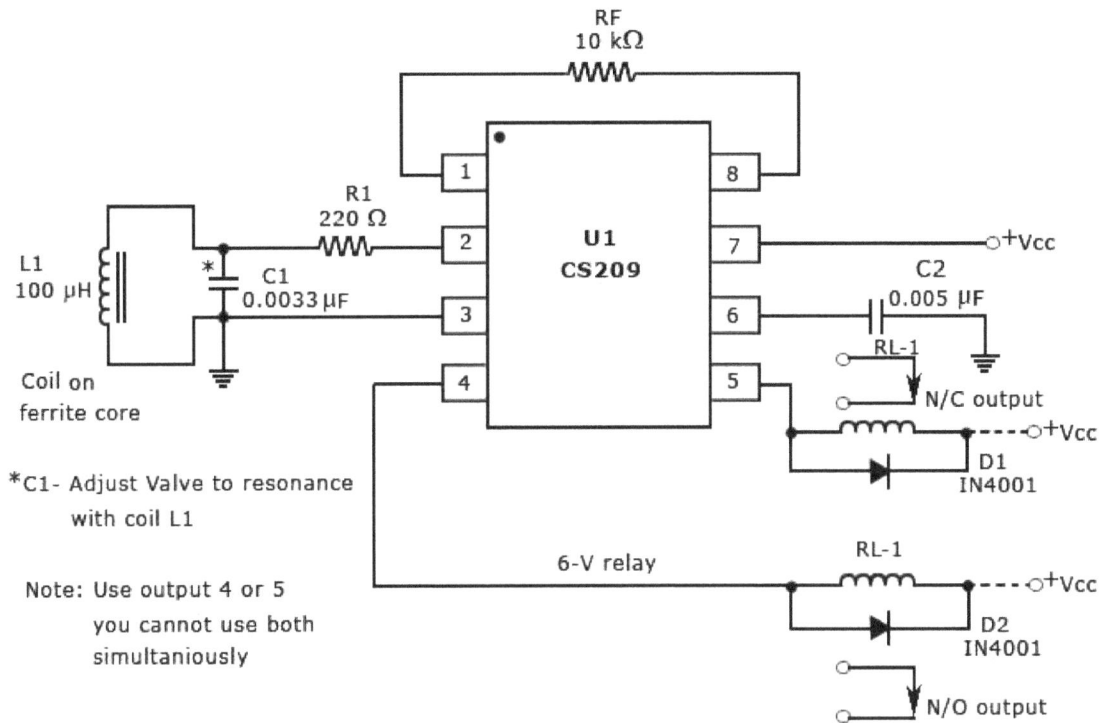

Figure 42 - Inductive proximity switch circuit (relay)

Figure 42 shows a circuit for an inductive proximity switch with complementary relay outputs. Note that part numbers (IN4001 diodes and CS209 integrated circuit) are listed, as well as values for the resistors, capacitors and inductor. This can be useful when repairing circuits, though in the case of sensors they usually cannot be repaired.

Components like the integrated circuit can easily be found online, the CS209A is a 24vdc circuit specifically designed for inductive (metal sensing) proximity switches. Internally it consists of current regulators, an oscillator, peak detection, a comparator, and two complementary output stages. These terms describe circuits made up of the same types of discrete electronic components described previously, all contained in a single integrated circuit "**chip**".

The most common factor in damaging electrical components is heat, often a consequence of too much voltage across a device which results in high current and breakdown of its materials.

Exercise 8

1. What 3 units of measure are used for classifying resistors?

_____

2. What term is used to describe resistance in AC circuits, varying in frequency?

_____

3. A capacitor opposes a change in _____ (voltage/current) while an inductor opposes a change in _____ (voltage/current).

4. What is the purpose of a transformer? _____

_____

_____

5. What is the difference between a circuit breaker and a fuse? _____

_____

_____

6. What are the names for devices that have a coil and contacts built into the same component? _____

7. What are three possible uses for semiconductor components?

   a. _____

   b. _____

   c. _____

8. What are the two different types of junction transistors?

    a. _____

    b. _____

9. What are two of the common functions transistors can be used for?

    _____

10. What is the most common cause of failure for electrical components?

    _____

11. What is the purpose of a "flywheel" diode?

    _____

    _____

12. A resistor has stripes on it in this order: Red, Red, Green, Brown, Gold. What is its resistance and tolerance? (Use Appendix C in the back of this book)_____

DC or Direct Current

Most low-level control devices such as sensors and valves are DC or direct current devices. Direct current is basically one-way flow of electrical current (electrons) through conductors (wires).

A good example of a DC circuit is a flashlight with batteries. In Figure 43, each battery provides 1.5 volts for a total of 3 volts. When the switch is closed completing the circuit, the light bulb will illuminate. The circuit diagram for a flashlight is also shown.

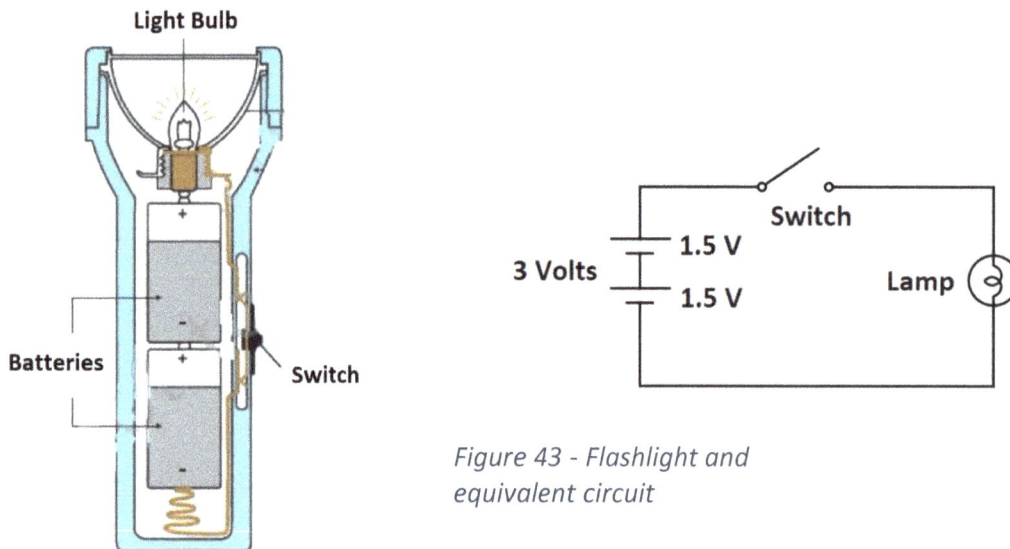

Figure 43 - Flashlight and equivalent circuit

Voltage (V) is the "pressure" that cause the electrons or **current** (I) to move through the circuit. The amount of current that flows is dependent on the **resistance** (R) of the circuit. In figure 42 the resistance is represented by the lamp or light bulb.

If two values are known for a circuit, the other value can be calculated using the formula

V = I x R. This can also be converted to **I = V/R**, or **R = V/I**. The measurement for voltage is expressed in **Volts**, for current is **Amperes** or **Amps**, and for resistance is **Ohms**. This set of formulas is called **Ohm's Law**.

If the resistance of the lamp in the circuit of figure 42 was measured as 300 Ohms, the current flowing through the circuit would be I = 3/300, or 0.010 Amps. This would be 10 milliamps, abbreviated mA (10 mA).

In industrial automation, 24 vdc is commonly used for sensors (inputs) and solenoid valves (outputs). Rather than a battery, a 24-volt power supply is used.

While this section is not intended to be a full treatment of electrical circuits and electricity, it is important to understand some of the basic concepts and terminology of components and circuit analysis.

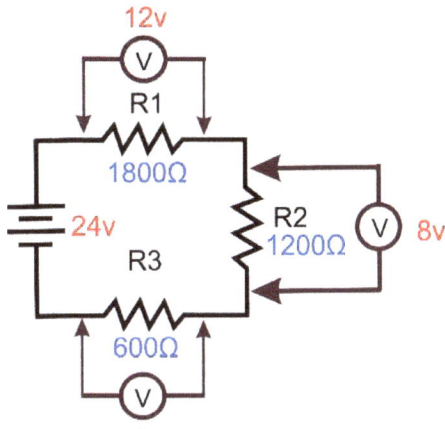

Figure 44 - Series circuit

Analyzing DC circuits requires an understanding of the concepts of **series** and **parallel**. Figure 44 shows several devices in series. The voltage is divided proportionally across the components based on their resistance, for this reason this can be thought of as a **voltage divider**.

Resistances can be added up to determine the total resistance of the circuit. If the voltage is known, and the total resistance can be determined, then the current can be solved for using Ohm's Law. Since the total resistance is 3600 ohms, the current is 24/3600, or 6.67mA. These resistances do not necessarily represent resistors, they could represent any device.

Figure 45 shows several devices in parallel with a constant voltage across them. Since the voltage is constant, the current will be divided proportionally through the devices. This is known as a **current divider**. Once again, if the voltage and the total current is known (by adding all the individual currents), the total resistance of the can be determined using Ohm's Law.

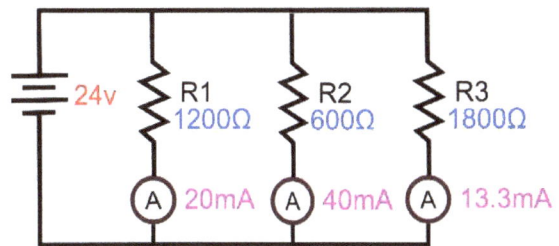

Figure 45 - Parallel circuit circuit

I (total current) = 73.33mA, R = 24 / 0.0733 = 327.28 Ohms! A much lower resistance than the same 3 resistances in series.

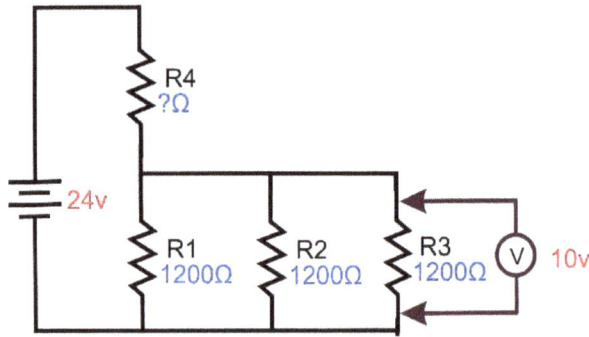

Figure 46 - Series Parallel circuit, calculate the resistor value

Devices can be placed in both series and parallel. To solve for voltages and currents, the individual sections must be evaluated. But how can an equivalent resistance be calculated in the parallel part of the circuit if the voltage or current is not known?

Figure 46 shows a series-parallel circuit. To calculate the total resistance, the formula 1/RT = 1/R1 + 1/R2 + 1/R3 + ... can be used for the parallel portion of the circuit. This gives a value of 400 ohms for the equivalent resistance. This circuit shows an example of using a resistor to create a 10-volt supply for the three devices R1, R2 and R3. What would the value of R4 have to be to accomplish this?

Again, here Ohms Law can be used. Since I =V/R for both series components of the circuit, and the resistance of the parallel equivalent is 400 ohms, 10/400 = 14/R4. This transposes to R4 = 14*(400/10) or 540 ohms. The total resistance would be 940 ohms, making the current 24/940 = 25.5mA.

An important reason for being able to calculate the total current in a circuit is to specify fusing and wire sizing for a given circuit. The resistances also generate heat, which can be represented by the wattage or power dissipation of the components. The formula for **Power** is **P = V x I**, this can be used to select the appropriate wattage rating of the resistor. 0.0253 amps x 14 volts = 0.357 Watts, so a ½ watt resistor should be selected. Additional formulas for power are $P = V^2/R$ and $P = I^2 * R$.

AC or Alternating Current

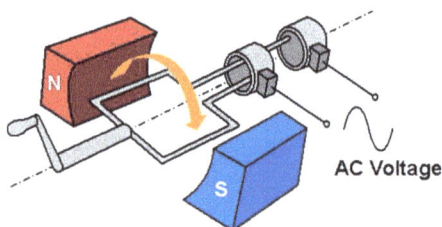

Figure 47 - AC Power Generation

Alternating current or AC is generated by rotating an electromagnetic coil (rotor) inside a stationary coil (stator). The same effect can be obtained by rotating a coil between magnets or rotating the magnets in front of coils. This induces a current that changes direction every half rotation.

Figure 47 shows wires being rotated in front of magnets with slip rings to transmit the current. While this is a simplistic view, it shows the concept well. The speed at which the rotor moves is called the **frequency** of the AC power. In North America and much of South America, the frequency is 60 cycles per second, or 60 Hertz (60Hz). In most of the rest of the world it is 50Hz.

Like frequency, commercial voltages delivered to buildings varies around the world. North America and parts of South America household voltages are 110-127V, while much of the rest of the world operates at 220-240V.

A major reason AC voltage is used across the world rather than DC is that it can easily be raised or lowered using a transformer. It is also more efficient to transfer power over longer distances with AC. Power is transferred at much higher voltages, typically 66,000 (66kV) or above, then stepped down using transformers at the point of use. It is also generated and transmitted in three different phases, each 120 degrees apart. Figure 48 shows how 3 phase power is generated. The alternating wave forms are **sinusoidal**, or **sine waves**.

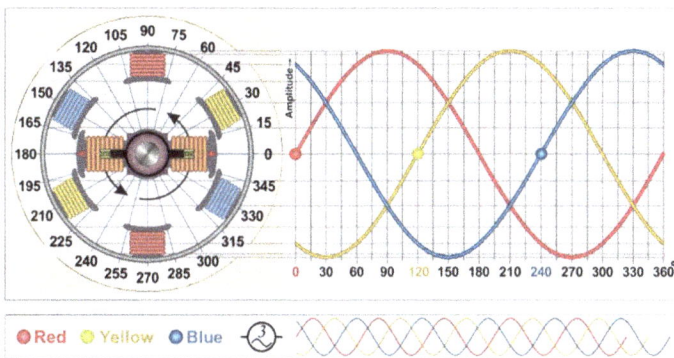

Three-phase power in most US manufacturing facilities is converted to 480VAC for machinery, in Europe this is usually 400V. Some higher power motors may use 4160VAC or even higher in process control applications such as gas pipelines.

Figure 48 - Three phase power generation

The frequency at power plants is controlled very accurately since timing circuits often depend on it in control systems. AC motors run at a multiple of the applied frequency minus a factor called "slip", which is due to the rotor of a motor always being behind the stator. If there was no slip factor, typical 4-pole motors would operate at 1800 revolutions per minute (RPM), but due to this factor the rated frequency of US motors is **1750-1780 RPM**. The no load (no slip) speed of a 4-pole motor at 50Hz is 1500 RPM, so the rated speed output would be slower.

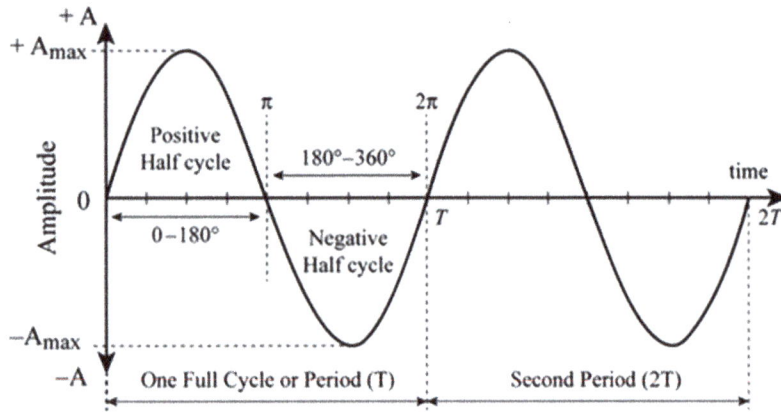

Figure 49 - Sine wave

If a single sinusoidal waveform as shown in Figure 49 is examined, the voltage moves in equal positive and negative cycles. This means that the voltage is constantly changing at the frequency of the generator, and the average is actually zero! So what is meant by "115 volts" being available from an electrical outlet?

115-120vac is actually the **effective** voltage across the whole waveform. Another term for this is **root-mean squared**, or **RMS** voltage. If the waveform was measured using an oscilloscope, it would show a peak voltage of 162.6 volts, which is $115 \times \sqrt{2}$. The negative peak would follow the same formula. The easiest way to calculate the effective voltage if you know the peak is to multiply **Peak x 0.707**.

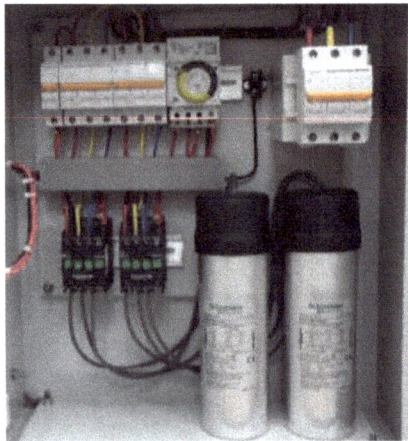

Figure 50 - Power Factor Correction Capacitors

Effective voltage across a purely resistive load can be calculated as described above, but most machines use motors that have an inductive load. Because the current lags the voltage in an inductor, the power that is apparent to the power company's meter is lower than what is calculated. The ratio of average actual power to apparent power is called the **power factor**.

The power company takes this into account when billing customers by charging a surcharge. Because of this, manufacturers will often use capacitors to change the power factor back to 1.0. Figure 50 shows a panel with Power Factor Correction Capacitors used to change the phase angle.

Digital and Analog Signals

When dealing with control systems for industrial machinery, it is important to understand the difference between digital and analog. These signals represent physical attributes and states of the system or environment to a controller.

Digital signals have two states, **on** and **off**, indicated by a **one (1)** or a **zero (0)**. An energized device provides an electrical output that can be used to indicate a piece of information such as the absence or presence of an object; this signal can also be referred to as being **True** or **False**.

Digital signals may be DC (commonly 24 volts) or AC (120 or 240 volts). For safety reasons, DC digital devices are more common.

Analog signals are used to provide varying voltage or current levels to a control system. These signals are scaled into numbers that represent physical attributes such as temperature, pressure or speed.

Analog signals usually vary from 0-10VDC (voltage) or 4-20mA (current). 0-5VDC or 0-20mA are also common. These signals are then often scaled into Signed Integer values. More on the topic of numbering systems will be discussed later.

Additional information on the processing of analog signals and their accuracy will be covered in the next section.

Exercise 9

1. If a 1.5 volt battery is used to illuminate a light bulb with a resistance of 10 ohms, what is the current?

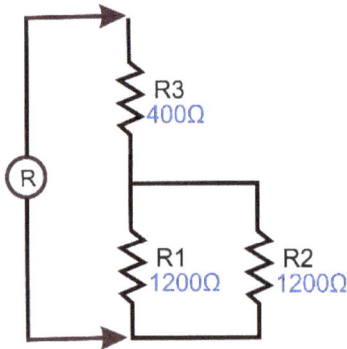

_____

2. Calculate the resistance of figure 51: _____

3. If 24 volts DC is applied across this circuit, what is the current?

_____

4. What is the power dissipated by R3 in this circuit at 24vdc? _____

5. What is the frequency of AC power in the United States? _____

Figure 51 - Resistive Circuit

(Circuit diagram: R3 400Ω, R1 1200Ω, R2 1200Ω, with a resistance meter R)

6. What do digital and analog signals represent? _____

_____

Analog Signal Processing:

Specifications for measuring equipment often includes data on the accuracy of an analog signal in representing the measured parameter. **Accuracy** is usually specified as a plus or minus percentage of how close a measured value is to its true value. **Repeatability** is the degree to which repeated measurements will come to the same value; sometimes this is referred to as precision. **Linearity** is how well the device's measurements approximate a straight line across its measuring range. Measuring devices are calibrated such that the most linear and accurate range should be right around the normal operating or "working" area of the scale.

Resolution is the smallest change that can be detected by the measuring system. While the sensor itself produces an analog signal that can theoretically produce an infinite number of values, measuring systems convert the signal into a finite number of values. An Analog to Digital Converter (ADC) performs this function.

ADCs are classified by their bit resolution; typically this is from 12 to 16 bits. The number that is produced by an ADC is typically a signed integer, which has 65,536 possible values, ranging from -32,768 to +32,767. This allows for negative values for a -10v to +10v range. If only positive values are used, this effectively produces a 15-bit resolution.

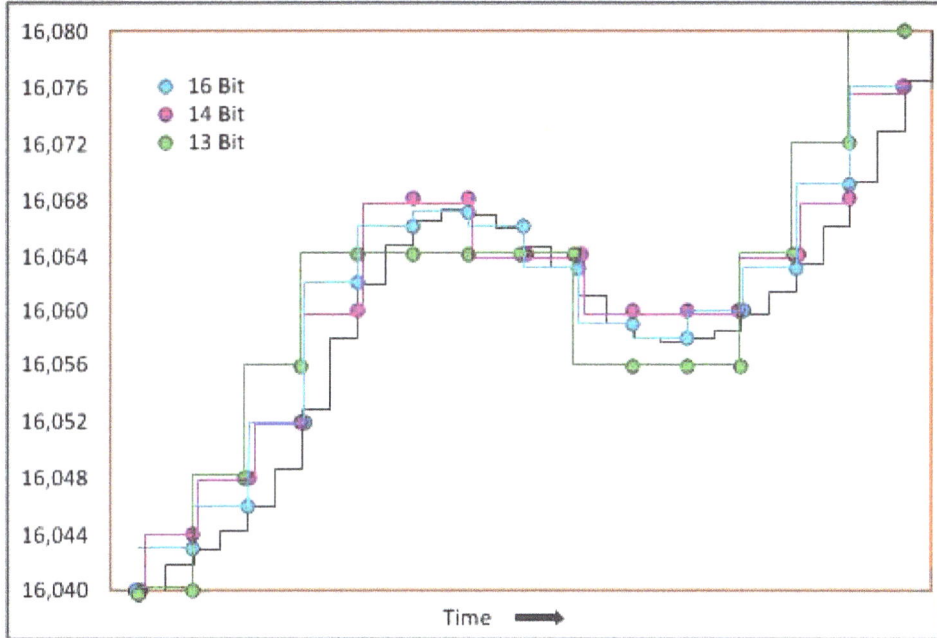

Figure 52 - Resolution

Figure 52 shows the effect of different resolutions on a 0-32,767 range. While a 0-10v signal at 16-bit resolution produces a value for every number, lower resolutions step by 4 or 8 values at a time. Some 12-bit resolution systems will produce values from 0-4095. These numbers then need to be scaled to Engineering Units, reflecting the actual value being measured. Usually this will be a REAL or floating-point value.

Exercise 10

1. In what area of a device's range should an instrument be calibrated?

_____

2. What format is the number produced by an ADC usually converted to?

_____

Communications

Electronic devices and controllers use a wide variety of different methods to communicate with each other, from serial communication protocols to ethernet, or even parallel connections between devices using I/O connections. Some methods are open protocols that are understood across many manufacturer's products, while others are specific to the brand.

Serial Communications

Serial means that bits are transmitted on one wire at a time in a series of high and low electrical signals, or "1's and 0's". This differs from parallel communications, such as printers, where the bits are transmitted and received on many wires in parallel with each other. Serial communication uses different physical formats, usually in the form of RS232, RS422, or RS485.

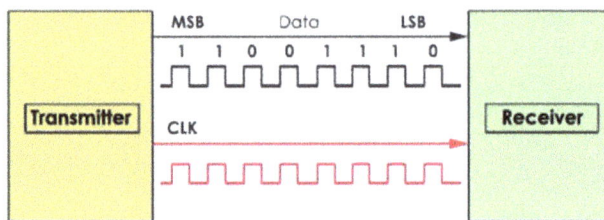

RS, or "Recommended Standard" communications include RS232, RS422, and RS485. RS defines the wiring and format of the message, but not its language or protocol.

Figure 53- Serial Communication

Controls manufacturers usually develop their own protocols and method of communication based on these recommended standards, making communications between the different vendors systems problematic. Each manufacturer will have its own driver for its language; for example, in Allen-Bradley's PLC the serial RS232 protocol is called **DF1**, while in Siemens it is called **MPI**. There are also protocols that have been widely used across different platforms such as **Modbus**, originally developed by Modicon in 1979, but generally the native protocol is specific to the manufacturer.

A couple of acronyms to be aware of when communicating between applications and devices:

OLE – Object Linking and Embedding. A proprietary technology developed by Microsoft that allows embedding and linking to documents and other objects.

OPC – Open Platform for Communications, formerly OLE for Process Control. A series of standards and specifications for industrial telecommunication. OPC was designed to provide a common bridge for Windows-based software applications and process control hardware.

RS232 protocol is used for many of the programming interfaces between laptop computers and PLCs or other controllers. Because serial ports are often not present on newer computers, a USB adapter may have to be used between the computer and cable. RS232 requires that parameters such as Baud rate (speed), bits, and parity be set the same on all stations for communication to take place. If the transmit and receive pins (TD and RD) are the same on both devices, a **null modem** adaptor will be needed. This cross-connects pins 2 and 3.

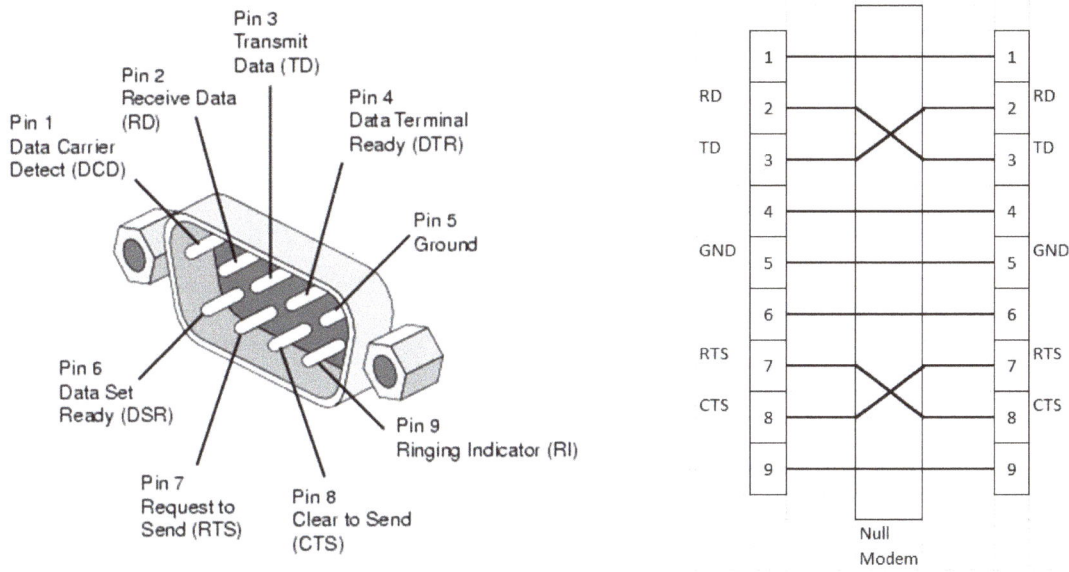

Figure 54 – RS232 pinout and null-modem

RS485 is another common standard for serial communications. It uses a single twisted pair of wires and can be "daisy-chained" between multiple computers or controllers.

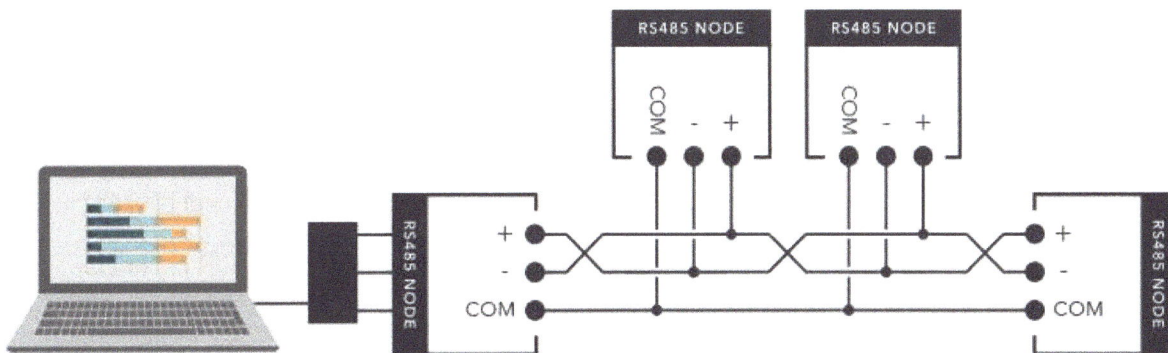

Figure 55 - RS-485 or "multi-drop" connections

RS485 is commonly used for remote I/O and multi-drop networks. Programming devices such as computers can be used on these networks through the use of an adapter. As with RS232, the Baud Rate and protocol must be set the same on all member stations or "nodes".

Examples of RS485 networks are Profibus, DeviceNet, Modbus and Data Highway/DH+.

RS422 is often called peer-to-peer, or **PtP**. It also uses a twisted pair of wires, but this protocol only supports communication from a single device to another.

The common thread to all of these protocols is that the language or interpretation of the bit patterns is not defined. The "language" spoken on these protocols will differ from device to device.

The interpreter for these different languages is called a **driver**. This is the software piece that is responsible for determining what the messages carried on these wires means. It connects a device and software, or two different software packages together.

A note on twisted pair communications like RS485 and RS422: Twisted pairs of wires reduce the amount of electrical interference from other signal wires that run in parallel. Noise can be further reduced by shielding the pairs as a whole or individually; it is important that the shields are only grounded on *one end*, otherwise even more noise may be introduced. This also applies to analog signals carried on twisted pair cable.

Figure 56 - USB symbol

USB or Universal Serial Bus is a standard developed in the 1990s to connect computer devices and peripherals such as keyboards, digital cameras and pointing devices to computers. It not only provides communications, but also can power devices. USB is much faster than standard serial connections.

As with other serial protocols, the language is not defined on a USB connection. There are also several physical configurations or shapes for the socket and plug.

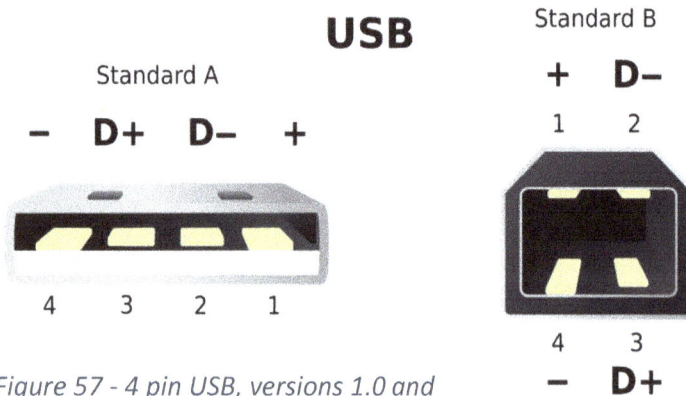

Figure 57 - 4 pin USB, versions 1.0 and 2.0 (Simon Eugster)

Data rates are generally classified by version, with newer versions being faster. Figure 57 shows the physical shape of standard A and B type USB ports, which have 4 pins. These apply to version 1.0 and 2.0, which operate at 1.5 or 12 MBits/sec. Version 2.0 can also operate at up to 480 MBits/sec for the high-speed version.

USB 3.0 and 3.1 also have an A and B style but have 9 pins. These versions can operate at 5 or 10 GBits/sec, which is similar to low ethernet speeds.

USB, like RS232, is limited in cable length. The standards were developed for items on the same tabletop, but gateways or signal boosters are available to extend the length.

Ethernet

Ethernet is a family of computer networking technologies consisting of a set of wiring and communications standards. Systems communicating over Ethernet divide a stream of data into shorter pieces called **frames**. Each frame contains source and destination addresses, and error-checking data so that damaged frames can be detected and discarded; most often, higher-layer protocols trigger retransmission of lost frames.

Cabling for Ethernet may consist of coaxial cable, several twisted pairs, or fiber-optics. Most PLC connections use standard twisted pair CAT 5 cable and RJ45 connectors.

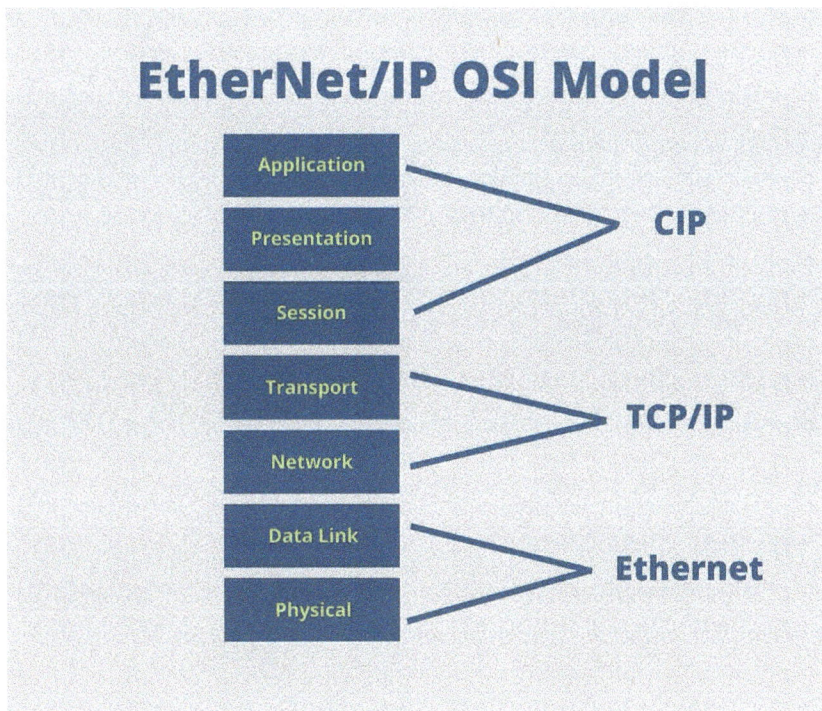

Ethernet follows a seven-layer structure defined by the Open Systems Interconnect (OSI) model as shown in Figure 58. These layers describe the lowest or physical layer, as well as various methods of interconnecting and networking between different areas, or domains.

Many different protocols are included on this definition, including TCP/IP (for connecting dissimilar devices across the internet), BOOTP (for setting initial addresses), and SMTP (for e-mail).

Figure 58 – Ethernet/IP OSI Model

PLC manufacturers generally define their own language and protocols for using Ethernet for both data communications and to connect to inputs and outputs (I/O).

Because it is important that I/O communications are **deterministic**, control device manufacturers follow the Common Industrial Protocol, or **CIP**. This ensures that deterministic signals are sent and received within a specified period of time and are therefore predictable. Examples of CIP include Ethernet/IP for Allen-Bradley and ProfiNet for Siemens.

Ethernet and Network Terms

Following is a list of terms that it is important to know when discussing an Ethernet network:

Client: A computer or device that initiates a request for data.

Server: A computer or device that responds to the client by providing or accepting data.

LAN (Local Area Network): a network that that connects computers or devices in one single location. Usually administered by a Server computer or a domain controller.

WAN (Wide Area Network): a network of LANs connected by Gateway or Router devices.

Workgroup: Computers in a workgroup can share files folders and printers.

Domain: A collection of host computers supervised by a domain controller computer.

Bridge: A device that interfaces between two similar networks.

Gateway: A device that allows modules on two different communication networks to interface.

Hub: A solid-state device that connects Ethernet modules in a star configuration.

Switch: A solid-state device that connects Ethernet capable devices. It also has buffering capability to reduce collisions.

Router: A solid-state device that has the capabilities of a switch but connects Ethernet modules on different networks. Often includes a **Firewall**.

TCP/IP (Transport Control Protocol/Internet Protocol): A common protocol that allows computers with different operating systems to exchange data.

CIP (Common Industrial Protocol): A method of communicating used in industrial automation. Provides deterministic communications for control, safety, synchronization, motion, configuration, and information.

Socket: A package of subroutines that provides access to TCP/IP and CIP functions.

NIC (Network Interface Card): A communication adaptor with an Ethernet port in a computer.

BootP (Bootstrap Protocol): A computer networking protocol that assigns a static IP address to a device from a configuration server.

DHCP (Dynamic Host Configuration Protocol): A network management protocol that automatically assigns a temporary IP address and other parameters to network devices.

IP Addressing

Part of the Network Layer protocol defined for TCP/IP is the IP addressing model. The original internet addressing scheme was proposed using IPv4 addressing that used 32-bit addresses. For human readability, IP addresses are expressed as four decimal numbers with a dot between each. This format is sometimes called "dotted decimal" notation. It divides the address into groups of 8 bits each, with a range of 0-255.

Addresses are divided into network (netid) and host (hostid) parts. The network part is also referred to as the **IP prefix**. All hosts attached to a network share the same network share the same IP prefix but must have a unique host part.

To support different sizes for the netid and hostid parts of the address, a class system was developed as a rule to partition the IP address space. The space was originally divided into three different classes, Class A, Class B, and Class C. This is shown in figure 59.

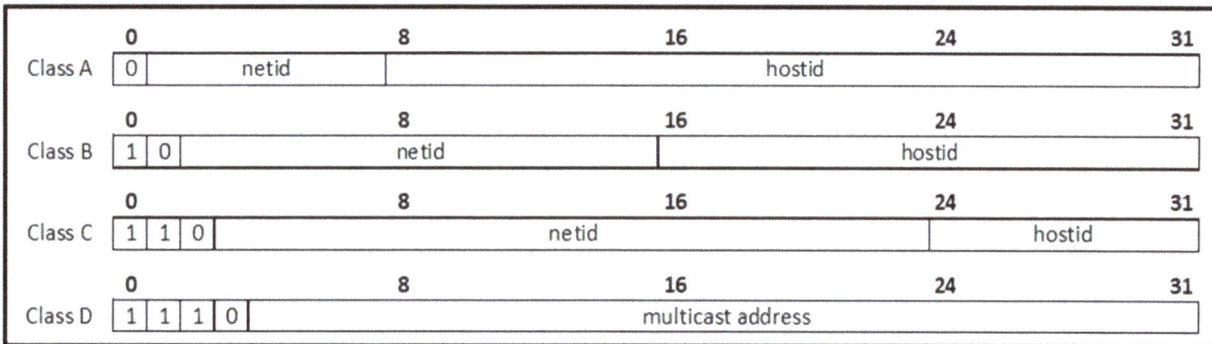

	0		8	16	24	31
Class A	0	netid			hostid	
Class B	1 0		netid		hostid	
Class C	1 1 0		netid			hostid
Class D	1 1 1 0		multicast address			

Figure 59 - Ethernet Class Addressing

The netid portion of the address is separated with a mask of ones, resulting in a mask of 255.0.0.0 for Class A, 255.255.0.0 for Class B, and 255.255.255.0 for Class C networks. Though this "classful" system is not commonly used anymore, it is common to see 192.168.xxx.xxx networks used in manufacturing facilities with a 255.255.255.0 mask (192 is 11000000 in binary). These masks are also sometimes abbreviated by the number of significant bits in the mask, so a Class C mask is compactly written as "/24". A bit-wise logical "AND" operation can then be done between the host address and the netmask to determine if two addresses can communicate. By enlarging the mask to /22 or /21, larger networks can be created.

The number 0 and 255 can't be used as a hostid in the last "octet" of the address, leaving a range of   1-254, or 254 devices possible in a class C network. With a mask of 255.255.252.0 (/22), 1016 devices can be placed on a network!

Communications Troubleshooting

There are both physical and software issues that can interfere with communications. Cables for RS232 devices may have molded or encapsulated ends, where the back side of the pin connections can't be accessed; in this case, it may be necessary to replace the entire cable. 9 Pin connectors can also be purchased and field wired, this usually involves soldering a multiconductor cable to the pins.

Using a known good cable to check against a suspected bad cable is the quickest way to determine if it needs to be replaced. The cable can also be checked pin by pin for continuity using a multimeter. Keep in mind: in the null-modem diagram in Figure 54, many cables have crossed pinouts.

If the cable is part of an installed system and was working yesterday, and it has no visible damage, it is unlikely that the cable is the problem. Ensure that the software parameters are set correctly; things like baud rate, protocol and the correct driver are important. These may have settings both on the computer and on the device. If no one has downloaded a program to the device recently, the computer side is likely where the problem is.

RS485 networks often require terminating resistors at the ends of the network. **Profibus** is a common protocol used for PLCs and devices; there are switches on the plugs for ON and OFF. The plugs at the ends of the network should be on, and for all of the intermediate devices the switch should be off. **DeviceNet** networks use actual resistors for some terminations, wired across the terminal blocks.

Ethernet networks often have switches and routers as connection points in the network. Ports on these devices have small yellow or green lights that blink as traffic occurs. These can help determine if the cable itself is good. To test whether an address is accessible, using "**ping**" at the DOS command prompt in Windows will tell you whether the communications are working:

C:\ ping 192.168.0.100 will tell you whether the destination address 192.168.0.100 is accessible by returning a "reply from…" message.

There are also utilities such as WireShark or AngryIP that will show network traffic for ethernet networks. A common cause of ethernet problems is the address and mask on the user's computer being incompatible with the destination address. Keep in mind also that all ethernet is not the same; an Ethernet/IP driver cannot communicate with a ProfiNet device.

Exercise 11

1. What does it mean to communicate "serially"?

_____
_____
_____
_____

2. What is the main difference between RS422 and RS485?

_____
_____
_____

3. What is "RS" an abbreviation for in RS232? _____

4. What is the purpose of a software driver for communications? _____

_____
_____

5. How many layers are in the OSI Ethernet structure? _____

6. Why do PLC manufacturers use CIP for control instead of TCP/IP? _____

_____

7. What is the difference between a router and a switch? _____

_____
_____

8. How many devices can be addressed on a network where the mask is 255.255.254.0?

_____

Data

Data Types and Formats

When dealing with data in software and computer-based controllers, there are various numerical data types and formats to be aware of.

BOOL or Bit: This is the simplest form of data. It only has two possible states, on and off, or one and zero. If the states are called "True" and "False", it is more properly called a BOOL, which implies a logical operation. A bit may be just one part of a larger element, such as a byte or word.

Byte or SINT: A Byte is a group of 8 bits. This is sometimes called a Single Integer, or SINT. A Byte's value ranges from 0 to 255, or 0000_0000 to 1111_1111. What is a half of a Byte, or four bits? A **Nibble** of course!

Integer (INT) or WORD: 16 bits make up an integer, or INT. While an Integer always implies a number that mathematical functions can be applied to, a WORD may sometimes only be used for logic functions, such as AND or OR. An Integer can have 65,536 possible values, from 0000_0000_0000_0000 to 1111_1111_1111_1111.

The bit on the far right of the string of 1's and 0's is known as the Least Significant Bit, or **LSB**, while the bit at the far left is the Most Significant Bit, or **MSB**. An **Unsigned Integer** ranges in value from 0 to 65,535. A **Signed Integer** uses the MSB as a "sign bit"; if it is a 1, the value is negative and if it is a 0, the value is positive. A Signed Integer ranges in value from -32,768 to +32,767. Most PLCs use Signed Integers.

Double Integer (DINT) or Double Word (DWORD): A Double Integer has 32 bits. Like an Integer, there are Signed and Unsigned versions, determined by the MSB. A DINT ranges in value from 0 to 4,294,967,295 (Unsigned) or – 2,147,483,648 to +2,147,483,647 (Signed).

REAL or Floating Point: A REAL number is a 32-bit number that can express fractional values, or decimal points. Because the decimal can be shifted to the right or left within the value to change its magnitude, the decimal point can be said to "float". Unlike Bytes, INTs and DINTs, it is not possible to look at individual bit values to determine the value of the number. A REAL number is composed of a Mantissa, an Exponent and a Significand, and the bits within the number are not used for anything else other than expression of the value. REALs have a range of 1.1754944e-38 to 3.40282347e+38.

Additional data types such as Long Integers (LINT, 64 bits), Long Reals, (LREAL, 64 Bits) and STRINGs (an array of Bytes signifying text characters) are also used in computers and PLCs.

Numbering Systems

Understanding How Bits Can Become Numbers:

0	0
0	1
1	0
1	1

If the previous talk of ones and zeros is confusing, think of it this way: for a single bit, there are only two possible values: 0 or 1, off or on.

For two bits, there are four possible values: a) Both off or 0,0 b) The first off and the second on or 0,1 c) The first on and the second off, or 1,0 and d) Both on or 1,1.

4's	2's	1's	
0	0	0	**0**
0	0	1	**1**
0	1	0	**2**
0	1	1	**3**
1	0	0	**4**
1	0	1	**5**
1	1	0	**6**
1	1	1	**7**

Figure 60 - Bit Combinations

For three bits, the possible number of combinations increases to 8 as shown: the values above the columns show the value of each "place" in the row. The string of ones and zeros is known as binary, a base 2 system. As a new position is added to this string, the value of the previous column is doubled.

For four bits, the possible number of combinations increases to 16, and the value of the next column becomes the "8's" position. The values then start at 0000, or zero decimal, and increase to 1111, or 15 decimal.

The value of each column then doubles as a placeholder (16, 32…), as does the possible number of combinations (32 @ 0-31, 64 @ 0-63). Remember, computers and PLCs can only "think" or process data in terms of ones and zeros since they are really just a collection of on-off switches, albeit very tiny ones.

In addition to the data types listed above, data can also be expressed in different ways. For instance, a Byte can be shown as a string of ones and zeros (Binary or Base 2): (0110_1011), as a decimal (Base 10) number: (107), or as a Hexadecimal (Base 16) number: (6B).

Integers: Computers and PLCs operate most efficiently when using numbers that are multiples of two; this is because at its heart, a microprocessor is just a collection of on-off switches. An Integer is therefore just a series of binary values signifying increasing values for each bit as shown below:

(MSB/Sign)															(LSB)
32,768	16,384	8192	4096	2048	1024	512	256	128	64	32	16	8	4	2	1
0	1	1	0	1	0	0	1	0	1	1	0	1	0	0	1

Figure 61 - 16 Bits = 26,985

To determine the decimal value of a Signed Integer for the binary value, it is necessary to add up all of the place values where there is a 1: 16,384 + 8,192 + 2,048 + 256 + 64 + 32 + 8 + 1 = **26,985**.

(MSB/Sign)															(LSB)
32,768	16,384	8192	4096	2048	1024	512	256	128	64	32	16	8	4	2	1
1	0	0	1	0	1	0	1	1	0	1	0	1	0	1	0

Figure 62 - Signed Integer = -27,222

This Signed Integer value has a 1 in the MSB. The decimal value is determined by adding all of the other place values, 4,096 + 1,024 + 256 + 128 + 32 + 8 + 2 = 5,546, and then subtracting the result from 32,768, or **-27,222**. This is also known as **2's Complement**.

BCD: Prior to the use of OITs (Operator Interface Terminals), digital devices such as thumbwheel switches and 7-segment displays were used for entering and displaying decimal values in the PLC.

To enter a number into a PLC's memory location, each thumbwheel switch required 4 digital inputs into the PLC. Decimal numbers from 0-9 could be set for each decimal digit; when the thumbwheel reached 9, it would then roll over to the zero position again. This meant that combinations such as 1010 (10) or 1011 (11) were not possible.

In the same way, outputs were used to illuminate the individual segments of a 7-segment display, each display requiring 4 digital outputs. It was important in this case that illegal combinations of outputs (decimal 10 and above) could not be sent to the device.

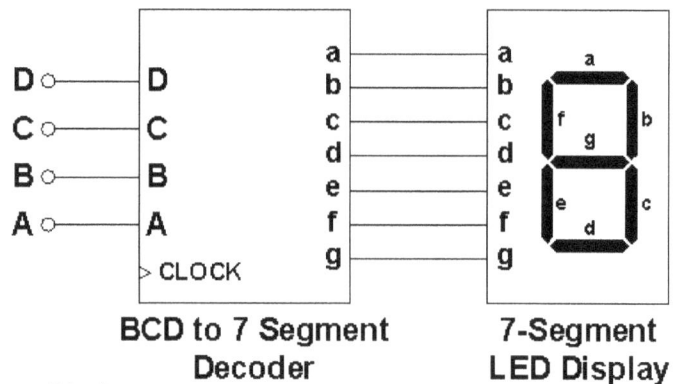

Figure 63 - Thumbwheel Switch and 7 Segment Display

This coding of bits into decimal equivalents is known as **Binary Coded Decimal**, or **BCD**. Some operator interfaces still require BCD for display of values, so conversion of integers into BCD format is sometimes needed. In addition, there are PLC platforms that use BCD for Timer and Counter values.

Binary Code A B C D	Decimal Number	BCD Code B_5 B_4 $B_3$$B_2$$B_1$
0 0 0 0	0	0 0 0 0 0
0 0 0 1	1	0 0 0 0 1
0 0 1 0	2	0 0 0 1 0
0 0 1 1	3	0 0 0 1 1
0 1 0 0	4	0 0 1 0 0
0 1 0 1	5	0 0 1 0 1
0 1 1 0	6	0 0 1 1 0
0 1 1 1	7	0 0 1 1 1
1 0 0 0	8	0 1 0 0 0
1 0 0 1	9	0 1 0 0 1
1 0 1 0	10	1 0 0 0 0
1 0 1 1	11	1 0 0 0 1
1 1 0 0	12	1 0 0 1 0
1 1 0 1	13	1 0 0 1 1
1 1 1 0	14	1 0 1 0 0
1 1 1 1	15	1 0 1 0 1

Figure 64 - Decimal and BCD

As the table in Figure 61 shows, BCD codes above 1001 require that another 4 digit value be used for the next decimal number. To display the number 9,999, the 16-bit pattern would read 1001_1001_1001_1001

The equivalent 9,999 expressed as a Signed Integer would be 0010_0111_0000_1111.

To express a Signed BCD number, the most significant four bits are used as a "sign" character, as with the Signed Integer. A negative number is signified by a 0001, while a positive number uses 0000.

The range of a 16 bit Signed BCD number is then only -999 to +999!

Binary	Decimal	HEX
0000	0	0
0001	1	1
0010	2	2
0011	3	3
0100	4	4
0101	5	5
0110	6	6
0111	7	7
1000	8	8
1001	9	9
1010	10	A
1011	11	B
1100	12	C
1101	13	D
1110	14	E
1111	15	F

Figure 65 - Hexadecimal

Hexadecimal: In order to display the full 65,536 possible values in sixteen bits in only four characters, a base 16 numbering system called Hexadecimal is used. The only reason the base 10 Decimal system is used is that humans calculate best in this format; all because we have 10 fingers and 10 toes. As mentioned before, computers are most efficient when they calculate in multiples of two.

After a group of four binary digits (a "**Nibble**", or half of a Byte) reaches 1001, the next value of 1010 can't be expressed as a numeral without using something outside of the values 0-9. In order to describe a base 16 number using 16 different symbols, the symbols use the letters A-F as shown to the left after the number 9.

Because Hexadecimal is base 16 and a multiple of 2, it is very easy to convert from Binary to Hexadecimal. Simply separate the binary into groups of four and convert each group.

Octal: Base 8 numbers are also commonly seen in computerized systems. For instance, Siemens I/O addressing is octal; this means that only the numbers 0-7 are used. The numbers after 7 would be 10, 11 …17, 20, 21 and so on.

Like Hexadecimal, because Octal is a multiple of 2 it is very easy to convert to and from Binary. Separate the binary into groups of three and convert each group; the highest value in any group of three is 111, or 7.

Exercise 12

1. Convert the Binary number "0110_1100_1011_0111" into…

Decimal _____   Hexadecimal _____   Octal_____

Can this number be converted to BCD? Why or why not?

_____

_____

2. How would you write the Binary number "1001_0101_1000_1001" as a Signed Integer?

_____

3. Convert the Decimal number 417 into:
   BCD _____   Binary _____   Hexadecimal _____

4. Write the number 2A9E in Binary :

   _____

   What is this number in Decimal?

   _____

5. How many Bytes are in a Double Integer?

   _____

Input Devices

Sensors

Sensors are devices that provide electrical signals to control systems to indicate some kind of status or event. They may provide these signals in digital or analog format.

One of the most basic type of sensors is a **pushbutton** or **switch**. Pushbuttons provide a digital one or zero signal and may be **toggled** on and off (**maintained**) or be **momentary**. Contacts may also be Normally Open (on when activated) or Normally Closed (off when activated).

Figure 66 - Pushbutton with NO (Green) and NC (Red) Contacts

Figure 67 - Normally Open and Normally Closed Pushbutton symbols

Switches may be as simple as an on/off signal or may have multiple positions, each corresponding to a different set of contacts. Like pushbuttons, they may be maintained or spring-return.

Industrial pushbuttons and switches are made of separate components; the **operator** or front part, and the **contact blocks** or electrical connections. Normally Open (NO) contact blocks are usually green, while Normally Closed (NC) contact blocks are red.

Operators may also have lights behind them that can be illuminated, this is very common with buttons.

Figure 68 - Switch and Symbol

Buttons and switches are available in standardized sizes. 30mm, 24mm, and 16mm are common sizes with the measurement indicating the diameter of the device. Hole punches designed for these are available from most electrical suppliers; Greenlee is a well-known brand. There is a picture of a set of Greenlee punches in the Tools section of this book.

Limit Switches are mechanical sensors that are used to detect objects by contact. These may have rollers or rods ("whiskers") attached that move when an object pushes on them. These also usually have both normally open and normally closed contacts.

Figure 69 - Limit Switches

Proximity switches are non-contact electronic sensors that operate in various ways. These are often called **"proxes"** for short.

Inductive proximity switches detect metal objects by using a magnetic field generated by a coil of wire. When a metal object enters the range of the sensor, an oscillator circuit inside the sensor stalls, activating a switching circuit. The best targets for these switches are steel or iron; aluminum objects reduce the range of the sensor by 50% or more.

Inductive proxes come in a variety of different shapes, from cylindrical threaded "barrel proxes" to flat packages. Barrel proxes are very common since they can be easily adjusted to come close to the target without touching it. Barrel proxes are available **shielded**, where the metal threads extend to the sensing face, or **unshielded**, leaving the side of the sensor coil uncovered. Unshielded proxes have a longer sensing range but are more susceptible to damage.

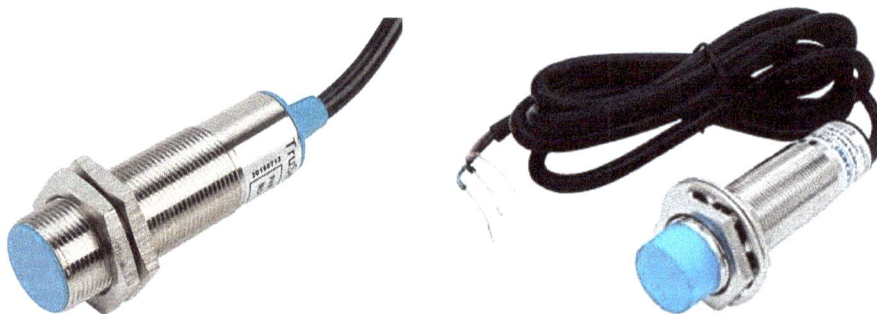

Figure 70 - Shielded (left) and Unshielded (right) inductive proximity switches

Hall Effect Sensors detect magnetic objects and are often used to detect the piston inside of a pneumatic cylinder. Usually these are mounted directly to the actuator.

Figure 71 -Cylinder mounted "proxes"

Like inductive proximity switches, hall effect sensors are commonly called "proxes" in industry. In addition to being used on air cylinders, they are often used in motor speed applications and current transducers. They are usually made of semiconductor materials and require threshold detection to act as a switch. Another device often used to sense cylinder position is a **Reed Switch**, which has a physical magnetized reed or lever that is actuated by the magnetic piston.

Inductive proxes and hall effect sensors can also provide analog outputs, this can be useful for short range distance and electric current measurement.

Capacitive proximity switches use capacitance to detect objects. Unlike inductive proxes, the object does not have to be metal. Capacitive proxes can be used to detect liquid through the side of a metal tank, or a finger placed on a capacitive pushbutton. In industrial applications they are common in level sensing and zero force ergonomic applications, i.e. "palm switches".

Touchscreens also use this capacitive principle to detect the presence and position of a finger or a stylus.

Photoelectric sensors use light to detect the presence or relative position of an object. There are three basic principles or types of these so called "**photoeyes**":

Through Beam: Light is emitted from one unit (emitter or transmitter) and received by the other (receiver). A switching circuit sends the output to the control system. While this is the most reliable method of sensing an object's presence with a photoeye, both units need power, making it the most expensive option. Outputs can be configured as "Light On" (output is on if light is detected, signifying the object is not present), or "Dark On" (output on indicates presence).

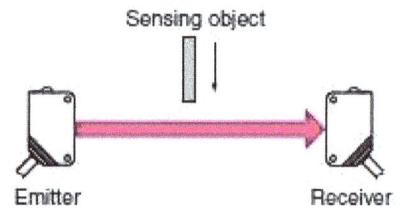

Figure 72 - Through-Beam

Retroreflective: Emitter and receiver are contained within the same unit. Light is reflected from a polarized reflector or reflective tape. Polarization shifts the light beam 90 degrees, which should ensure that light reflected from the object itself does not trigger the sensor. This type of sensor is not quite as reliable as a through beam photoeye but nearly so. It is less expensive since everything is contained in the same unit. Like the through beam sensor, it can be set as Light On or Dark On.

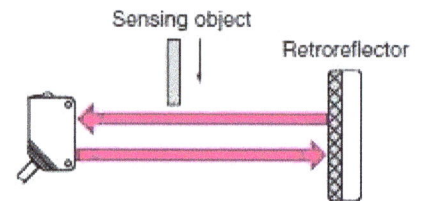

Figure 73 - Retroreflective

Diffuse: Like the retroreflective sensor, the emitter and receiver are built into the same unit, but unlike it, there is no polarized filter on the receiver. Light detected is reflected from the object itself, signifying its presence. This makes the result of the Light On/Dark On switch the opposite, Light On indicates presence. Diffuse photoeyes are not as reliable as the other two configurations, the amount of light reflected back from the object varies depending on the size and color of the target.

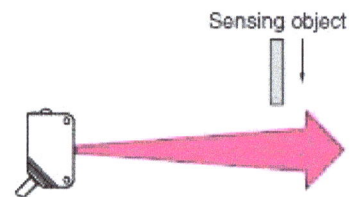

Figure 74 - Diffuse

The color of the light transmitted by the emitter can affect the range of the sensor, especially with a diffuse sensor. Red targets appear as white and reflect more light than green targets with a red light. Infrared light is also often used to provide a longer sensing range, this can be problematic since it is not visible.

Diffuse sensors may also provide an analog output that varies with the amount of light received. This can be used for color or distance detection.

The sensors listed here may have individual wires that can be connected to terminals (**"flying leads"**), or they may have standardized plugs that connect to cables. These plugs are referred to as quick disconnect cables and sensors, and come in specific sizes and configurations.

Micro quick disconnect plugs are also referred to as **M12** connectors as they have a 12mm locking thread. They are designed for washdown and corrosive environments and usually have 3-5 pins for standard sensors. They are made in both straight (axial) or right angle (90 degree) connections.

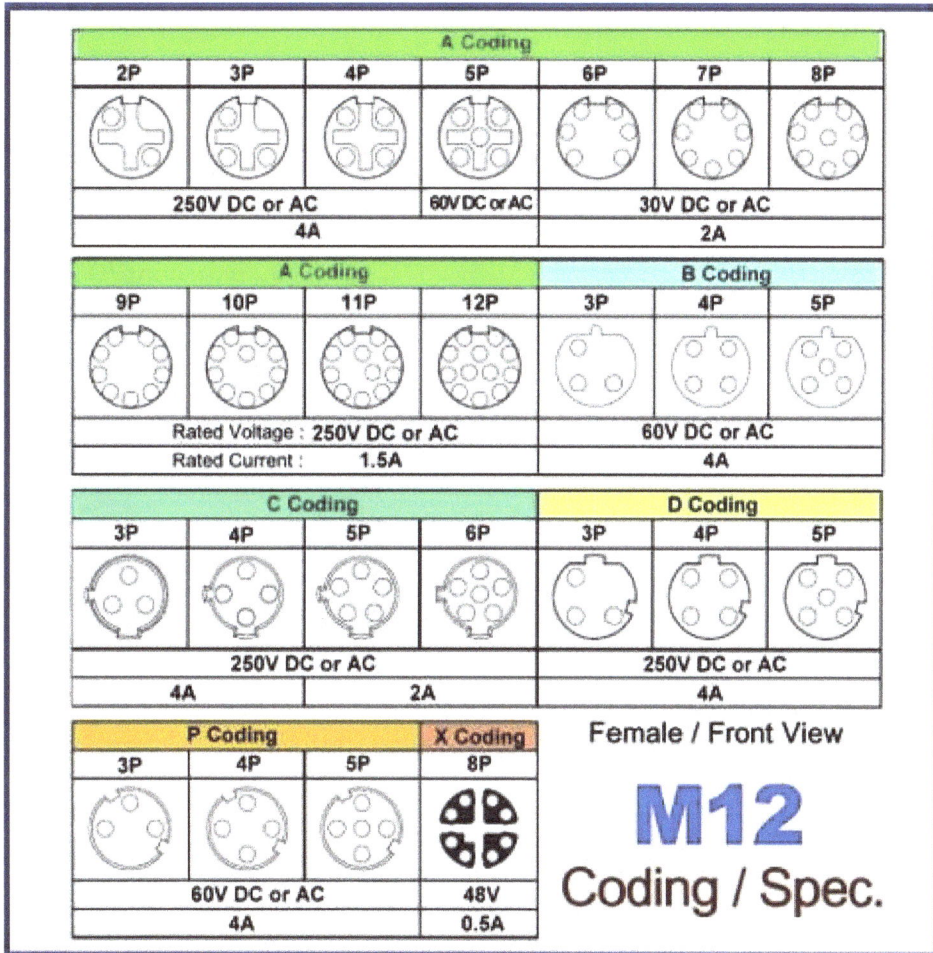

Figure 75 - M12 connectors (Courtesy of Taiwan Trade)

Mini quick disconnect plugs have an **M8** connection and are smaller than the micro or M12 size. They also usually have 3-5 pins.

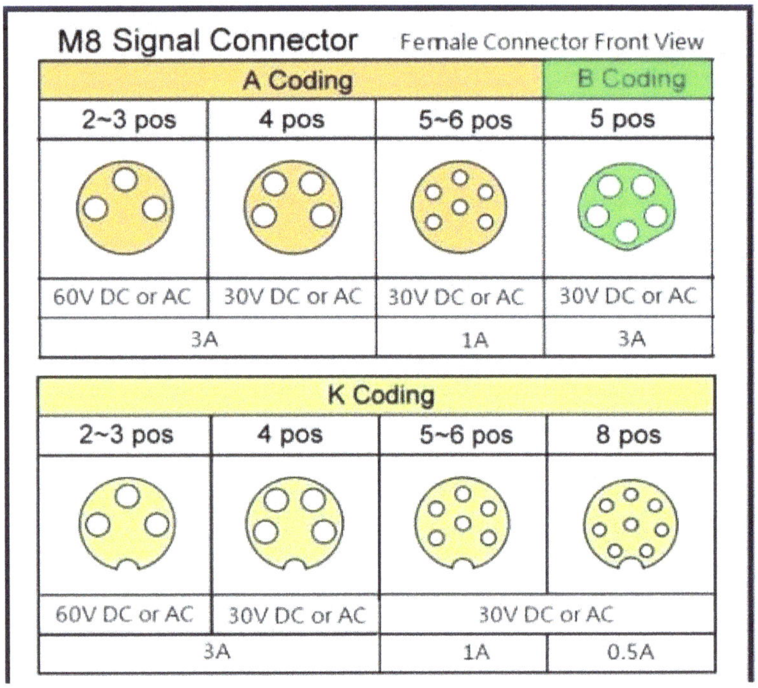

Pin arrangements can differ and cables may also be keyed. Cables may also have more pins for special purpose devices.

Wire colors may vary depending on the manufacturer, but typically brown wires connect to 24VDC and blue wires connect to 0V or ground/DC common. Signal wires can be black or white depending on the sensor.

Figure 76 - M8 Connectors (Courtesy of Holin)

Exercise 13

1. What is the simplest form of discrete input device?

_____

2. What type of sensor must make physical contact with a part to detect it?

_____

3. Why would an unshielded prox be used instead of a shielded one?

_____

4. What is the most reliable type of photoelectric sensor?

_____

5. What color is the wire that usually connects to +DC voltage in a sensor cable?

_____

Discrete Sensor Wiring – Sink/Source

Sensors and devices may be AC or DC, or may provide a contact closure (relay) so the user can switch the voltage of their choice.

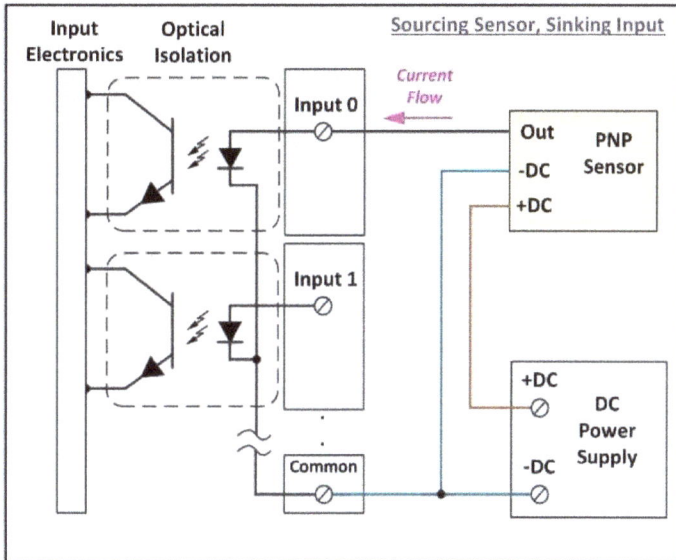

Figure 77 - Sourcing Sensor

An important concept in sensor selection and interfacing is that of **sinking** and **sourcing**. When using 24VDC sensors, the device can either provide a positive DC voltage (source) to the control device, which then sinks the current, or the control input can provide a positive DC voltage back to the sensor, which sinks the current. The important factor in these connections is that sourcing devices interface with sinking control inputs, and sinking devices work with sourcing control inputs. Sourcing devices are PNP, while sinking devices are NPN.

Some devices have both types of electrical interface (black and white wires), and others only have one, so it is important to ensure that the sensor is correct.

Most machines use sourcing sensors, but many machines from Japan use sinking sensors with sourcing PLC cards. PLC cards can usually be wired as sinking or sourcing by applying the correct polarity to the common terminal on the card, but some cards are dedicated to one configuration.

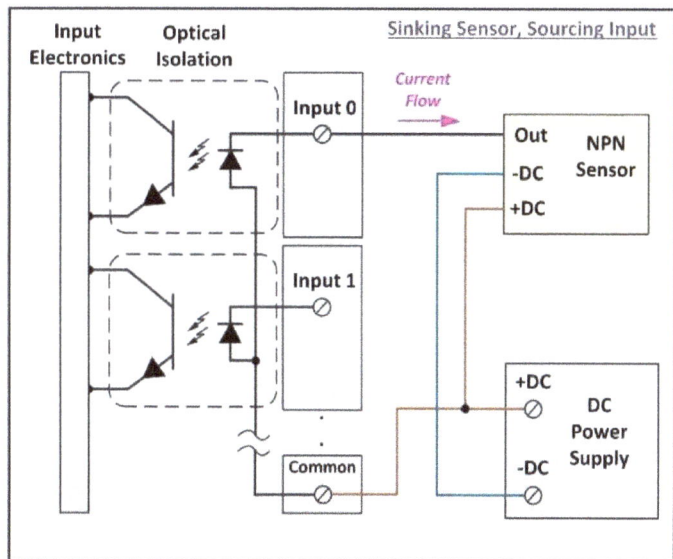

Figure 78 - Sinking Sensor

Special Purpose Sensors

There are some types of sensors that don't fit the standard categories listed previously, either because they don't use standard digital or analog outputs, or they require signal conditioning.

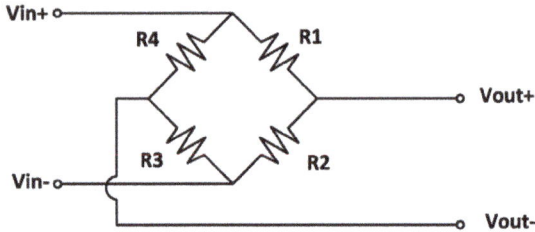

Figure 79 - Wheatstone Bridge load cell circuit

Load cells and **strain gauges** are made of materials that change their resistance when deflected. These are usually made of a metal alloy, but quartz crystal is sometimes used. An electrical excitation voltage is applied and a resistive network called a Wheatstone bridge is used to create a voltage change. R1-R4 in Figure 78 are strain gauges interconnected in a loop as a load cell so that the Vout signal changes as force is applied. The mV level voltage change is then amplified to a linear 0-10V signal or to 4-20mA.

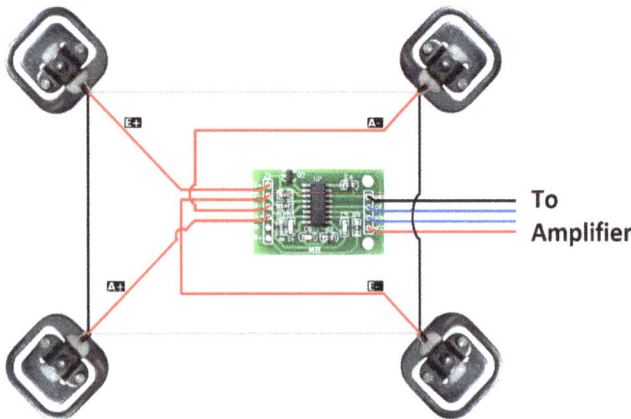

Figure 80 - Load Cell Scale

Load cells are used to weigh objects and are often connected to a summing circuit that accepts multiple inputs. The summing circuit is then connected to an amplifier. In addition to weighing objects, load cells can be used to determine fluid tank levels if the density of the liquid is known. 3 or 4 load cells are often used at the base of the object to be weighed as shown in Figure 80.

Thermocouples are non-linear junctions made of different types of metal. There are several combinations of dissimilar metals that create a characteristic temperature curve that are commonly used in industrial automation; these are used based on the desired measuring range and the environment in which they are located. Thermocouples create a micro-voltage that matches a curve built into the device that is used to amplify the signal. Because of this, the device needs to be set to the thermocouple type.

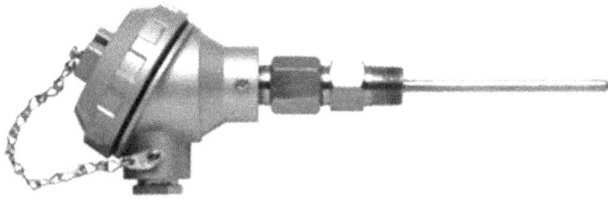

Thermocouples are often made in the form of a probe with a terminal head or "thermowell" as shown in Figure 81. The probe serves as a protective sleeve for the thermocouple junction.

Figure 81 - Thermocouple Probe

The most commonly-used thermocouples are made of nickel alloys. Listed below are some of the common types and their temperature ranges:

Type E: (chromel-constantan) -270 °C to +740 °C. Often used in cryogenic applications.

Type J: (iron-constantan) -40 °C to +750 °C. Very common in industrial use

Type K: (chromel-alumel) -200 °C to +1350 °C. Very common in industrial use.

Type T: (copper-constantan) -200 °C to +350 °C.

Because thermocouples are often used in high temperature and reactive atmospheres, they can deteriorate over time. It is important to monitor the device for errors on a regular basis.

Connecting thermocouples to the controller or amplifier should be done using terminal blocks specific to the thermocouple type or by using extension wire of the same category.

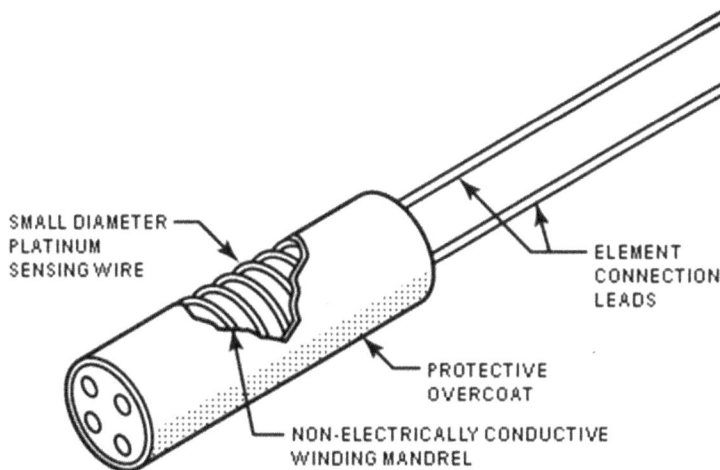

RTDs (Resistance Temperature Detectors) are also used for temperature detection at temperatures below 600 °C. These are usually composed of platinum, nickel or copper wire and require an amplifier with a power source.

Figure 82 - Platinum RTD (RdF Corp.)

RTDs are considered to be linear within their specified range but must still be calibrated. Calibration is usually done with ice baths or secondary instruments.

RTDs are more accurate than thermocouples within the same range but react more slowly. If process temperatures are between -200 °C and 500 °C, industrial RTDs are usually the preferred option over thermocouples.

Calibration of temperature sensors requires a known, good reference device that can measure the temperature at the typical setpoint of the process. It is also sometimes done at the triple, freezing, or boiling points of N2, water, or oils, or at the melting point of metals. An offset is then often applied to the scaling of the device.

Scaling is the process of converting a signal into a number that describes a physical property such as temperature, pressure, speed or weight. Raw signals are generally based on voltage, current, or pulse counts, and are brought into a controller in their raw form. Raw data is then converted mathematically to user units, often called **Engineering Units**.

Signals are often linear and follow the equation for a line, $Y=(m*X)+B$, where **Y** is the units of the Y axis and **X** is the units of the X axis. **B** is known as the **Offset**, while **M** is the **Scalar**, determined by dividing the "rise", or increase of the Y axis, by the "run", or increase of the X axis.

As an example, look at the graph in Figure 83:

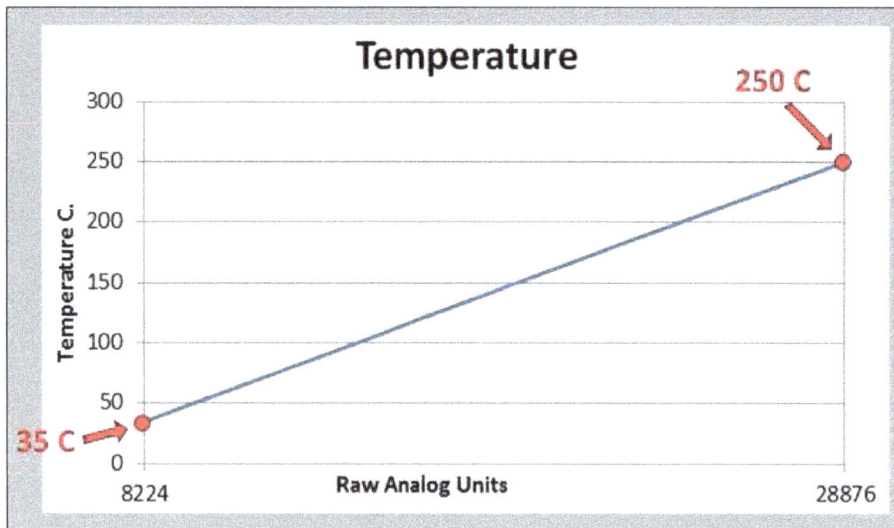

A temperature sensor produces a 0-10v signal. This is wired into an analog card which produces a signal that ranges from 0-32,767, a signed integer.

Figure 83 - Scaling Graph Example

A thermometer is used to measure the actual temperature at two different points and the raw value is recorded for each measurement; the first point P1 is recorded as 8,224 on the analog card at 35 degrees C, while the second (P2) is recorded as 28,876 at 250 degrees C.

The first step in scaling the raw measurements into degrees is to calculate M, the Scalar. The "rise" or difference in Y values is Y2-Y1, 250-35, or 215. The "run" or difference in X values is X2-X1, 28,876-8224, or 20,652. Dividing the rise by the run produces a Scalar M of 0.01041061.

The next step is to calculate the offset B. Since Y=mX+B, the B factor can be calculated as B=Y-MX. Substituting the values for P1, which were Y1 and X1, the calculation becomes B = (35-(0.01041061*8,224)), which yields an offset of <u>-50.6168894</u>. These two constants, M and B, can be used to calculate Y for any inserted value of X.

As an example of how to use this formula in calculating a temperature, assume that the temperature sensor reads a value of **10,512** into the analog card. If the formula y=mx+b is used, the temperature is (0.01041061*16,512) – 50.6168894, or **121.28 degrees C**.

Another way to describe this formula that is often used in PLCs is to use the following:

Scaled Output = Raw Input * ((EU Max-EU Min)/(Raw Max-Raw Min)) – (Raw Max * (EU Max – EU Min/(Raw Max-Raw Min)) – EU Max)

Where EU Min and EU Max are the minimum and maximum values for engineering units, and Raw Min and Raw Max are the minimum and maximum of the raw signal.

For example: The raw input of a signal is 12 mA, the range raw min and raw max are 4mA and 20mA. We want to convert the signal to a percentage, so the engineering min and max are 0% and 100%.

12 * ((100-0)/(20-4))-(20*((100-0)/(20-4))-100) = 50, indicating 50%

Not all sensors produce a linear output. Thermocouples, as mentioned previously, are non-linear and the output is in micro or millivolts. There is a characteristic curve that is used in the controller to produce the output.

Encoders and resolvers produce a linear relationship between pulse count or frequency and speed, so the offset B will be zero in the formula.

Linear Variable Differential Transformers, or **LVDTs** are used to measure position or distance. They use a series of three solenoidal coils wrapped around a tube; this slides along a rod that can be up to a few inches or several hundred millimeters long. An alternating current at 1-10kHz frequency is applied, which generates a linear voltage change along its path. Like load cells and RTDs, this requires an amplifier to generate a 0-10V or 4-20mA signal proportional to

Figure 84 - LVDT Probes

distance. LVDTs are very accurate over their usable range; they can be used as individual measurement probes or for position feedback in servomechanisms. Figure 84 shows a cluster of LVDT probes used in a gauging application.

Because the sliding core does not touch the tube, there is no friction. LVDTs are also **Absolute** devices; when the power is removed and reapplied, the same voltage will be present so the positional information is not lost.

Encoders produce pulses relating to distance or rotation of a motor. Because the pulses switch very rapidly, standard PLC input cards react too slowly to capture them. Special cards or devices called **High Speed Counters** are used to accumulate counts.

Rotary encoders usually use a metal or glass disc with slits and a photodiode to create pulses as the disc rotates. There are two versions of these used in industrial applications: **absolute**, which maintains position information when power is removed, and **incremental**, which simply send a fixed number of pulses for every rotation.

Encoders are made with different resolutions or counts per rotation. Resolutions may be anywhere from 10 to 5000 pulses per revolution, abbreviated as ppr. Voltage and output requirements may be anywhere from 5 to 30 volts DC.

Quadrature

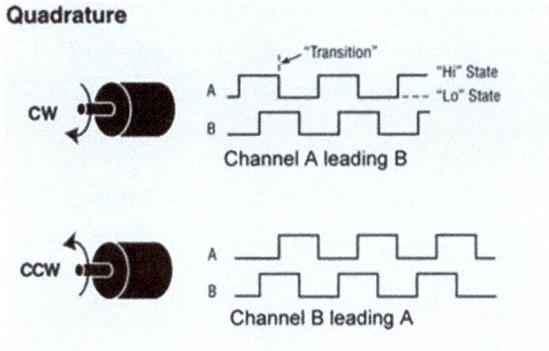

Figure 85 - Quadrature Encoder

Incremental encoders have two output signals, **A** and **B**, which are out of phase by 90 degrees. The direction of rotation or the encoder shaft can then be determined by whether A leads B, or B leads A. This arrangement is called **Quadrature**. A single index pulse is often present for homing purposes, this is labeled **Z**, also known as the **reference**.

An inverse A and B pulse is also sometimes provided, referred to as /A and /B. This is sometimes called a **differential pair** or **complementary outputs**. This can be useful for error checking with higher speed applications, where switching can be fast.

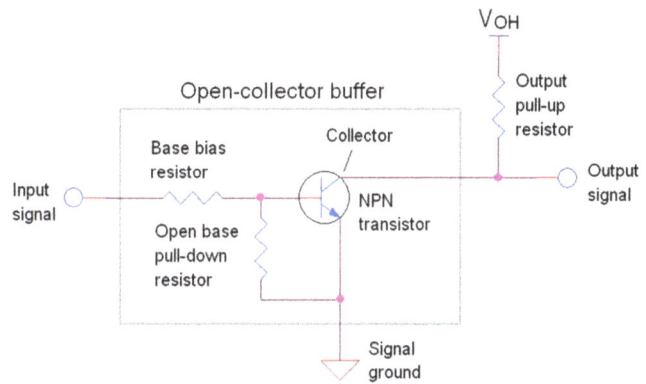

Figure 86 - Open Collector encoder interface

Decimal	Binary	Gray	Gray Decimal
0	0000	0000	0
1	0001	0001	1
2	0010	0011	3
3	0011	0010	2
4	0100	0110	6
5	0101	0111	7
6	0110	0101	5
7	0111	0100	4
8	1000	1100	12
9	1001	1101	13
10	1010	1111	15
11	1011	1110	14
12	1100	1010	10
13	1101	1011	11
14	1110	1001	9
15	1111	1000	8

Figure 87 - Binary Gray Code

Other types of outputs employed by encoders include **open collector**, which sinks current from the high-speed counter. A pull-up resistor is used with this type of output, some encoders provide this internally.

Push-pull outputs are TTL or transistor level signals that interface directly with solid state logic circuitry. This can only be used when the encoder is located very close to the circuit or uses a short, shielded cable.

Absolute encoders produce combinations of bits that represent absolute positions on the rotation. This requires more wires as resolution increases, up to 12 signals for a 4096-count encoder. These may produce all of the positions in a single rotation (single turn) or over the course of multiple rotations (multi-turn). Multiturn encoders use an internal gearing system to do this.

Absolute encoders have multiple code rings with different binary weights that provide a data word representing the position of the rotation. A problem with using standard binary number combinations to represent the position is that more than one signal can change at the same time. If the slots or pulses don't line up correctly the position can be represented incorrectly and be off by many counts. For this reason, **Gray code** is sometimes used, ensuring that only one digit changes at a time, as shown in Figure 87.

The advantage of absolute encoders over incremental is that if power is lost, the positional information is maintained. There are also battery-backed versions that do this electronically.

Linear encoders use a transducer or reader paired with a scale that encodes position. There are both incremental and absolute versions of these; technologies include optical, magnetic, inductive, capacitive, and eddy current. In addition to producing pulses like rotary encoders, the output may be an analog signal.

Resolvers are rotary sensors that produce a sinusoidal signal representing the position of the shaft. An AC signal is applied to the stator (R), and two two-phase windings (S1 and S2) 90 degrees apart are used to receive a sinusoidal signal from the rotor. The rotor acts as the secondary of a transformer; as it revolves it induces a signal back into the S1 and S2 windings.

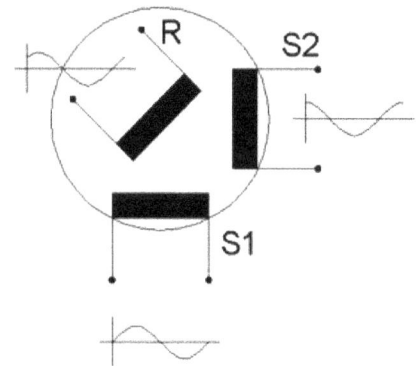

Figure 88 - Resolver Windings

Since the S1 and S2 windings are 90 degrees out of phase, the signals can be compared at any point to determine the position of the rotor.

$$a = \frac{VS1}{VR}$$

$$a = \frac{VS2}{VR}$$

0°   180°   360°

Rotation angle $\alpha \rightarrow$

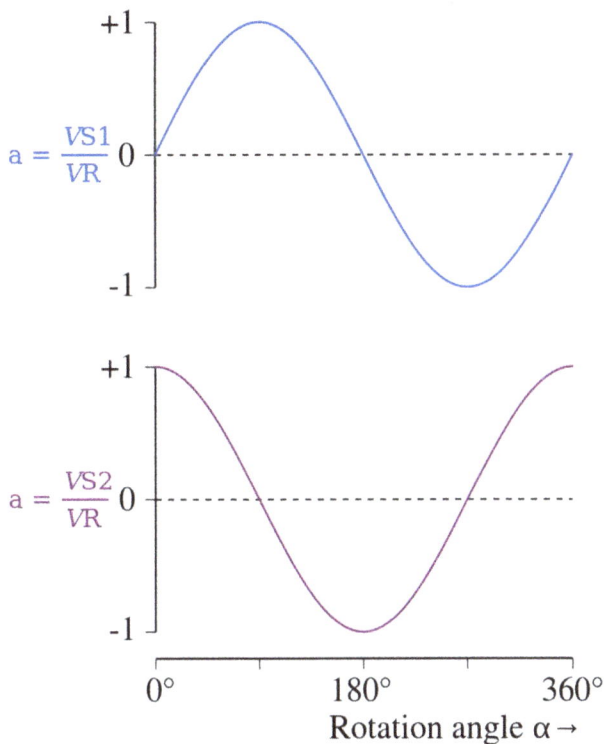

Figure 89 - Resolver Signals

The signals resulting from the S1 and S2 windings act in a sine-cosine relationship. These signals are converted to digital values using Analog to Digital Converters (ADCs) from 10 to 16 bits in resolution.

Resolvers are considered to be less accurate than encoders but are also generally less expensive to manufacture. They are commonly used for speed control of AC and DC servo motors. They also survive shock and vibration better than encoders.

Other measurement methods: Linear encoders are sometimes used to count pulses on a printed scale. Other techniques include magnetic, inductive, capacitive, and eddy current. They are used in motion systems, high precision machining tools and CNC machines.

Laser measurement is sometimes used for measuring distance but is not accurate to sub-millimeter distances like some of the other methods listed.

Measurement of physical distances and rotation is pretty simple to troubleshoot and calibrate. Since the measuring device can be moved physically, simply compare the devices movement with the physical distance or rotation to determine if it is functioning correctly.

Electrical noise is a common cause of inaccuracy in measurement. Ensure cables are properly shielded (terminated on one end only!) and that the shield is grounded. Loose or broken physical linkages between the instrument and tooling is another possible source of errors. Check rotational couplings carefully.

Barcode readers and scanners are optical devices used to read printed or stamped codes, decode the data, and send it to a computer. They consist of a light source, a lens, and a light sensor that captures the image for analysis.

Laser scanners use a laser beam as a light source and either a reciprocating mirror or a rotating prism to scan the beam into a line across the barcode. These are used to read single dimensional (1D) or line-type barcodes. The line and space thicknesses are encoded into different formats, so the decoding software needs to know which code type is being used.

Common 1D barcode formats include Pharmacode (pharmaceutical packaging), Code 39 (common in industrial use), UPC or Universal Product Code (common in retail), and ITF-6, used to encode additional data such as item quantity or container weight. There are many other formats, including various postal codes for different countries, ISDN numbers on books, and railroad equipment.

Figure 90 - 1D UPC Barcode

There are many factors that affect the readability of barcodes. Contrast and size of the bars and spaces, distance of the reader from the code, and scratches or marks on the barcode are just a few of these. It is common for producers and users of barcodes to have a quality management system that includes verification and validation of the code. This can include rating the readability of the code and comparing the data against a database.

Figure 91 - Standard Data Matrix (left) and QR code (right)

2D or **data matrix** barcodes are also common in industry. **QR codes** are a common data matrix type barcode used in industrial applications.

2D barcodes are read by using an optical scanner, or array of pixels. Unlike 1D scanners, optical scanners have no moving parts. This makes them more robust, and they can also read 1D codes.

Machine vision systems can also read both 1D and 2D barcodes; more will be covered on this subject in a later section. The cost of "smart sensors" that do simple inspections is often lower than that of bar code readers.

Scanners can either be hand-held or machine mountable. They can be classified into three categories of communication: the older RS-232 types, which needed special programming to transfer input data to the application, PS2/AT keyboard types which plugged into a keyboard port using an adapter, and USB scanners, which plug into a USB port. The last two use a **keyboard wedge**, which transfers data as if it was typed into a computer keyboard.

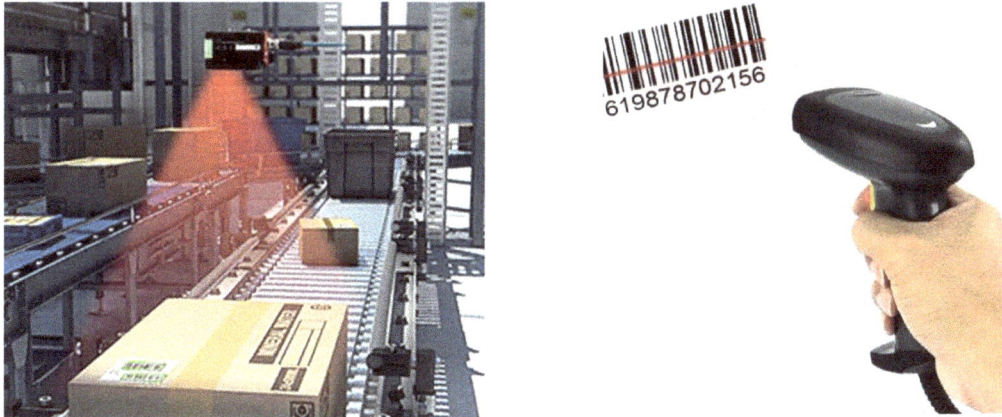

Figure 92 - Mounted and Handheld Barcode Scanners

Radio-frequency identification (RFID) scanners use electromagnetic fields to automatically identify and track tags attached to objects. An RFID system consists of a tiny radio transponder, a radio receiver, and a transmitter. When triggered by an electromagnetic interrogation pulse from a nearby RFID reader device, the tag transmits digital data, usually an identifying inventory number, back to the reader.

Passive tags are powered by energy from the RFID reader's interrogating radio waves. Active tags are powered by a battery and thus can be read at a greater range from the RFID reader, up to hundreds of meters.

Unlike a barcode, the tag does not need to be within the line of sight of the reader, so it may be embedded in objects like pallets or painted over.

Sensor Troubleshooting

Discrete sensors provide electrical signals to some type of control device; usually there is an indicator both on the sensor and on the input of the controller. The one on the controller is the most important, since the light going on and off indicates both the status of the sensor and the integrity of the wiring between the sensor and the controller.

Devices such as photoeyes and limit switches can be tested by placing an object in front of the eye or manually moving the lever on the switch. Inductive proximity switches require a piece of metal (preferably iron or steel) to be placed in front of the sensor face, while Hall Effect sensors require a magnet. The cylinder piston itself has a magnet inside, so the sensor should switch as the piston is moved.

Capacitive sensors can be triggered with moisture containing objects such as a hand or, if reading through the side of a tank, by placing it where the liquid level is.

Analog sensors provide signals as voltage or current. Portable testers, sometimes battery powered, can be used to read signals from sensors. Reading the value from the sensor in the software of the controller or from an HMI is often the quickest way to test it, but this is often not possible without the system running. 0-10vdc and 4-20mA signals are scaled inside of the controller to provide meaningful values such as pressure, position or temperature, incorrect values could indicate a need for calibration.

Encoder pulses can be read with a multimeter, but there are also lights on the highspeed counter cards that show the status of the channels.

Troubleshooting of temperature measuring devices usually involves comparison of the displayed temperature with a known reference and checking the connectivity of the wiring. When replacing wiring terminals for thermocouples, be sure to uses thermocouple wire and junctions! Using different metals can create a new thermal junction that disrupts the curve.

Thermocouple wire can be identified by the jacket or insulation color. Unlike many wiring schemes, the red wire is often the negative side.

Testing barcode scanners involve the following:

- Checking the code for legibility.
- Checking the cable for damage.
- Checking scanner indicators for power.
- Checking handheld scanner charge or batteries.
- Having a known good sample to test the reader with.

Exercise 14

1. What type of device is a PNP sensor? (sinking/sourcing)

_____

2. Can load cells and LVDTs be connected to PLC analog inputs?

_____

3. What is required to connect an encoder to a PLC?

_____

4. Do incremental encoders hold their count value when power is removed?

_____

5. Can copper wire be used to extend the length of a thermocouple? _____

6. What are common causes of error when measuring distance or rotation?

_____

_____

7. What are two types of 2D bar codes?

_____

8. What are the two types of RFID tag?

_____

Instrumentation

In the process control industry, devices that are used for indicating, measuring, and recording physical quantities are collectively known as **instrumentation**. These can be as simple as level gauges on fluid tanks and direct-reading thermometers or complex multi-sensor systems. Sensors may be connected to simple **Transmitters**, which condition the signal and pass it to larger control systems, or to self-contained control units such as **Loop Controllers**.

Often instrumentation is used in **Closed Loop** control systems.

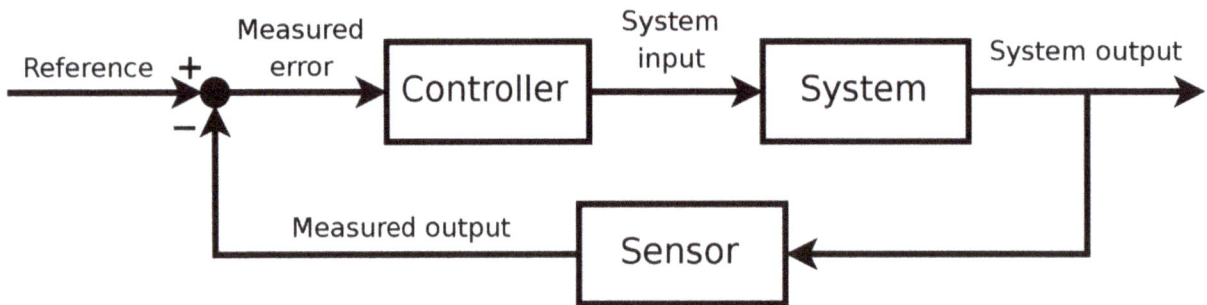

Figure 93 - Closed Loop Feedback

A closed loop system uses feedback from a sensor to maintain the setpoint of a physical variable, such as temperature, flow, or pressure. As the feedback value increases, it subtracts from the output of the system. This feedback value is proportional to the error and may be applied positively or negatively. As an example, suppose a gas burner is used to heat an oven; as the temperature sensor value increases, the gas quantity is lowered until when the oven is at the correct temperature and that the gas is at the correct level to maintain it.

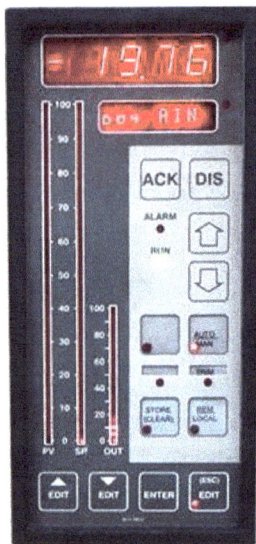

There are a wide variety of loop controllers available for this purpose. Some have analog outputs, to control devices like proportional valves, while others may pulse discrete outputs, to provide energy to electric heater coils. Some may be dedicated to a specific type of variable, such as temperature or fluid flow, while others may be generic, such as the Preferred Utilities unit shown in figure 94.

Figure 94 – PCC-III Loop Controller (Preferred Utilities)

Figure 95 - Temperature Controller (West)

Loop controllers can accommodate multiple signals for complex systems such as boilers, where air and fuel flow, temperature and water quality may all have to be controlled and monitored in the same controller. The PC III can be expanded to up to 15 analog inputs by means of added option boards.

Controllers like this can also communicate with the larger control system by using open protocols like Modbus or ethernet. They are programmed by using the buttons on the faceplate or by using software on a computer, making them very flexible.

Typical signals received from industrial transmitters are 4-20mA current loops, but various other data signaling methods are also used. Some sensors such as thermocouples are non-linear and must be connected to special devices like the single loop temperature controller in Figure 95.

Figure 97 - Flow Transmitter

Instruments like the **Flow Transmitter** in figure 97 can have more than one type of output: the typical analog 0-10vdc or 4-20mA signal and a high-speed pulse output signal like an encoder. Pulse outputs are more useful for determining quantities rather than continuous flow values.

HART (Highway Addressable Remote Transducer) communication protocol is sometimes used to transmit data, directly overlaying communications over the 4-20mA loop. This allows simple data transmission using the existing analog wiring of a system. Commands to the instrument and data such as the device address and status can be communicated in this way. **IO-Link** is another point-to-point technology used to transmit additional sensor or actuator data over a 3-wire sensor cable through a gateway.

Closed loop systems use more complex algorithms than the simple proportional feedback signal shown in Figure 93. An important concept to understand in closed loop systems is that of **PID (Proportional Integral Derivative)** control.

PID Control

Proportional control, as mentioned previously, calculates an error based on the difference between the setpoint (**SP**) and a measured process variable (**PV**), which might be temperature, flow or speed, or any other variable that requires control. This error is often referred to as e(t), a varying value over time that corresponds to SP-PV. The proportional term is often referred to as **Kp** in formulas.

The proportional control value or control variable (**CV**) is applied to a control device, such as a valve or heater. If the error is large, the control value applied to the device will be correspondingly (proportionally) large.

PID is sometimes called three-term control, because in addition to the proportional term P, there are two other terms: I (integral) and D (derivative).

The **Integral** or I term accounts for previous error values and "integrates" them over time. The I term contributes to reducing the error and is not always needed; usually it is used for unusual disturbances in the error. It is referred to as **Ki** in formulas.

The **Derivative** or D term estimates the future trend of the error by looking at the derivative or rate of change. It is sometimes called anticipatory control, the more rapidly the error changes, the greater the damping effect on the control variable. As with the I term, D is not always required, so PI or PD control is sometimes sufficient. The derivative component is referred to as **Kd** in formulas.

Balancing the three terms for the control system is called **loop tuning**. The values calculated will depend on the response of the whole system: the behavior of the measuring sensor, the control element (valve or heater, etc.), control system delays, and the process itself. Initial settings may be calculated or approximated, but the values are usually refined by "bumping" the values slightly to obtain the optimal response.

There are well known formulas for calculating parameters for PID loops: Zeigler-Nichols method, "good Gain", and Skogestad's method are a few. Below is a trial-and-error version that is sometimes used:

1. Predefine and set up a PV, CV, and SP. For example, assume PV is a temperature input 4-20mA signal (an RTD) downstream of the CV. The CV is a modulating analog steam control valve. The set point would then be the desired water temperature the PV would need to achieve by adding steam.

2. Set integral and derivative values to zero.

3. Start the process and adjust the proportional/gain tuning parameter until the PV starts modulating above and below the SP

4. Time a cycle or period of this oscillation. Record this time as the natural period or cycle time.

5. After timing and recording the cycle, reduce the P value to half of the setting needed to achieve the natural cycle.

6. Set the I parameter to the natural cycle. This will decrease the amount of time it takes for the PV to reach SP than with the P setting alone.

7. The D value may usually be safely set to approximately one-eighth of the I setting. This value helps with "damping" or controlling the overshoot of the process. If the process is noisy or dynamics are fast enough, the D value may often be left at zero.

While moving from the P setup to adding the I value, it is important to stop and restart the process. If the process remains running and attempts are made to adjust variables, the system will be looking at previous cycles and will not be a true indication of PID performance in a start-up scenario.

There are many other possible modifications to parameters in hardware and software controllers; here are just a few:

- Set point softening

- Rate before reset

- Proportional band

- Reset rate

- Velocity mode

- Parallel gains

- P on PV

Users should start with the standard PID form before setting these values. It is also usual to make sure the loop update time is 5 to 10 times faster than the natural period.

Servo systems and software programs usually have auto-tuning algorithms that use results from the input information, known system characteristics, and detected loads to approximate the initial PID settings. Many manufacturers preset the algorithm variables based on the hardware selected. Process control and servo system loops act very differently though, and a general knowledge of PID principles can be useful when setting initial parameters.

Exercise 15

1. What is a "closed loop" system?

_____

_____

_____

2. What types of output signal does a flow transmitter have?

_____

3. What do the letters P, I and D stand for in PID Control?

_____

4. What are the P, I and D terms referred to in formulas? _____

5. What should be done after setting up the P variable and before setting up the Integral?

_____

P&ID (Piping and Instrumentation Diagrams)

In the instrumentation and controls field, technicians use these diagrams to identify devices. Diagrams indicate the mechanical equipment such as pressure vessels, tanks, cooling towers, and piping, as well as control devices such as valves and pumps. Letters indicate the type of device, while numbers are used for locations.

Though used primarily in process control such as oil and gas or chemical plants, P&IDs are often part of the documentation in manufacturing equipment.

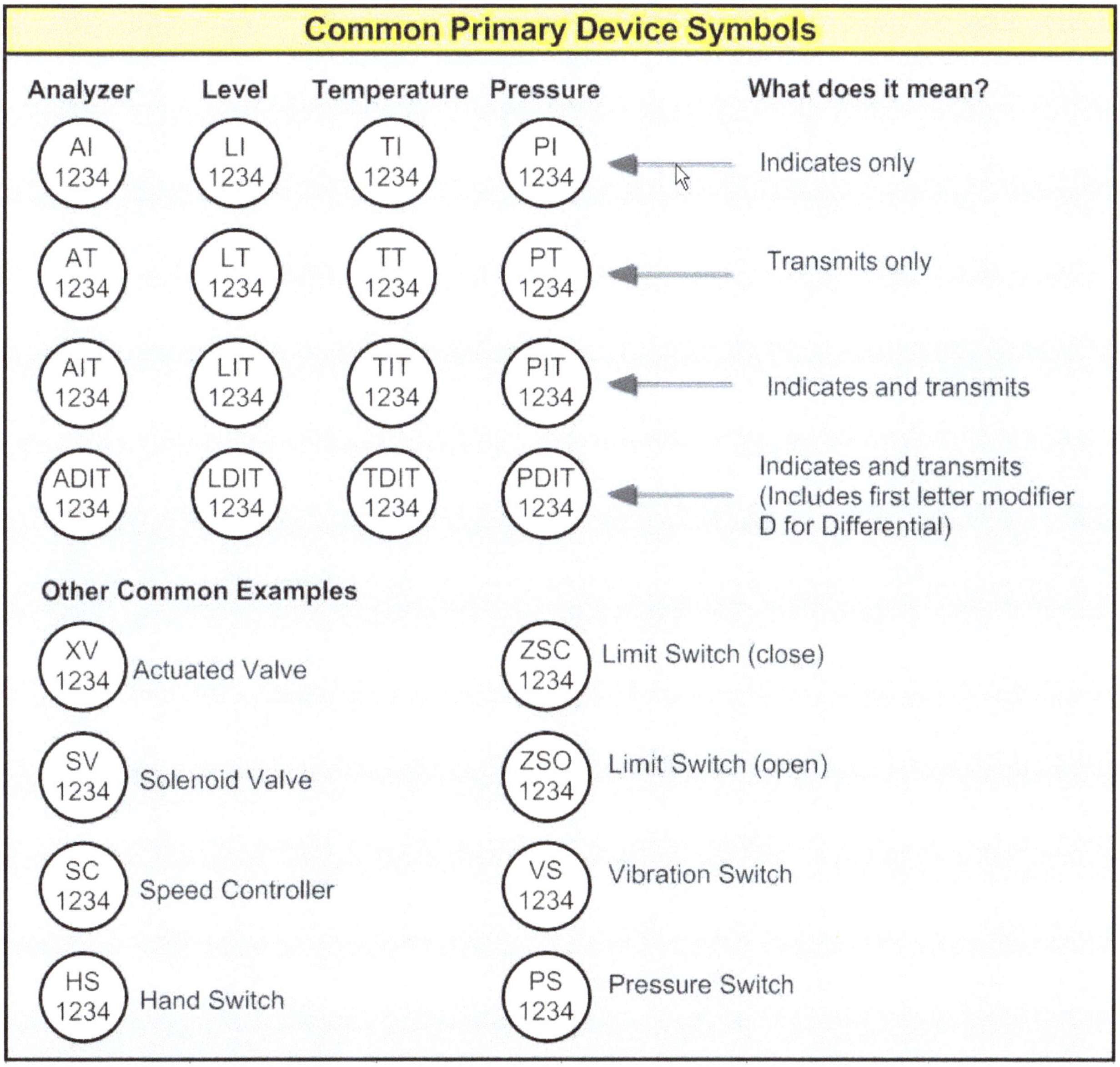

Figure 98 - P&ID Letter Designations

Symbols contain designations for the different types of instruments and devices in a process system as shown in figure 98.

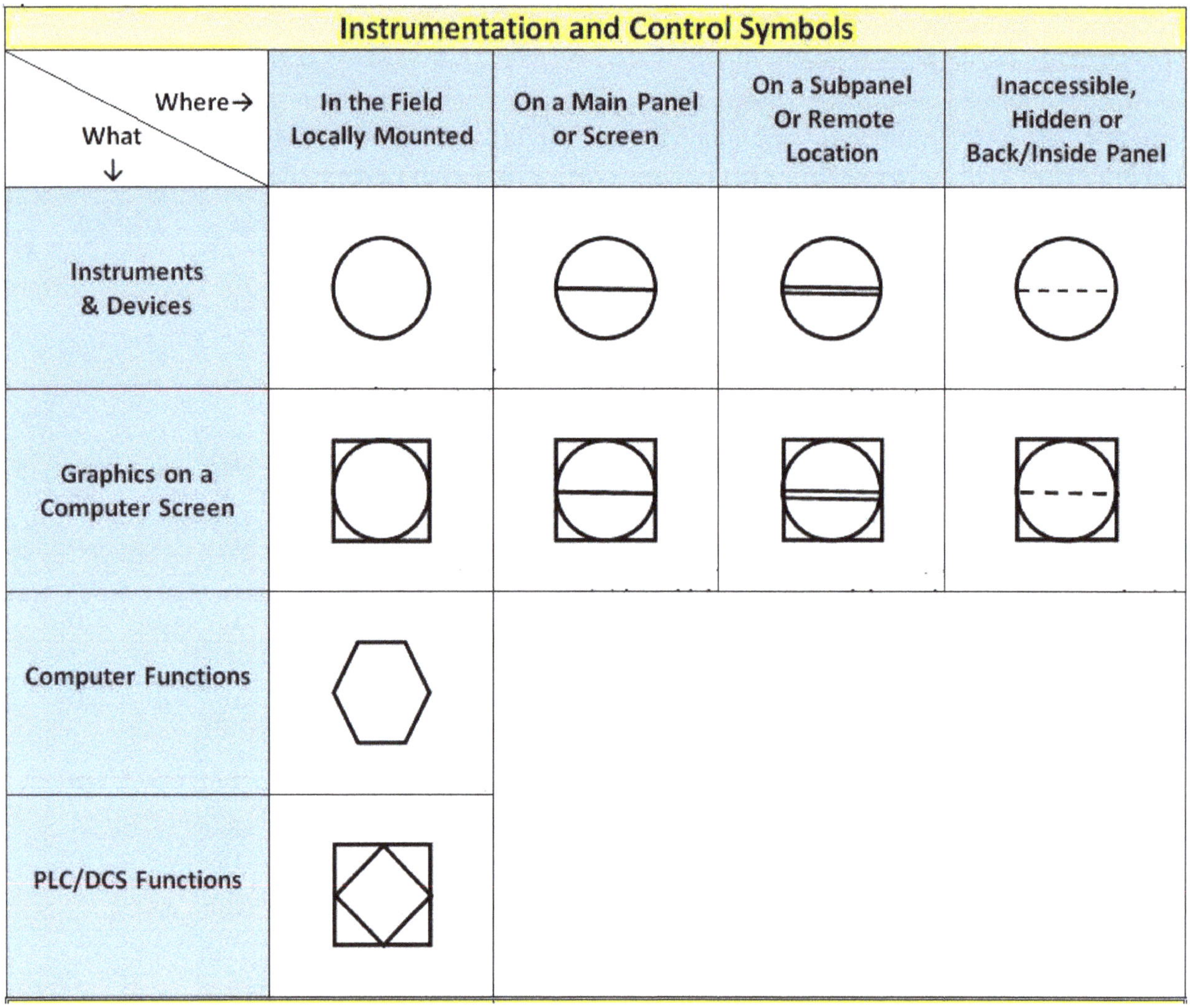

Instrumentation and Control Symbols

Where→ ⟍ What ↓	In the Field Locally Mounted	On a Main Panel or Screen	On a Subpanel Or Remote Location	Inaccessible, Hidden or Back/Inside Panel
Instruments & Devices	○	⊖	⊖	(⚬ dashed)
Graphics on a Computer Screen	⊡	⊡	⊡	⊡
Computer Functions	⬡			
PLC/DCS Functions	◇ in □			

Figure 99 – P&ID Symbol Designations (Modified from AIChE document)

The shape of the designation indicates the type of symbol and the type of area it is located in. As shown in Figure 99, some of these symbols are virtual devices or graphics.

The number designations identify the "loop number". They also often indicate the area, building, or skid where the device is located. This can differ based on the designer's decisions and company template. For example, **FIT8235** might be a **flow indicating transmitter** in **building 82**, **control loop 35**.

Additional shapes and standard ISO symbols are also used in P&IDs. An example of a Piping and Instrumentation Diagram is shown in figure 100.

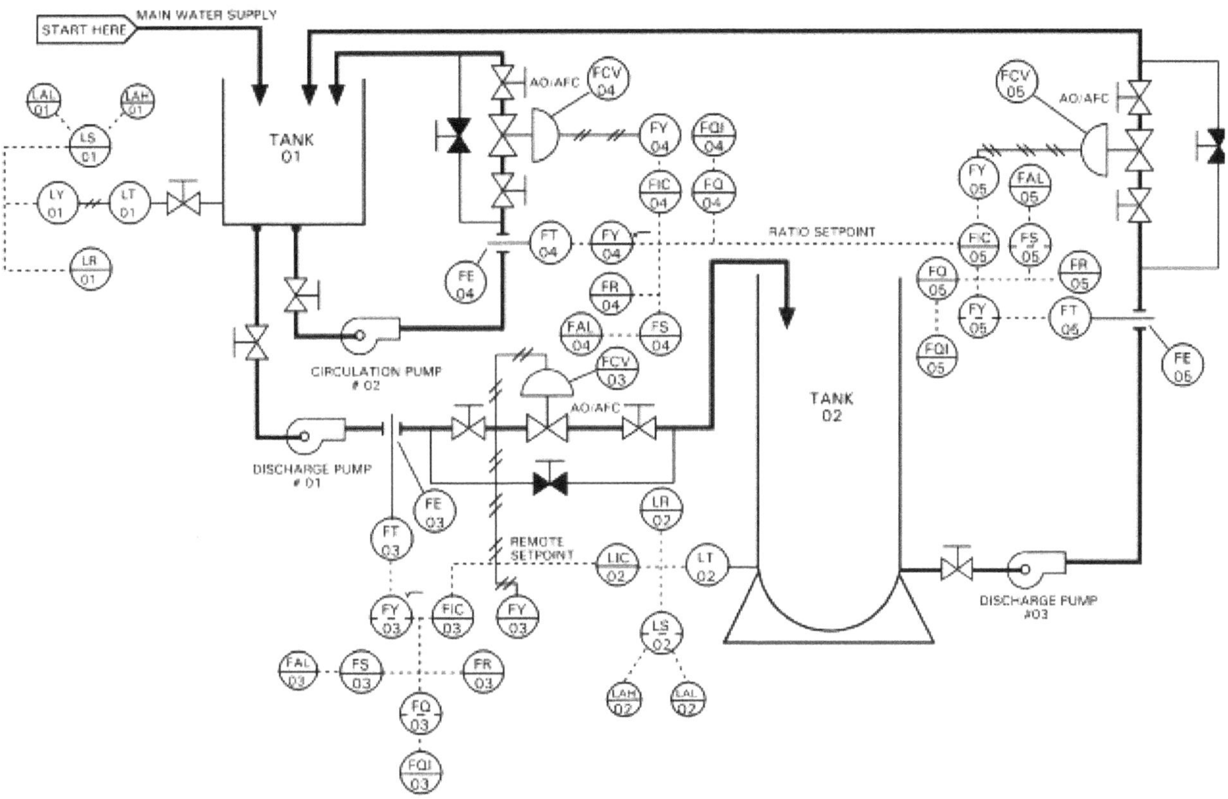

Figure 100 - Simple process P&ID

While this is a simple example of circulation between tanks, it shows detail of pumps, valves, and instruments in the system. Additional information on pipe sizes would also usually be listed here.

Figure 101 shows just some of the valve symbols used in P&IDs. There are hundreds more symbols and designations that may be found in these diagrams, and they can differ widely by company and process.

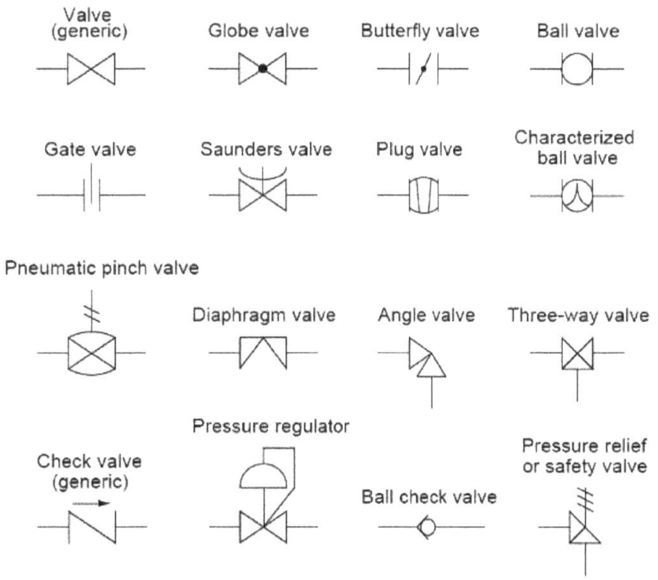

Figure 101 - Valve Symbols

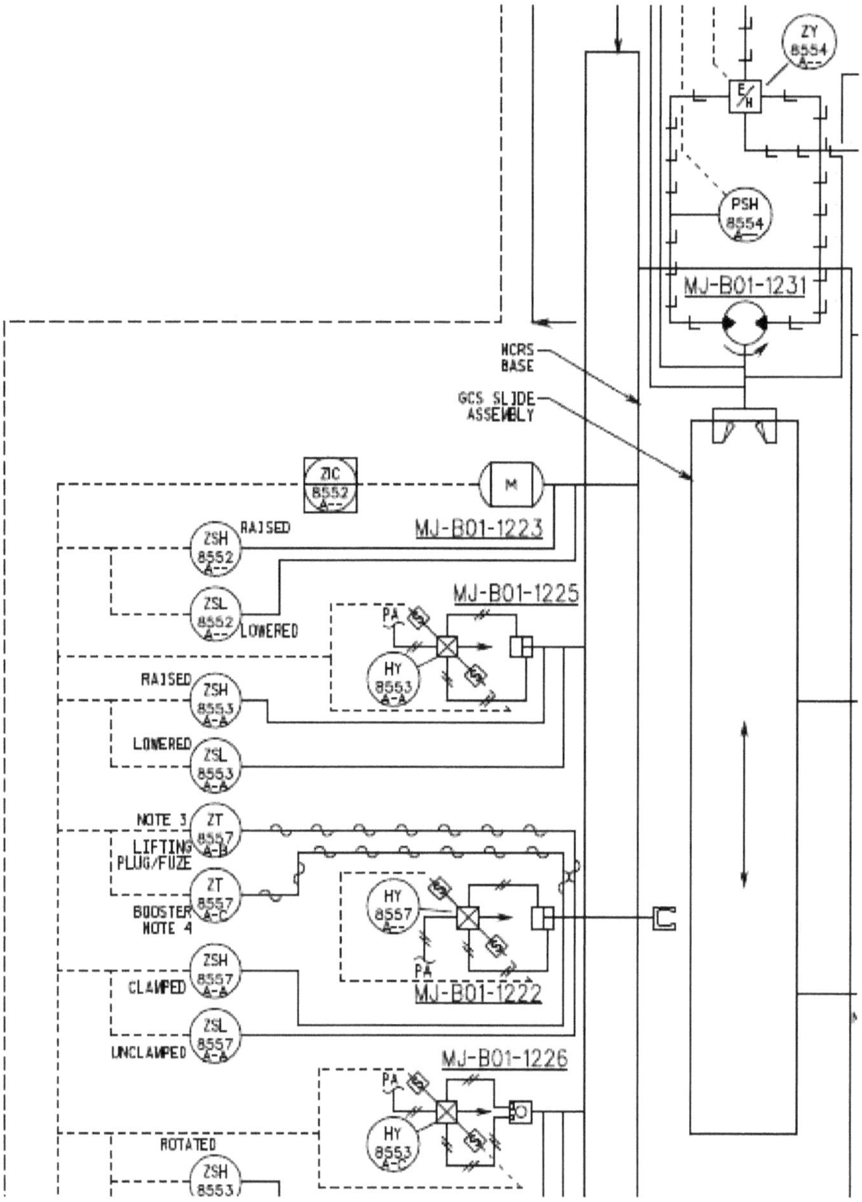

Figure 102 - Machine P&ID

The diagram in Figure 102 is a section of a mechanical machine rather than a process system. It shows the difficulty of representing actuators and assemblies with P&IDs. Some companies and government agencies require these as part of their documentation package, so learning to read and use them could be important, even if you are not involved in process control.

Safety Systems

Machine and process control isn't complete without a discussion of machine safety systems that prevent damage to personnel or equipment. These can address standards on procedures meant to prevent injury, regulations on machine guarding, and the circuits and components used in building a compliant safety system.

Machine Safety Standards

There are a wide range of standards directed at machine safety. A **technical standard** is an established norm or requirement regarding technical systems. It is usually a formal document that establishes uniform engineering or technical criteria, methods, processes, and practices.

Most standards are voluntary in the sense that they are offered for adoption by people or industry without being mandated in law. Some standards become mandatory when they are adopted by regulators as legal requirements in particular domains.

In general, ANSI standards are directed to manufacturers, integrators, and users of machinery.

ISO and EN standards are for suppliers of machinery, not users, unless users also have the role of supplier of industrial machinery. These standards allow movement of like goods into and within Europe. Most of these standards have also been adopted in North America.

OSHA standards provide requirements only to users (employers) for occupational safety, but can include responsibilities to employees (example, Lock-out Tag-Out).

Following are some of the definitions used in safety standards:

CCF – Common Cause Failure – Failures of different items due to the same event

Control Reliable – A concept stating that the safety system be designed, constructed, and installed such that the failure of a single component within the device or system should not prevent normal machine stopping action from taking place, but shall prevent a successive machine cycle from being initiated until the failure is corrected. To achieve "Control Reliability", a device should feature both redundancy and fault detection.

Dangerous Failure – A Failure with the potential to put the SRP/CS in a hazardous or fail to function state.

DC – Diagnostic Coverage – The extent of fault detection mechanisms

Failure – Termination of the ability of an item to perform its required function, i.e. "After the device failed, it was faulted". Failure is an event; fault is a state.

Fault – State of an item characterized by inability to perform its desired function, i.e. "Device is Faulted"

Harm – Physical injury or damage to health

Hazard – Potential source of Harm

MTTFd - Mean Time to Dangerous Failure – Reliability of components

Muting – Temporary automatic suspension of a safety function by the SRP/CS.

PL – Performance Level – defined in terms of probability of dangerous failure per hour

PLr – Required Performance Level

Risk Assessment - The combined effort of 1. identifying and analyzing potential (future) events that may negatively impact individuals, assets, and/or the environment (i.e., risk analysis); and 2. making judgments "on the tolerability of the risk on the basis of a risk analysis" while considering influencing factors (i.e., risk evaluation). A risk assessment analyzes what can go wrong, how likely it is to happen, what the potential consequences are, and how tolerable the identified risk is. As part of this process, the resulting determination of risk may be expressed in a quantitative or qualitative fashion. The risk assessment is an inherent part of an overall risk management strategy, which attempts to, after a risk assessment, "introduce control measures to eliminate or reduce" any potential risk-related consequences.

SIL – Safety Integrity Level – Numbered 1-4, 1 is the lowest SIL and 4 is the highest level of safety integrity

SRP/CS – Safety Rated Part of a Control System. The part that responds to safety-related input signals and generates safety-related output signals.

Structure Categories – 5 levels: B, 1, 2, 3, 4. Classification in terms of resistance to faults and subsequent behavior.

Switched Outputs – PLC or other controller outputs to actuators that are disconnected by the safety circuit. This is done by removing power from the output card's power terminal such as VDC.

Systematic Failure – Failure related to a cause that can't be eliminated without modifying the design, manufacturing process, operational procedures, documentation, or other relevant factors.

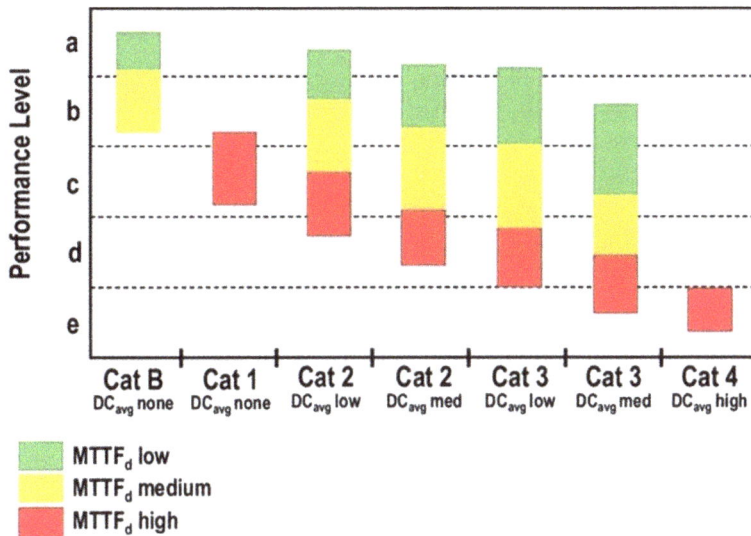

Figure 103 shows a performance level chart with safety categories. After performing a risk assessment of the machine or system, the requirements of the safety system can be determined, which then leads to selection of safety components.

Figure 103 - Safety Categories and Performance

Machine Guarding

There are many standards related to machine guarding that discuss a shield or device covering hazardous areas of a machine to prevent contact with body parts or to control hazards like metal chips or parts from exiting the machine. Physical guarding is the first line of defense to protect operators from injury.

Point guarding refers to the guarding of moving parts on a machine that present a hazard to the machine operator or others who may encounter the hazard. OSHA 1910.212 requires these guards to be "affixed to the machine where possible" and requires a tool to be used to remove it.

Fixed perimeter guarding refers to a barrier placed around a work area where an automated piece of equipment performs a function. This generally includes fencing or wire partitions, but may also take the form of monitored gates, pressure sensitive mats or light curtains.

Safety Devices and Architectures

Standards address types of guarding used on machines including electronic monitoring and control of safety circuits. This includes electrical interlocks and devices that stop equipment operation when activated.

Before 1995 and the implementation of the Machinery Directive, control architectures were categorized as SIMPLE, SINGLE CHANNEL, SINGLE CHANNEL – MONITORED and CONTROL RELIABLE. The general idea behind control reliable circuits before then was that the circuit would default to a safe state and that components were safety rated. Since then, systems have evolved to include regulations that categorize degrees of hazards and preclude any single point of failure in a system.

It is useful to examine safety circuits beginning with the original "Emergency Stop" or E-Stop circuit.

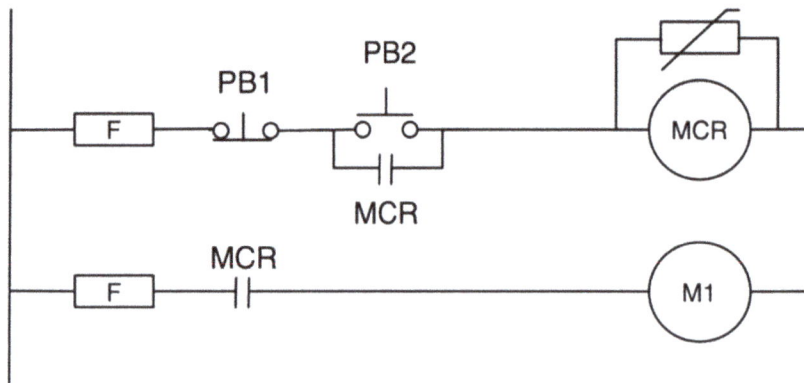

Figure 104 shows an early "master-control-relay" circuit with a single contactor. This architecture was designed to maintain the contactor coil circuit once the Start or Power On button (PB2) was pressed.

Figure 104 – Basic Start-Stop Circuit

Power to the output elements of the machine controls was supplied via contacts on the contactor, which is why it was called the Master Control Relay or 'MCR'. The Power Off button (PB1) could be labeled that way, or you could make the same circuit into an Emergency Stop by simply replacing the operator with a red mushroom-head push button. These devices were usually spring-return, so to restore power, all that was needed was to push the Power On button again.

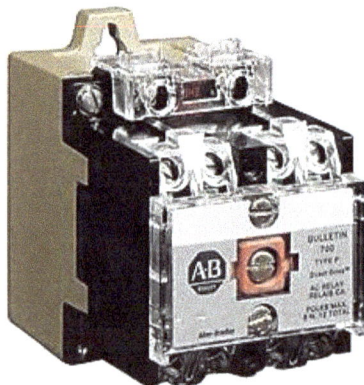

Typically, the components used in these circuits were specified to meet the circuit conditions, but not more. Controls manufacturers brought out over-dimensioned versions, such as Allen-Bradley's Bulletin 700-PK contactor which had 20 A rated contacts instead of the standard Bulletin 700's 10 A contacts.

Figure 105 - Allen-Bradley Bulletin 700-PK Contactor

When interlocked guards began to show up, they were integrated into the original MCR circuit by adding a basic control relay (CR1 in Figure 106) whose coil was controlled by the interlock switch(es) (LS1), and whose output contacts were in series with the coil circuit of the MCR contactor. Opening the guard interlock would open the MCR coil circuit and drop power to the machine controls.

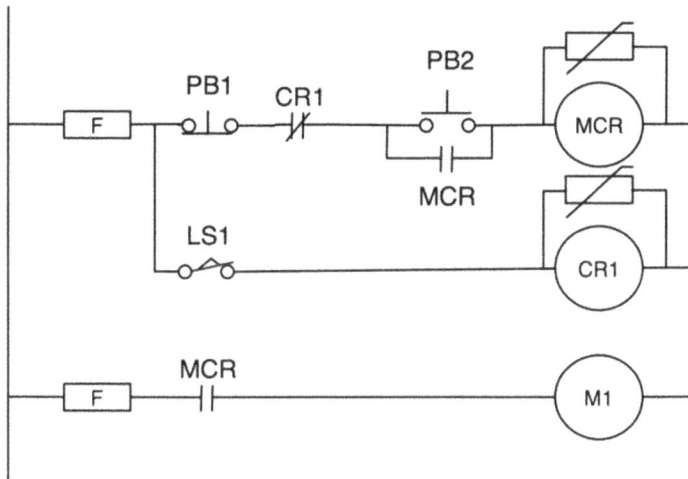

Figure 106 - Basic Start Stop Circuit with Interlock

'Ice-cube' style plug-in relays were often chosen for CR1. These devices did not have 'force-guided' contacts in them, so it was possible to have one contact in the relay fail while the other continued to operate properly. Force-guided contacts are mechanically linked and can't switch independently.

The MCR contacts in this circuit could also be placed in series with the voltage supply terminal of output cards on a PLC. This often required extra contacts on the contactor, both for output card power and for input card monitoring.

LS1 could be any kind of switch. Frequently a 'micro-switch' style of limit switch was chosen. These snap-action switches could fail internally shorted, or weld closed, and the actuator would continue to work normally even though the switch itself had failed.

These switches are also ridiculously easy to bypass. All that is required is a piece of tape or an elastic band and the switch is no longer doing its job.

The problem with these circuits is that they can fail in several ways that aren't obvious to the user, with the result being that the interlock might not work as expected, or the Emergency Stop might fail just when you need it most.

These original circuits are the basis for what became known as 'Category B' ('B' for 'Basic') circuits. There was no diagnostic coverage for the system.

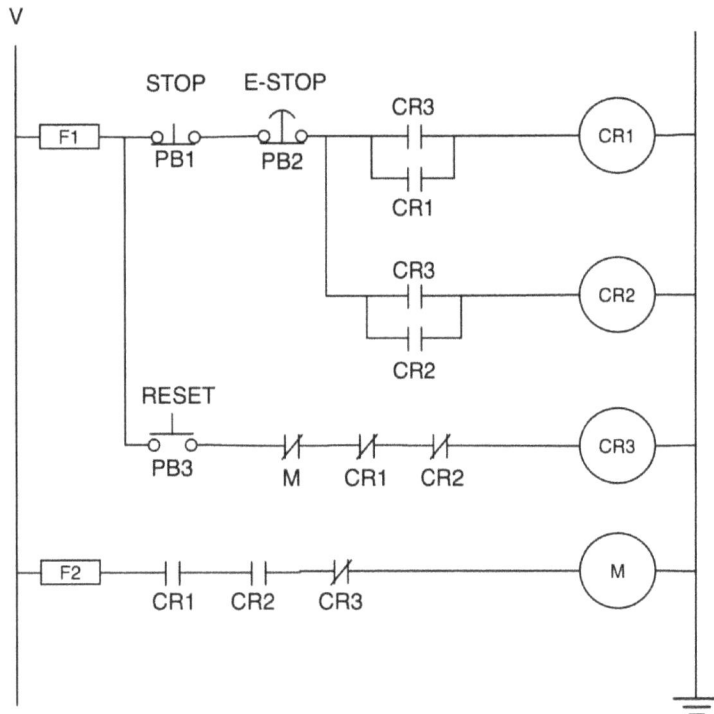

Figure 107 - Sample Category 2 Safety Circuit

Modern safety circuit design is based on risk analysis that calculates the possible frequency and hazard level for personnel. It involves diagnostics that detect faults in the system and requires components rated for safety. **Category 2** is the first classification requiring monitoring. Checks of the safety system are performed at machine start-up and periodically during operation if the risk assessment and type of operation shows that it is necessary.

Figure 107 is an example of what a simple Category 2 circuit constructed from discrete components might look like. Note that PB1 and PB2 could just as easily be interlock switches on guard doors as push buttons on a control panel. For the sake of simplicity, surge suppression is not shown on the relays, but you should include MOV's or RC suppressors across all relay coils. All relays should be constructed with 'force-guided' designs and meet the requirements for well-tried components.

How the circuit works:

1. The machine is stopped with power off. CR1, CR2, and M are off. CR3 is off until the reset button is pressed, since the NC monitoring contacts on CR1, CR2 and M are all closed, but the NO reset push button contact is open.
2. The reset push button, PB3, is pressed. If CR1, CR2 and M are off, their normally closed contacts will be closed, so pressing PB3 will result in CR3 turning on.
3. CR3 closes its contacts, energizing CR1 and CR2 which seal their contact circuits in and de-energize CR3. The time delays inherent in relays permit this to work.
4. With CR1 and CR2 closed and CR3 held off because its coil circuit opened when CR1 and CR2 turned on, M energizes, and motion can start.

In this circuit the monitoring function is provided by CR3. If any of CR1, CR2 or M were to weld closed, CR3 could not energize, and so a single fault is detected, and the machine is prevented from re-starting. If the machine is stopped by pressing either PB1 or PB2, the machine will stop since CR1 and CR2 are redundant. If CR3 fails with welded contacts, then the M rung is held open because CR3 has not de-energized, and if it fails with an open coil, the reset function will not work. Therefore, both failure modes will prevent the machine from starting with a failed monitoring system, if a "force-guided" type of relay is used for CR3. If CR1 or CR2 fail with an open coil, then M cannot energize because of the redundant contacts on the M rung.

This circuit cannot detect a failure in PB1, PB2, or PB3. Testing is conducted each time the circuit is reset.

Category 3 system architecture is the first category that could be considered to have similarity to "Control Reliable" circuits or systems as defined in the North American standards. It is also the first level with dual channel or redundancy. Self-monitoring relays certified by the manufacturer include features that prevent application of DC voltage from outside of the system being used to bypass components in the circuit. The signals passing through the safety devices use solid-state circuits that use a special type of output known as **OSSD**, for **O**utput **S**ignal **S**witching **D**evice. OSSD is often used in devices like light curtains, safety scanners, or safety mats. It provides a pulsed self-monitoring function to continuously test the circuit.

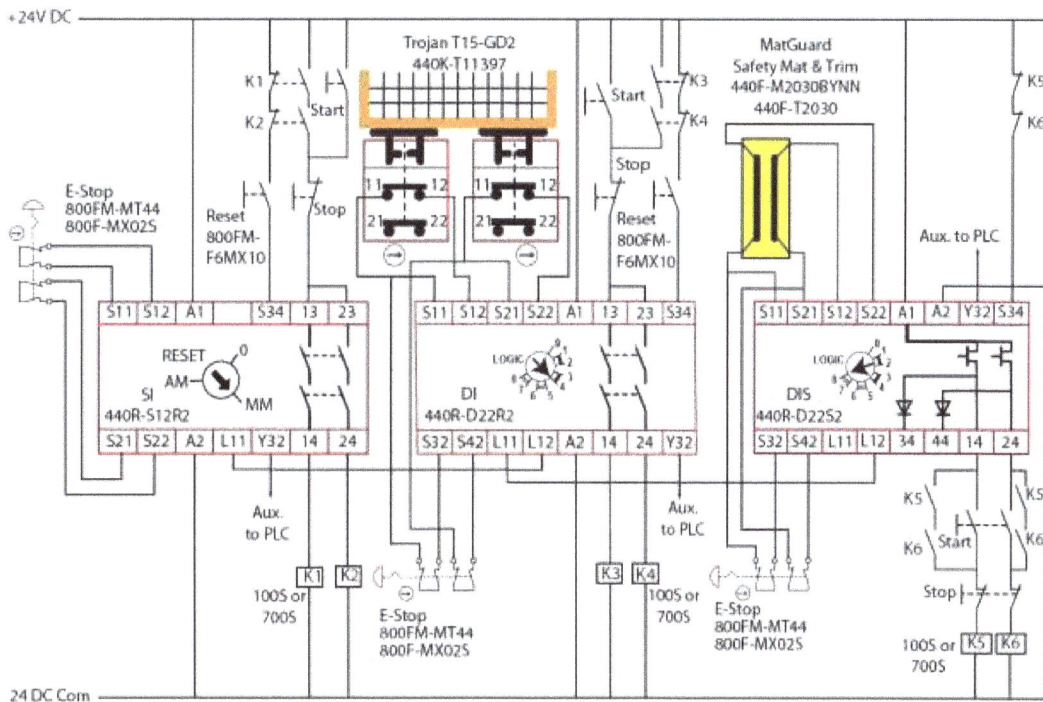

Figure 108 - Category 3 Safety Circuit (Allen-Bradley SAFETY-WD001A-EN-P - June 2011, P. 6)

Figure 108 shows the application of safety relays in a complete system that includes the emergency stop, a gate interlock, and a safety mat.

Here are the subsystems:

The **emergency stop circuit** uses the 440R-512R2 relay on the left side of the diagram. This particular system uses Category 3 architecture in the e-stop system, which may be more than is required. A risk assessment and a start-stop analysis are required to determine what performance level is needed for this subsystem.

The **gate interlock circuit** is in the center of the diagram and uses the 440R-D22R2 relay. As you can see, there are two physically separate gate interlock switches. Only one contact from each switch is used; one switch is connected to Channel 1, and the other to Channel 2. Notice that there is no other monitoring of these devices (i.e., no second connection to either switch). The secondary contacts on these switches could be connected to the PLC for annunciation purposes. This would allow the PLC to display the open/closed status of the gate on the machine HMI.

The output contactors, K3 and K4, are monitored by the reset loop connected to S34 and the +V rail.

One more interesting point – did you notice that there is a "zone e-stop" included in the gate interlock? If you look immediately below the central safety relay and a little to the left, you will find an emergency stop device. This device is wired in series with the gate interlock, so activating it will drop out K3 and K4 but not disturb the operation of the rest of the machine. The safety relay can't distinguish between the e-stop button and the gate interlocks, so if annunciation is needed, you may want to use a third contact on the e-stop device to connect to a PLC input for this purpose.

The **safety mat subsystem** is located on the right side of the diagram and uses a second 440R-D22R2 relay. Safety mats can be either single or dual channel in design. The mat shown in this drawing is a dual-channel type. Stepping on the mat causes the conductive layers in the mat to touch, shorting Channel 1 to Channel 2. This creates an input fault that will be detected by the 440R relay. The fault condition will cause the output of the relay to open, stopping the machine.

Safety mats can be damaged reasonably easily, and the circuit design shown will detect shorts or opens within the mat and will prevent the hazardous motion from starting or continuing.

The output contactors, K5 and K6 are monitored by the relay reset loop connected to S34 and the +V rail.

This circuit also includes a conventional start-stop circuit that doesn't rely on the safety relay.

Just like the gate interlock circuit, this circuit also includes a "zone e-stop". Look below and to the left of the safety mat relay. As with the gate interlock, pressing this button will drop out K5 and K6, stopping the same motions protected by the safety mat. Since the relay can't tell the difference between the e-stop button and the mat being activated, you may want to use the same approach and add a third contact to the e-stop button, connecting it to the PLC for annunciation.

Category 4 safety systems build on both Category B and Category 3 systems. In addition to redundancy and "well-tried safety principles", the following is required:

- a single fault in any of these safety-related parts does not lead to a loss of the safety function, and

- the single fault is detected at or before the next demand upon the safety functions, e.g., immediately, at switch on, or at end of a machine operating cycle, but if this detection is not possible, then an accumulation of undetected faults shall not lead to the loss of the safety function.

The circuitry for Category 4 safety systems looks like that of Category 3, the difference is in the components and the self-monitoring functions.

Safety System Components

Modern safety components used in electrical circuits include protective devices to detect personnel, power interlock guards that prevent entry into hazardous areas, logic units that ensure safety functions, emergency stop devices, and two-hand control devices. These are wired together following recommendations from manufacturers and the Machinery Directive 2006/42/EC. They are designed and tested by the manufacturer to meet designated safety categories as described previously.

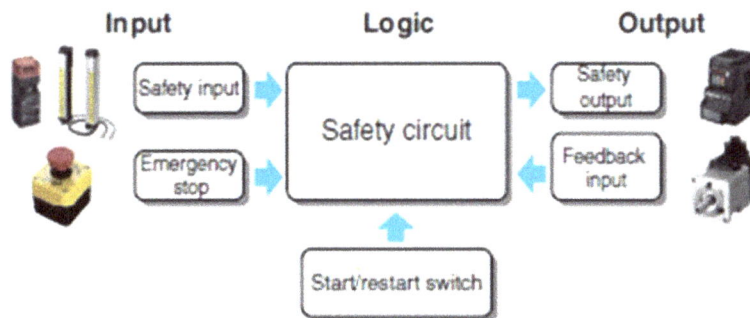

Figure 109 - Safety System Components (Omron)

Input components detect the state of the device to ensure safety. They include items such as door switches, safety rated limit switches for axis overtravel, light curtains, pressure sensing safety mats, area laser scanners and enabling switches for robots. They are often dual-channel devices and include features like ensuring safe operation even when there is a failure.

Direct opening action ensures that contacts open through the pressing action even if a contact is welded. Safety switches are also designed not to be easily defeated, meaning intentionally disabling the safety effects. This includes using screws, needles, or objects in daily use such as keys or coins to defeat the mechanism.

Interlocking devices with locking capability are used when the stopping time of an actuator is greater than the time it takes for a person to access the hazardous area.

Input Components:

Safety input devices include components such as emergency stop pushbuttons or "E-Stops", light curtains, door or gate switches, area scanners, and pressure sensitive safety mats.

E-Stops are maintained red pushbuttons that must be pulled out to reset. Most E-Stops have two NC contacts used in the safety circuit for redundancy and one NO contact for feedback to a control system. They also may have a light, used to show when it has been activated.

Door and gate switches also have two NC and one NO contacts. They may also have an interlock that prevents it from being opened when the hazard is active as mentioned previously.

Some machine builders will wire these components in series without bringing the terminations back to the control panel, which can make troubleshooting the system difficult.

Light curtains are arrays of LEDs that transmit light to an opposing array or receivers. They are self-monitoring to ensure that other light sources can't interfere with the signals. Controllers can be programmed to "mute" sections, allowing items to pass through them without tripping if desired. Like other safety devices, they are dual channel.

Area scanners and **safety mats** are used to monitor personnel entering a hazardous area. Area scanners are programmable to ignore objects that are permanent. Both devices (and light curtains) generally have separate **safety controllers** that monitor the channels and allow them to reset.

Logic:

Safety controllers monitor the channel signals of the safety input devices. If the channels don't switch within a defined period, the device will fault and need to be reset. In this case both channels will need to see a properly timed on-to-off followed by an off-to-on signal before allowing the controller to reset.

Controllers may be separate units wired together as shown in Figures 108 and 109, or be part of a "**Safety PLC**" system with programmable logic and modular I/O.

The OSSD circuits wired to the safety channels are not standard DC voltage but are clocked pulses of different frequencies on each channel to prevent bypassing or cheating the signal with external voltage. There are also feedback inputs to the controller to monitor the function of the contactors or safety relays (Category 4).

Safety Output:

The safety relays or safety PLCs usually control redundant safety contactors, which in turn supply control power to output cards ("switched" outputs) on PLCs and enable signals on motion control and VFD systems. This ensures that motion can't occur when the system is tripped.

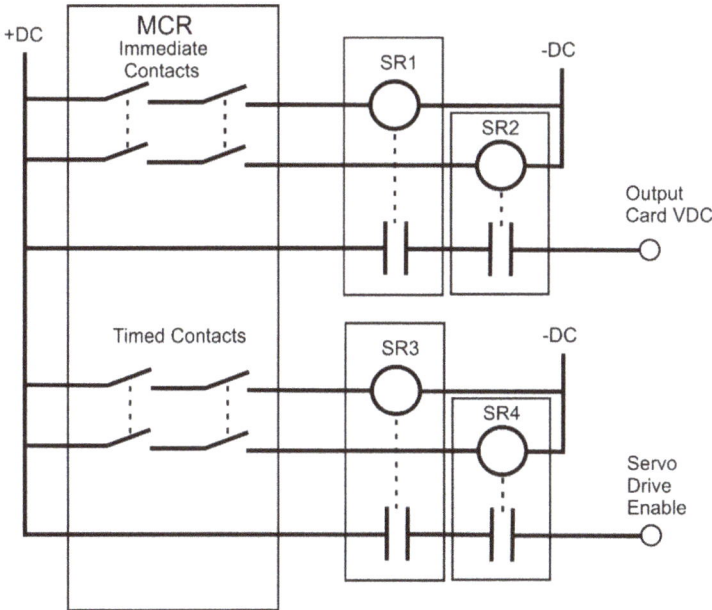

Figure 110 - Dual Channel Immediate and Timed Contacts

Figure 110 shows a simplified diagram of safety relays in series with power supplied to output actuating devices. The immediate and timed contacts can be used without safety contactors, but if multiple output cards or servos must be controlled then additional devices have to be wired in series for redundancy as shown.

The timed contacts allow a stop command to be issued to the servo controller before removing the enable system. If the enable is removed immediately, holding torque is released allowing the movement of an actuator to continue. The delay allows the servo actuator to come to a controlled stop before releasing holding torque. The time delay is adjustable by programming or using a rotary selector.

Figure 111 shows an Allen-Bradley GuardMaster safety relay. With proper wiring this relay is rated for Category 4 applications. The time delay can be adjusted from 0.5-10 seconds.

The indicator lights on the front show the status of the input circuits and both the instantaneous and timed contact outputs. These lights are quick aids to troubleshooting the circuit.

When configured for monitored manual reset, the output monitoring circuit is checked every time the reset switch is pressed. The circuit needs to see a successful on-off and off-on transition of both channels within prescribed time limits before resetting.

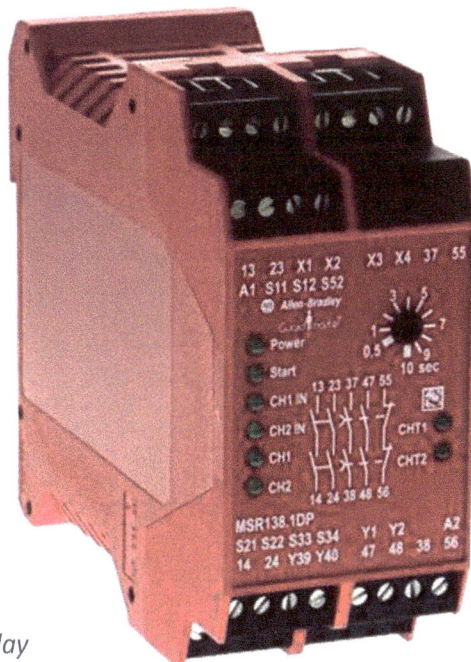

Figure 111 - Allen-Bradley MSR138DP Safety Relay

Exercise 16

1. Who are ANSI standards directed towards?

_____

_____

_____

2. What document should be completed before specifying safety components for machinery?

_____

3. What are "force-guided" contacts?

_____

_____

4. What is the lowest number of safety classification category that requires redundancy?

_____

5. Why can't standard DC voltage signals be applied to the safety input channels to "cheat" the system?

_____

_____

6. What are the timed contacts of a safety relay signal used for?

_____

_____

7. What are "switched" outputs?

_____

_____

Motor Control

Electric motors are used to drive actuators and machines in industry. Most motors used in industrial applications are AC, which means they operate off of the frequency of the power supplied. In the US this is 60Hz, while in Europe it is 50Hz.

Starters and Overloads

Motor starters are used to start and run AC motors. Smaller motors may only use a single phase of electrical voltage, but three-phase motors are most common in industrial applications, usually at 240 or 480VAC in the US. Starting a motor "across the line" will run the motor at a speed related to the applied frequency times the number of poles in the rotor, minus a "slip". In the US most motors run at about 1750 RPM.

A motor starter is composed of a contactor with a coil and an overload device for circuit protection. The direction of the motor's rotation is determined by the order of the applied voltage phases for a three-phase motor; reversing any two of the phases will change the direction.

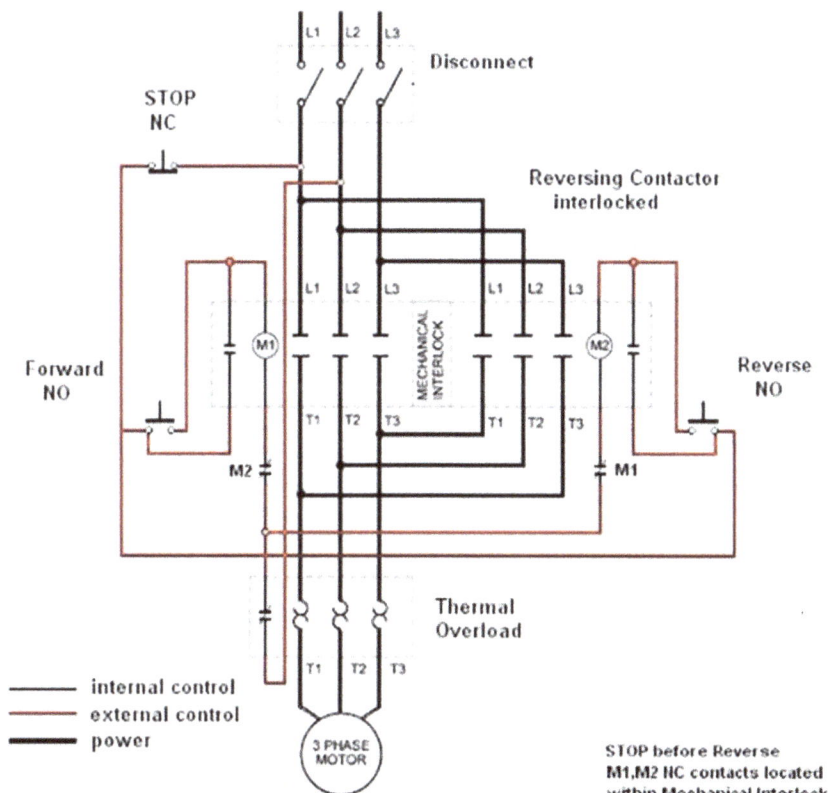

Figure 112 shows the wiring of a reversing motor starter with an overload. The coils are both wired through the overload's NC contact, which will de-energize the contactors if the overload is tripped. Both coils are also wired through the opposite coil's NC contact, preventing both coils from being energized simultaneously. The contactors are mechanically interlocked to prevent them from energizing at the same time. Pushbuttons are shown as a means of actuating the starter, but outputs from a PLC or other controller can also be used. In this case, logic would be used to prevent the two outputs from energizing at the same time.

Figure 112 - Reversing Motor Starter

The disconnect is also often fused, or a circuit breaker is used for branch circuit protection.

Overloads are typically bimetallic strips that uses a heating element to mimic the temperature of the motor based on current. They are made with different adjustable ranges that are chosen based on the full-load current of the motor and the required time delay.

Figure 113 shows a three-phase motor starter. The red button at the bottom is used to reset the overload.

Figure 113 - IEC Motor Starter

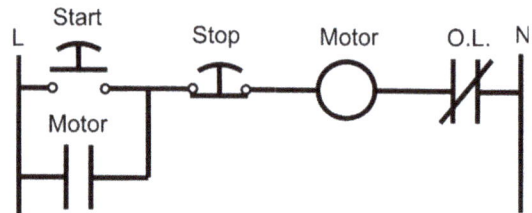

Figure 114 - Three-wire "hold-in" circuit

Three wire control circuits are used in motor starter pushbutton enclosures to energize the coil of the starter. The neutral side of the coil is often wired through the NC overload contacts as shown. If both pushbuttons are momentary, the start button must be pressed again after a stop to re-energize the coil. The NO contact labeled "Motor" is called the hold-in contact.

Soft Starters

A soft starter is a device that can be used to control the starting and stopping profile of an AC motor. The traditional method of applying full voltage "across the line" creates a high inrush current, often 6 to 7 times the motor's rated current. A soft starter temporarily reduces the electric current and torque during start-up, reducing mechanical stress and the effects of high current on the wiring. Soft starters are solid-state devices, but clutches, fluid couplings and autotransformers are also sometimes used. Unlike Variable Frequency Drives, they do not change the speed of the motor.

Figure 115 – Weg brand Soft Starter

The electrical circuit for a soft starter is shown in Figure 116. **SCRs** (Silicon Controlled Rectifiers), also known as thyristors, are solid state current-controlling devices. The SCRs are engaged during ramp up, and the bypass or run contactor is pulled in after maximum speed is achieved. This helps to significantly reduce motor heating.

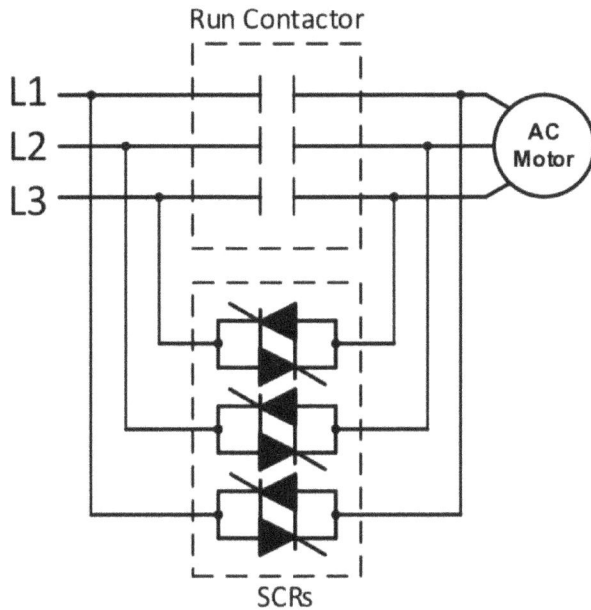

Figure 116 - Soft Starter Circuit

Soft starters are simpler devices than variable frequency drives, requiring fewer adjustments. In pump applications, soft starters can be used to reduce pressure surges, also called "water hammering". They also take up less space and are less expensive than variable frequency drives.

Variable Frequency Drives (VFDs)

A Variable Frequency Drive, or **VFD**, controls the speed of a motor by varying the applied frequency (and voltage). In industrial automation, these are sometimes known as simply "**drives**".

VFDs are used for both AC and DC motors, and AC drives can be used with single or three phase induction motors. An AC drive operates by first converting the applied AC voltage into DC, then using solid state electronic circuits to rebuild a sinusoidal waveform at the desired frequency to apply to the motor. This is called "Pulse Width Modulation", or **PWM**, where the average power delivered to a motor is controlled by effectively chopping up the waveform into discrete parts or pulses of varying durations as shown in Figure 117.

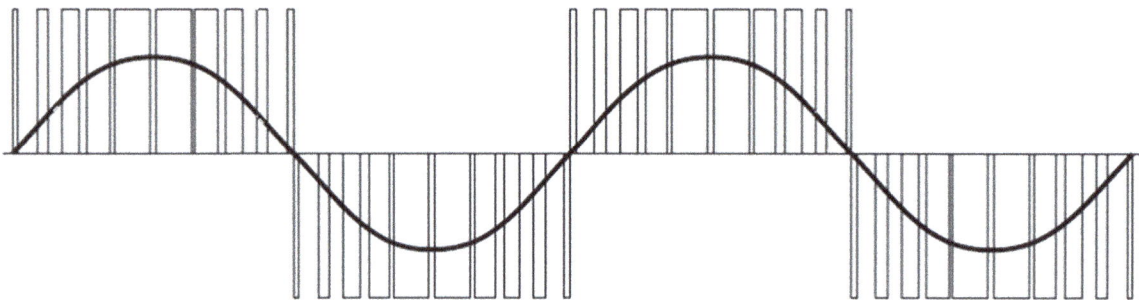

Figure 117 - Pulse Width Modulation for an AC VFD

Variable frequency drives are made up of three distinct sections; a rectifier bridge converter, a DC link that creates the bus voltage, and an inverter that converts the DC back into AC. Most inverter switching devices are Insulated Gate Bipolar Transistors (IGBTs).

VFDs are usually operated at frequencies at or below the rated nameplate frequency of the motor. Operation above the rated speed is possible, but the V/Hz factor of the nameplate is derated, reducing the available power.

By varying the voltage and angle from reference as well as the frequency, torque can also be controlled. Some drive manufacturers refer to these as vector or Direct Torque Control (DTC) drives.

An embedded microprocessor controls the overall operation of the VFD. Parameters such as acceleration, deceleration and preset speeds can be programmed, and feedback like bus voltage, current, and speed can be obtained. These parameters can be set and displayed by means of an operator interface built into the drive or exchanged with a controller by communications like Modbus or ethernet.

In addition to setting and displaying parameters from the operator interface, physical discrete and analog signals can be used to start and stop the drive, select preset speeds, and control the motor speed with a reference (usually 0-10VDC). Most discrete signals are 24VDC.

Starting and stopping a drive can be done using two different wiring schemes. **Two-wire control** directly energizes a run command input; if the signal is removed, the drive will stop. An enable signal can be used in this configuration to disable the run signal, but if the enable signal returns, the drive could start if the run signal is present. For this reason, two-wire control is not considered as safe as the alternative described next.

Three-wire control as shown in Figure 114 uses a stop signal to enable the drive. Usually, the stop signal is required to be on for the drive to run; that way, if the wire is cut, the drive will stop. A separate enable and STO (Safe-Torque Off) signal is still often used for safety purposes.

Two and three-wire control is often one of the selectable parameters for VFDs.

Digital inputs can be assigned functions such as selecting a preset speed, clearing alarms or faults, and placing the drive under automatic or manual control. **Digital Output** signals may be assigned to indicate that the drive is at speed, running, or faulted. Outputs may be normally open (form A) or form C with both NO and NC contacts. Form C allows dual connections for running/not running or faulted/not faulted. Using relay type outputs allow connections to external pilot lights for drive status.

A safety disable (**Safe Torque Off** or **STO**) input may also be used when required to meet industrial safety standards. This is in addition to the enable signal or stop signal for 3-wire control.

The **main circuit** wiring for a VFD accepts line voltage in and provides load power to the motor. VFDs may be single or three-phase, so it is important to select the correct drive for your motor. Three phase drive power input terminals are generally designated as R, S, and T or L1, L2, and L3, while output terminals to the motor are usually labeled U, V, and W or T1, T2, and T3.

Less commonly used are the **DC bus** terminals. Components such as DC reactors, power regenerative units, and dynamic braking modules can be interfaced with these terminals, which are usually labeled + and -. Even more uncommon uses for these terminals include the addition of capacitance or common bussing.

Besides wiring, parameter changes or dip switch settings are usually required for setup of a VFD. Manufacturers use different features and labels, so it is very important to carefully read the manual.

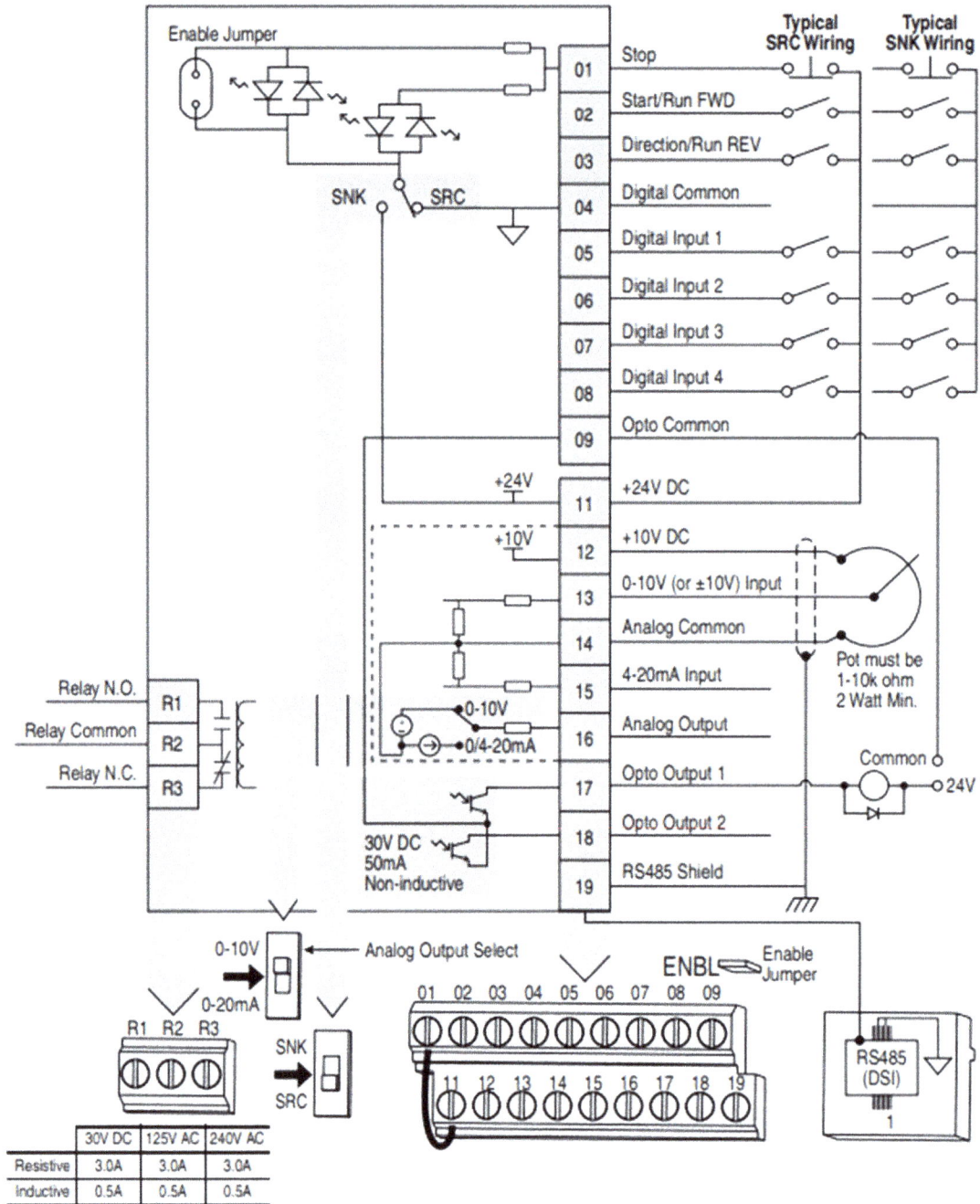

Figure 118 - Allen-Bradley Powerflex 40 Control Wiring

Figure 118 shows typical control wiring connections for a VFD. There are a lot of options shown in this diagram; sink vs. source wiring for discrete signals, 0-10V vs. 4-20mA for the analog I/O, and two wire vs. three wire control as indicated by the jumper from terminal 1 to 11. Digital inputs 1-4 are also configurable for preset speed selects or jogging and stopping profiles.

The output relay is also programmable as a fault or running/at frequency output, or feedback for direction or logical functions.

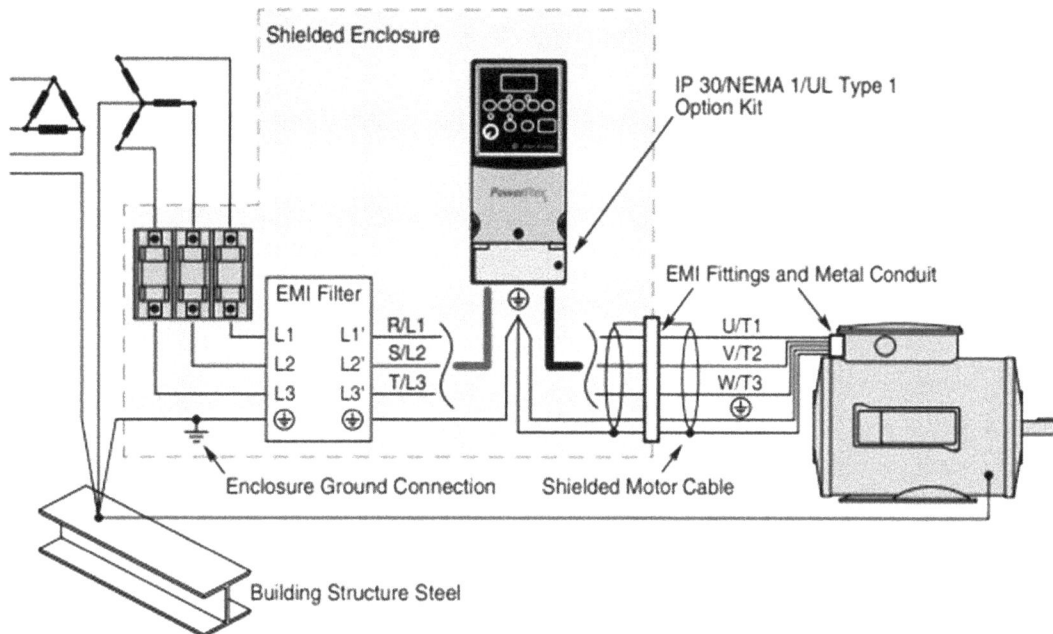

Figure 119 - VFD Power Wiring

Figure 119 shows power wiring for the same PowerFlex 40 drive. EMI is an acronym for Electro-Magnetic Interference. Because VFDs create powerful radio frequency signals, filters are required to minimize the effects on other components.

Proper grounding of the drive, motor and enclosure are also important considerations in VFD installations.

There are many internal diagnostic functions to detect problems with VFDs. Fault codes can be displayed on the operator interface on the front of the drive or sent via communications.

Other troubleshooting techniques involve measuring voltage on the control signal wiring and at both the infeed and outfeed of the drive. Because there are many auxiliary components in a VFD system, it is necessary to use electrical schematics and the drive user manual to guide the technician through the system.

VFD_Drive1:I		AB:PowerFlex40_...	VFD_Drive1:O		AB:PowerFlex40_...
+ VFD_Drive1:I.DriveStatus		INT	+ VFD_Drive1:O.LogicCommand		INT
VFD_Drive1:I.Ready		BOOL	VFD_Drive1:O.Stop		BOOL
VFD_Drive1:I.Active		BOOL	VFD_Drive1:O.Start		BOOL
VFD_Drive1:I.CommandDir		BOOL	VFD_Drive1:O.Jog		BOOL
VFD_Drive1:I.ActualDir		BOOL	VFD_Drive1:O.ClearFaults		BOOL
VFD_Drive1:I.Accelerating		BOOL	VFD_Drive1:O.Forward		BOOL
VFD_Drive1:I.Decelerating		BOOL	VFD_Drive1:O.Reverse		BOOL
VFD_Drive1:I.Alarm		BOOL	VFD_Drive1:O.OptoOutput1		BOOL
VFD_Drive1:I.Faulted		BOOL	VFD_Drive1:O.OptoOutput2		BOOL
VFD_Drive1:I.AtReference		BOOL	VFD_Drive1:O.AccelRate1		BOOL
VFD_Drive1:I.CommFreqCnt		BOOL	VFD_Drive1:O.AccelRate2		BOOL
VFD_Drive1:I.CommLogicCnt		BOOL	VFD_Drive1:O.DecelRate1		BOOL
VFD_Drive1:I.ParmsLocked		BOOL	VFD_Drive1:O.DecelRate2		BOOL
VFD_Drive1:I.DigIn1Active		BOOL	VFD_Drive1:O.FreqSel01		BOOL
VFD_Drive1:I.DigIn2Active		BOOL	VFD_Drive1:O.FreqSel02		BOOL
VFD_Drive1:I.DigIn3Active		BOOL	VFD_Drive1:O.FreqSel03		BOOL
VFD_Drive1:I.DigIn4Active		BOOL	VFD_Drive1:O.RelayOutput		BOOL
+ VFD_Drive1:I.OutputFreq		INT	+ VFD_Drive1:O.FreqCommand		INT

Figure 120 - VFD (PowerFlex 40) Parameters in PLC

Figure 120 shows the parameters that are automatically exchanged with an Allen-Bradley ControlLogix PLC when the PowerFlex 40 VFD shown previously is added to the ethernet I/O network. The I parameters show feedback from the device, and the O parameters show the commands to the drive. These addresses and parameters can be used just like physical I/O in the PLC to operate the drive!

Figure 121 shows the front of the PowerFlex 40. All of the parameters listed in Figure 120 are available for programming and display from the controls shown via menu selection. The drive can be operated via discrete wiring as shown in Figure 114, from PLC commands over ethernet, or by pressing buttons and adjusting speed from the drive controls shown here.

The bottom cover of the VFD is removed to connect the wiring. Voltage and current in VFD power wiring can be dangerous, follow proper safety precautions!

Figure 121 - PowerFlex 40 VFD with front accessible controls

Exercise 17

1.   What is added to a contactor to make it a motor starter?

_____

2. What is the typical speed of a motor started "across the line" with a motor starter in the United States? _____

3. How can the direction of a three-phase motor be changed on a motor starter?

_____

4. Can a "soft starter" change the speed of a motor? _____

5. What three methods can be used to start, stop, and change the speed of a PowerFlex 40 VFD? _____

_____

6. What are some of the functions that can be assigned to the digital inputs of a VFD?

_____

_____

7. What are power input terminals usually labeled on a 3 phase VFD?

_____

How are motor output terminals usually labeled?

_____

8. Do all VFD manufacturers use the same type of I/O and parameter arrangement?

_____

Motion Control

Motion control involves the use of motors for positioning and precise movement of actuators. While it is not always closed loop, it differs from the previous section on motor control in that the main goal is to achieve and verify a known position or movement.

Steppers

Stepper motors are brushless DC motors with multiple electromagnets arranged as a stator around a gear-shaped rotor. The circular arrangement of magnets are divided into groups called phases, each phase is energized together to make the motor "step" to the next position.

Microcontroller-based stepper drives are used to activate the drive transistors in the correct order. Typical stepper motor resolution is 200 steps per revolution, but with "microstepping" drives, up to 1600 steps per revolution can be achieved. Stepper drives are also sometimes called "choppers".

Stepper motors are typically operated without feedback devices such as encoders or resolvers, making them a less expensive method of positioning than with servo motors, but they also don't have much holding torque.

In addition to the motor and drive, some type of indexer is required. This may be built into the drive and communicate to a master controller, or a controller such as a PLC can send pulses to index the drive.

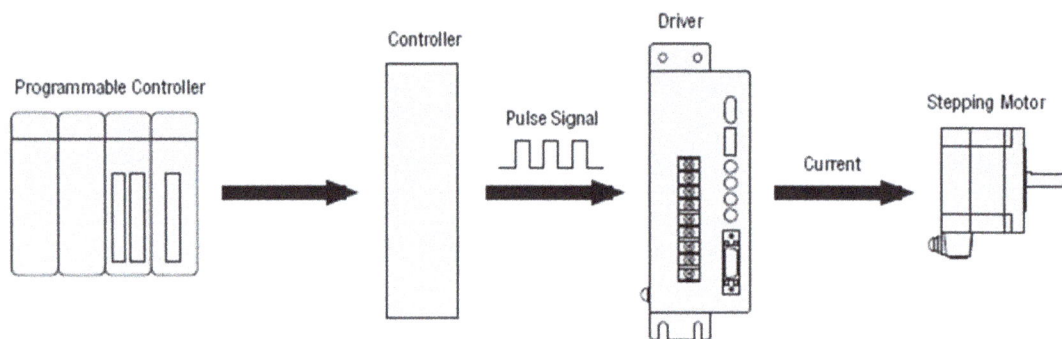

Figure 122 - Stepper Motor System

Troubleshooting a stepper system can involve checking voltages and communications in the control circuit, or even viewing the pulse train using an oscilloscope.

Servos

A servo or servomechanism is a device that uses feedback to control position and torque. They can be electric, hydraulic, or pneumatic, but most servos used in industrial automation applications are motor driven.

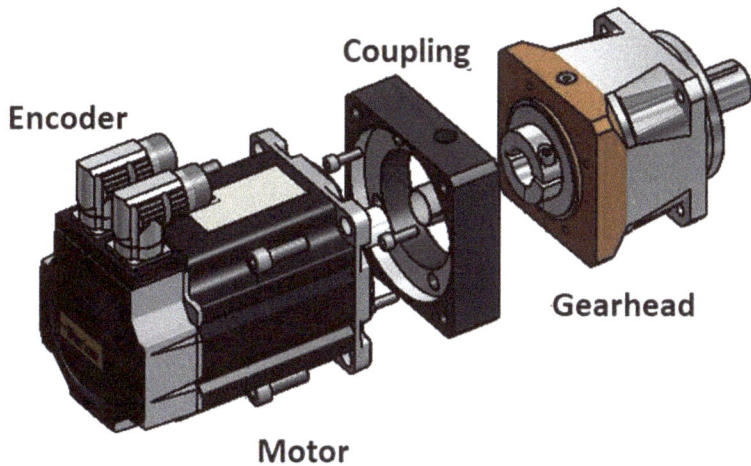

Figure 123 - Servo Motor Assembly

Servo motors can be brushed permanent magnet DC motors, brushless AC motors with permanent magnets or AC induction motors. They usually have a built-in encoder or resolver. They are also often integrated with a gearhead. There are two cable connections attached to the motor assembly to separate the signals from the encoder and sensors (feedback cable) from the motor power wiring.

Servo drives accept pulse inputs from the encoder as well as monitoring torque with current. Temperature sensor and brake control signals are sometimes included in the control cable. Servo drives are generally more sophisticated than VFDs, and often have logic capability built in. Modern controllers also almost always have high-speed communication ports that can be interfaced with other controllers to coordinate motion. Usually this is an ethernet-based communication protocol, but fiber optics are sometimes used.

Servo control algorithms are PID-based for position or torque control. Motors need to be tuned to the characteristics of the motor and load to ensure peak performance. For this reason, the motor and drive are often specified and sold as a set from the same manufacturer.

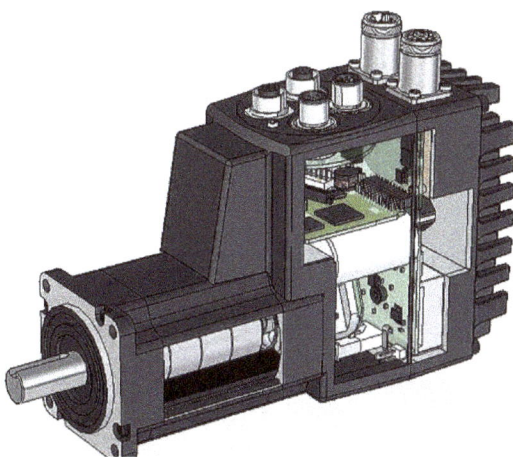

Figure 124 - Integrated Servo

Some motors have the drive and controller built in with the motor. These "**integrated servo**" motors can be networked together to perform complex tasks or serve as stand-alone positioners.

An important difference between a servo motor and a typical AC induction motor controlled by a VFD is that a servo has holding torque at zero speed. If the motor shaft is moved off its position while under control power it will try to correct itself, faulting the controller if the correct position is not achieved.

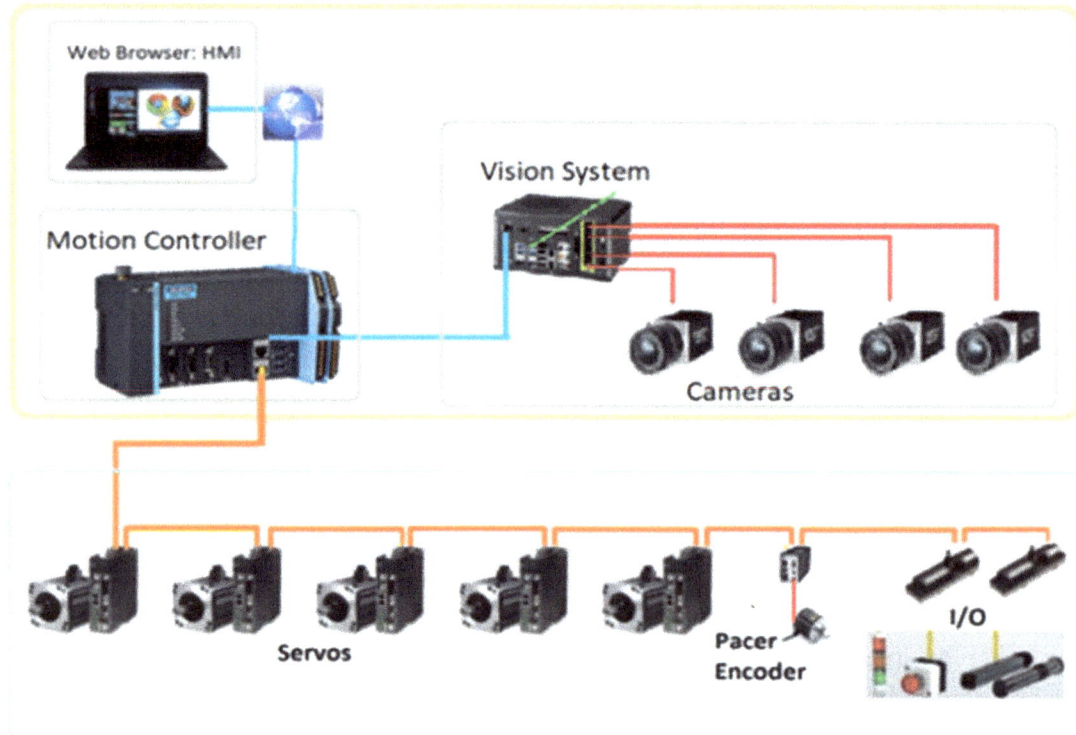

Figure 125 - Motion Control System

Coordinated Motion

When coordinating motion, a "master" controller or position is often used to pace the other controllers. Movement of one axis depends on the changing position of another, or on a virtual axis. Because of this, it is important that a fast communication network is used dedicated to the motion system. A dedicated **motion controller** may be used to coordinate servo axes. As illustrated in Figure 125, machine vision may be integrated to guide manipulators to the right location. Motion controllers may be integrated into PLC racks or be separate systems. Many of them have separate I/O modules and can be programmed in IEC 61131 PLC languages.

Troubleshooting servo systems usually requires knowledge of the platform's software in addition to typical methods of electrical diagnostics. Drives and controllers generally have built-in diagnostic functions to detect problems with the motor or attached load. Mechanical elements such as couplings are also subject to failure. As always, read your documentation!

Exercise 18

1. What is the typical resolution of a stepper motor?

_____

2. In addition to the encoder, what parameters are often monitored related to the motor?

_____

3. What is a motion device with a motor, drive and controller built into one unit called?

_____

4. Why does a coordinated motion control system require a dedicated controller and fast communications?

_____

5. What is an important difference between a servo motor and a standard AC induction motor?

_____

6. How many cables are usually used to connect a servo motor to its drive? _____

   What are they used for? _____

Robotics

There are many types of robots, some of which blend into the previous section on motion controllers. Coordinated motion systems often control three-axis **gantry**-type "pick-and-place" systems, but these aren't properly known as robots, since motion of axes are not necessarily coordinated simultaneously.

Industrial robots are used in manufacturing and material handling tasks, and their physical configuration depends on the function that is required. Payload and speed requirements help determine the type used in a particular application.

Robots can have up to 6-7 axes of motion, or as few as 3. Two axes are required to reach any point in an X-Y plane, and 3 are required to reach any point in X-Y-Z space. To completely control the position of tooling at the end of the "arm", three more axes are required in addition to X, Y and Z, these are pitch, roll and yaw.

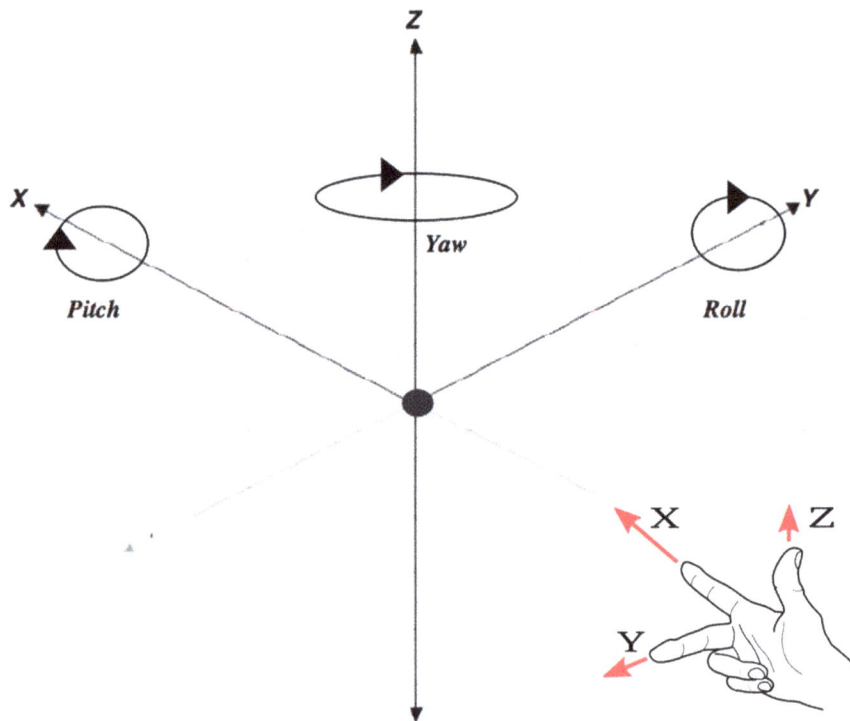

Figure 126 - Robot Coordinate Axes and Right Hand Rule

Figure 126 shows the six axes required to reach any point and orientation in three dimensional space, but robots use different coordinate systems, and joint configurations are different. The origin and directions are defined differently depending on the brand and can often be changed in the software. Fanuc robots use a "right-hand rule" based system, where Y is opposite to the figure shown above.

The X, Y, and Z locations are referred to as Cartesian coordinates, but can be defined from different reference points. If defined from the base of the robot or reference point of the environment, they are referred to as **World Coordinates**. In this case the origin's reference is stationary. When addressed from the view of the manipulator, they are referred to as **Tool Coordinates,** where the origin moves with the end effector. This may also include an **offset**, from the point where the tooling is attached to where it contacts the part.

Local Coordinates can also be defined, usually with origins within a working area. This allows references to be duplicated for pallets or other local systems.

Individual joints can also be controlled independently, usually defined in degrees. Distances are generally defined in metric measurements (millimeters) but can also be scaled to user defined units in the software.

In addition to X, Y and Z, roll, pitch, and yaw may be defined by other letters like U, V and W.

The area which a robot can reach is called the work envelope. Planes and boxes can be defined within the envelope to prevent collisions or for safety purposes, safety devices such as light curtains can also be integrated into a robot workcell.

Robot controllers constantly perform calculations to ensure they know where the robot is relative to defined points and paths. Axes must work together when maintaining positions along a defined path, so a robot is the ultimate form of coordinated motion control. This is why controllers are generally dedicated to the task of achieving and maintaining position.

An important problem to understand when dealing with robots is that of a **singularity**. In this case, the robot cannot move its end effector along a certain path due to either physical or mathematical constraints. Robots can end up in positions where it can't rotate the tooling around certain positions, this is sometimes referred to as **gimbal lock**. There are also other physical configurations where moving joints through certain orientations can cause damage to associated cables or hoses, so care is needed when moving robots close to singularity points or rotating axes too far. There are usually multiple joint configurations that can achieve the same tool position and orientation, this is often referred to as **redundant degrees of freedom**.

Robot controllers are often able to perform logic functions and operate external equipment, but usually they are built into workcells and connected to a "master" controller such as a programmable logic controller, or PLC. The controller may be a separate unit connected to the robot with power and signal cables or be built into the robot base. Connections may be physical 24vdc, communication links, or "pass through" ports and connectors with internal routing to the end effector or tooling. Pass through ports often include pneumatic hoses.

Robots can be categorized by their physical configuration. Figure 127 shows some of the common types of robots used in industrial applications.

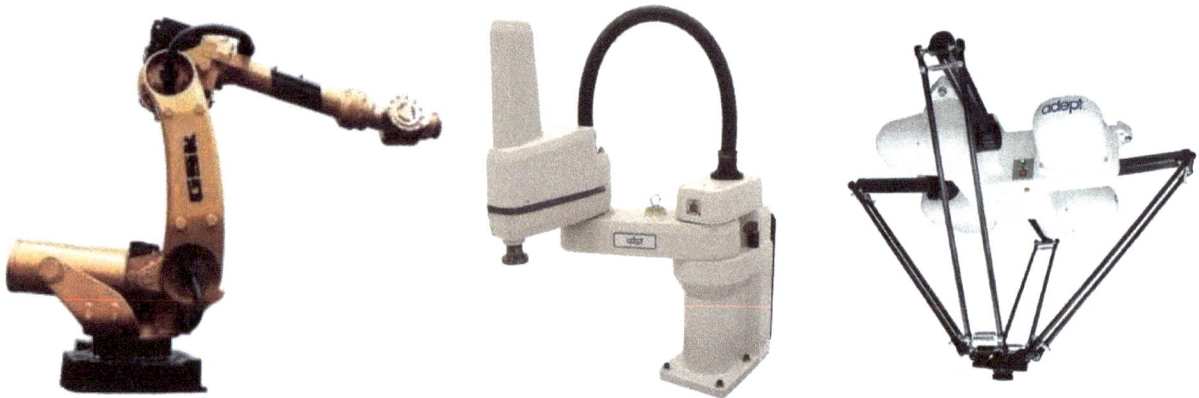

Figure 127 - Common Robot Types - 6 Axis Articulating Arm (Left), SCARA (Center), Delta (Right)

The **6-axis articulated arm** is very common in applications with heavy payloads, whereas the 4 axis SCARA is often used for oriented pick-and-place. The Delta configuration is very fast and often used in the electronics industry for component placement. An additional term to be aware of is **Cobot**, or **Co**llaborative Ro**bot**. These are robots that are designed for direct human interaction within a shared space and have a different configuration than those shown here.

Robots can be programmed using a computer or by means of a teach pendant. There are two types of code that need to be programmed, **procedures** and **positional data**. For a task where a robot end effector needs to move from one location to another, the starting and end points need to be defined, then a procedure needs to be written on how to get there. This may involve additional positions, along with external signals telling the robot an object is present or to begin the move.

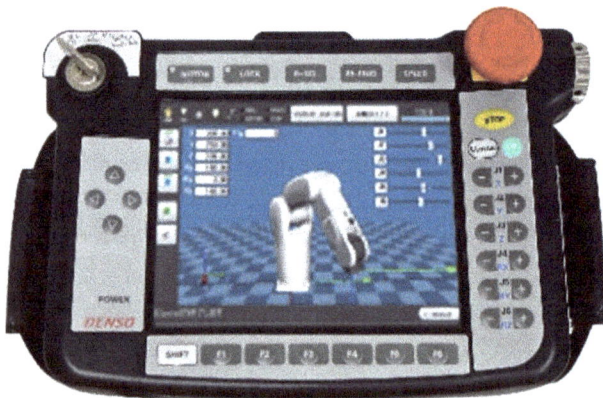

Figure 128 - Robot Teach Pendant (Denso)

Positions can be defined by listing them in the software, but using a pendant is easier. A teach pendant allows an operator to move individual axes, "driving" the robot to the desired location. A low speed is usually used for precision and safety. A 3 position "dead man switch" also needs to be pressed while the robot is being maneuvered. The spring-loaded switch needs to remain in the middle position, if it is depressed all the way or released, the robot can't move.

Procedures are series of movements to different positions. They can be triggered individually or linked together. There are a variety of languages used in robotics, generally proprietary to the manufacturer. They often resemble languages such as Basic or Assembly, with JUMP and MOVE statements.

Following are some of the programming languages used by different robot manufacturers:

- ABB – RAPID
- Comau – PDL2
- Epson – SPEL+
- Fanuc – Karel
- Kawasaki – AS
- Kuka – KRL
- Staubli – VAL3
- Yaskawa – Inform

Additional high-level scripting languages are also used to build data structures or create mathematical algorithms such as calculating paths or positions. Some languages allow parallel processing, allowing the robot to perform more than one action at a time, such as calculating movement vectors while a camera follows a moving object.

Positional data tables and programmed procedures reside in different memory areas, so one can be changed without affecting the other. This allows positions to be changed or "touched up" by editing the table with a computer or teach pendant.

Positions are often defined in world coordinates, but the positions of individual axes of a six-axis robot can differ with the tooling of the end effector being in the same position. Positions can be taught by driving the robot to a location with a specific axis configuration and selecting "teach", or by using a technique called "lead-by-the-nose". This technique allows the user to manually push the axes to a specific series of positions while the axes are relaxed, describing a path.

Troubleshooting and maintenance involves using the software or pendant to touch up (slightly correct) positions, replace tooling on the end effector, and maintain electrical or pneumatic connections. As with motion controllers and VFDs, robot controllers will indicate problems with the system by providing fault data. Most faults will cause the robot to stop moving and may require the operator to move the robot to a "safe" position after correcting a fault.

Robotic workcells interface with the robot, often with a PLC and HMI. The PLC communicates with the robot, displaying received fault codes and other data on the HMI. This involves two communication links, (robot-PLC and PLC-HMI) so ensuring they are working properly is important.

End effectors may have M8 or M12 cable connections, junction boxes with terminals, communication interfaces such as **ASI** (**A**ctuator-**S**ensor **I**nterface) or ethernet remote I/O. If the sensor terminations are under a cover, knowing what kind they are ahead of time can be helpful. Check your documentation or examine the gripper or tooling area to see these connections.

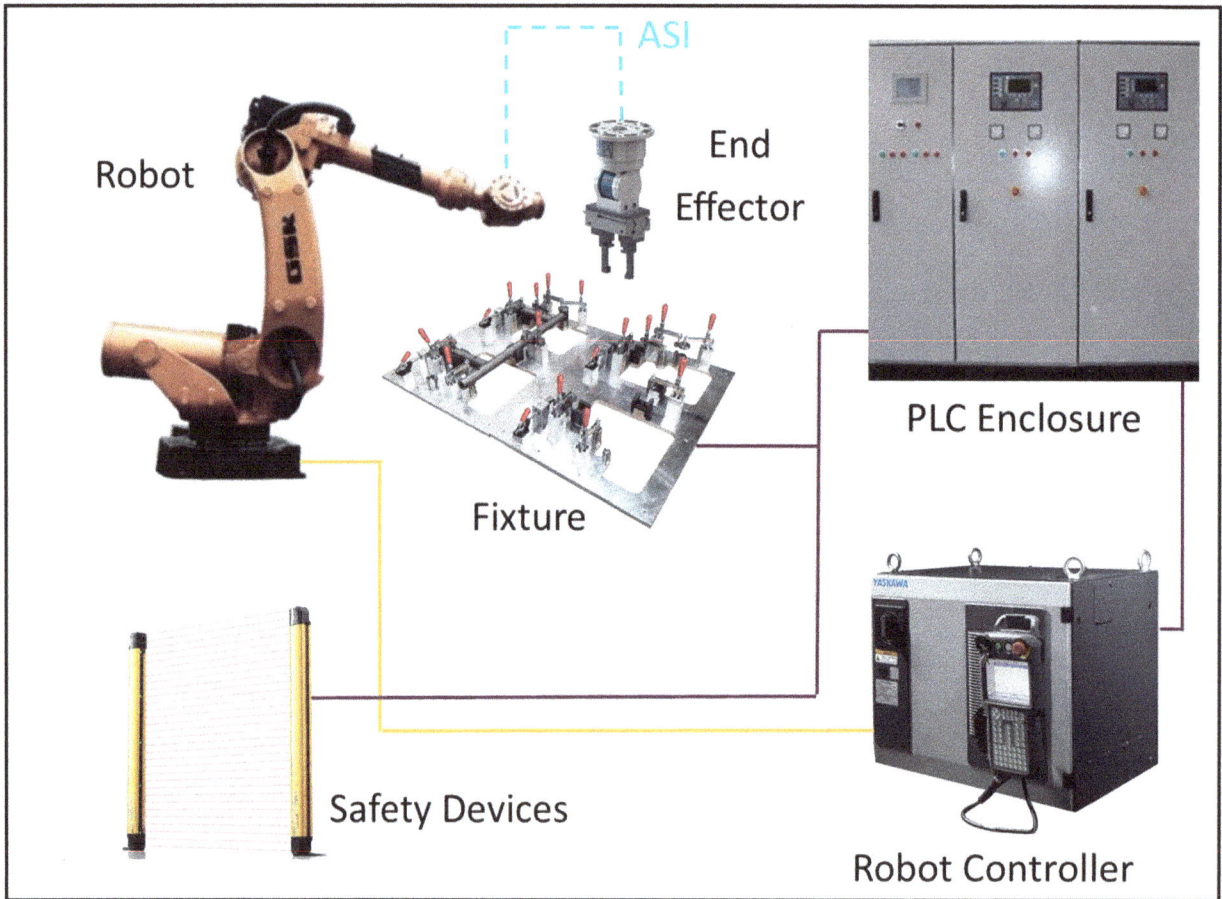

Figure 129 - Robotic Workcell

Figure 129 above shows a typical layout for a robotic workcell. The different colors of the lines illustrate that the connections between the different elements may be discrete wiring, communications, pneumatics, or a combination of power and feedback wiring in the case of the robot to controller connection. This can make troubleshooting complicated, as a wide range of knowledge is required in mechanical, electrical and control disciplines.

There are often actuators that are not controlled by the robot controller in such a system, such as the workpiece fixturing. This requires "handshaking" signals from the PLC and the robot controller. External systems for material handling and conveyor systems may also interface with the PLC, and multiple robots may also be present. Preventing collisions between multiple robots and tooling can be very complicated!

Safety devices such as light curtains, floor scanners and gate switches may interface with both the robot controller and the PLC.

Machine vision is also sometimes used to help locate parts for the robot, introducing another level of complexity to the system.

Exercise 19

1.   How many axes of movement are required to orient tooling to any position? _____

What are these axes called? _____

_____

2. Can positions be modified without changing the procedures in a program? _____

3. What is the easiest method to "touch-up" points on a robot? _____

_____

4. What does the acronym ASI stand for? _____

5. List some of the elements that may be found in a robotic workcell:

_____

_____

_____

_____

_____

_____

_____

Machine Vision

An important element of quality control in industrial automation is the use of cameras to examine products. This section describes the basic elements of a machine vision system.

What Is Machine Vision?

Machine vision is the use of a camera or multiple cameras to automatically inspect objects, usually in an industrial or production environment. A typical application might be on an assembly line; after an operation is performed on a part, the camera is triggered to capture and process an image. The camera may be programmed to check the position of something, its color, size or shape, or simply whether it is there or not. It can look at and decipher a standard or 2D Matrix barcode or even read printed characters.

After the product has been inspected, a signal is usually generated to determine what to do with it. The part might be rejected into a container or an offshoot conveyor, or passed on through more assembly operations, tracking its inspection results through the system. In any case, machine vision systems can provide a whole lot more information about an object than simple absence/presence type sensors.

Photoeyes and Pixel Arrays

Most people are familiar with a discrete photoeye. This is one of the most basic sensors in industrial automation; the reason we call it "discrete" or digital is that it only has two states: on or off.

The principal idea behind a diffuse photoeye is that it emits a beam of light and detects whether that light is being reflected from some object. If the object is not present, no light reflects into the photoeye's receiver. An electrical signal, usually 24 volts, is connected to the receiver. If an object is present, the signal turns on and can be used in a control system to make something happen.

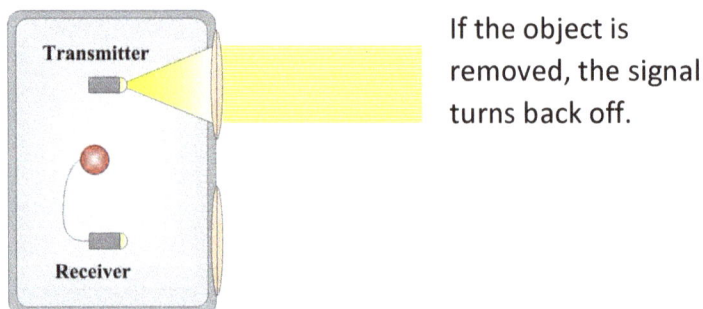

If the object is removed, the signal turns back off.

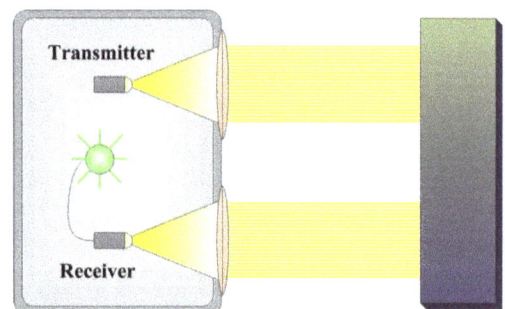

Figure 130 - Diffuse Photoeye, no target

Figure 131 - Diffuse Photoeye, ON state

Figure 132 shows a more advanced type of photoeye. This photoeye is analog; rather than only having two states, off and on, it can return a number signifying *how much* light is returning into its receiver. In the case of this particular photoeye, it can return 256 different values, from 0 (signifying no light, dark or black) to 255 (signifying lots of light, or white). It is returning a value of 76, or dark gray. This is about 30% of the maximum value of 255.

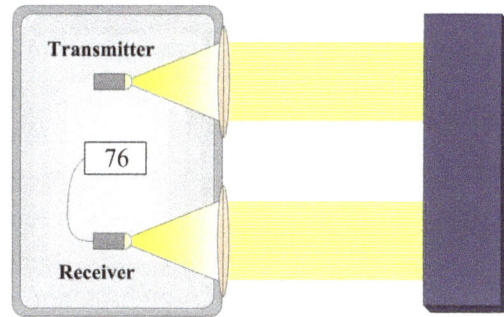

Figure 132 - Analog Photoeye, dark target

If a lighter object is placed in front of the sensor, it will return a higher number. If it produces 217, which is about 85% of the full range of 255, this indicates a much lighter shade of gray.

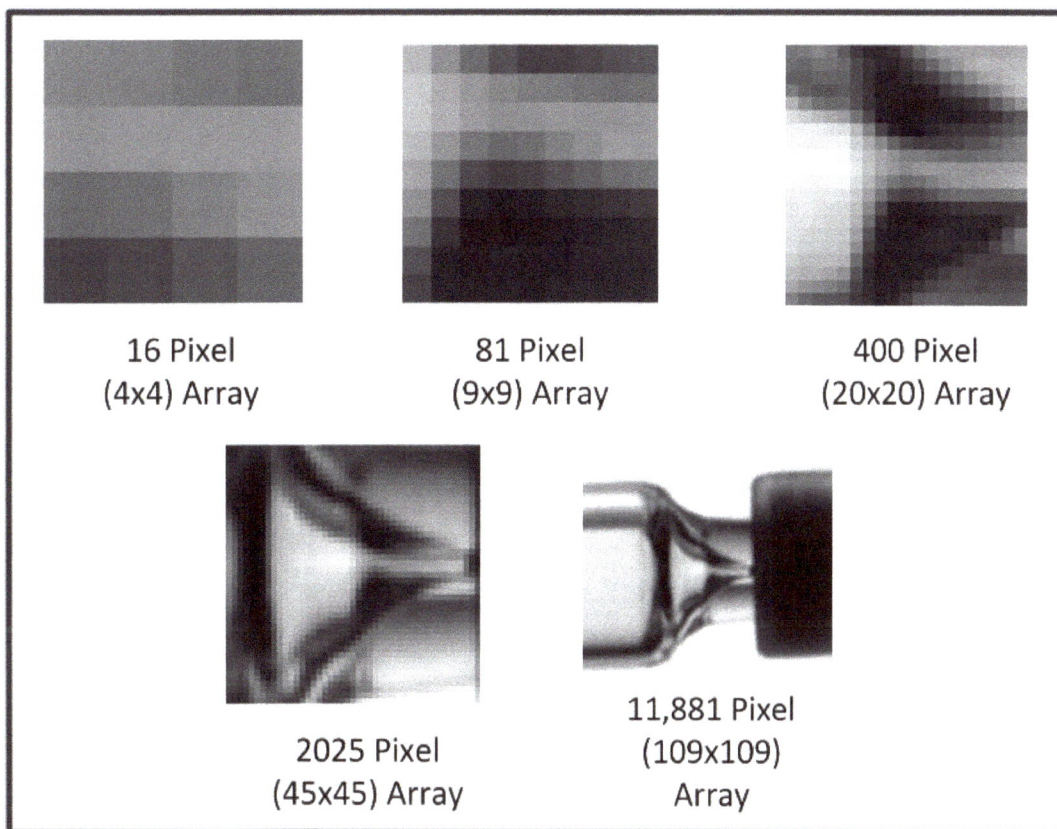

| 16 Pixel (4x4) Array | 81 Pixel (9x9) Array | 400 Pixel (20x20) Array |
| 2025 Pixel (45x45) Array | 11,881 Pixel (109x109) Array | |

Figure 133 - Ascending Arrays of Pixels making up an image

Now imagine what it would be like if we could arrange lots of tiny analog photoeyes in a square or rectangular array and point them at some object. This would create an image of the object in black and white, based on the reflectivity of wherever the sensor was aimed.

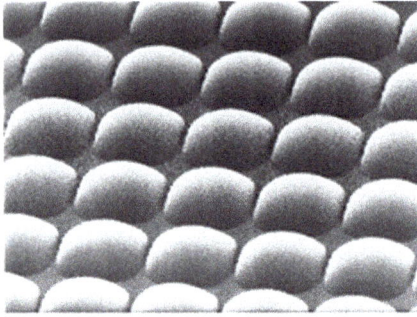

Figure 134 - Microscopic view of a black and white CCD

The individual sensed points in these images are referred to as "**pixels**". Of course, we don't use thousands of tiny photoeyes to create an image, instead a lens focuses the image onto a solid-state matrix of light detectors. A "CCD", (Charge Coupled Device) or CMOS (Complementary Metal-Oxide-Semiconductor) array of light sensitive solid-state devices are used in the matrix. Figure 134 shows a close-up view of what a black and white CCD looks like under a microscope.

Color Detection

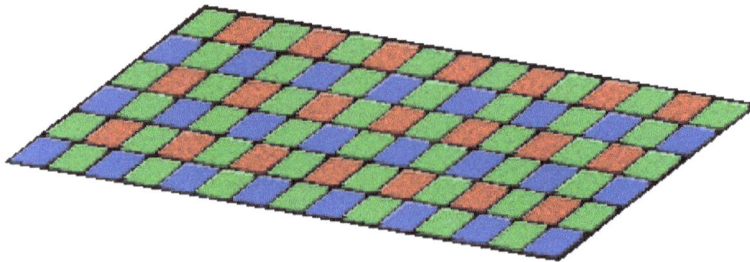

Figure 135 - "Bayer Filter" CCD pattern

Color detecting arrays are also often used in machine vision. A filter is placed over the grid of pixels and the amount of color reflected onto each sensor is turned into a numerical signal representing the intensity of each color. Notice that there are twice as many green filters as there are either red or blue ones.

Figure 136 shows a microscopic view of the surface of a RGB CCD. Of course, color detecting machine vision systems are more expensive than gray scale ones, so many machine vision systems use detectors without color filtering and processing.

Figure 136 - RGB CCD - Microscopic View

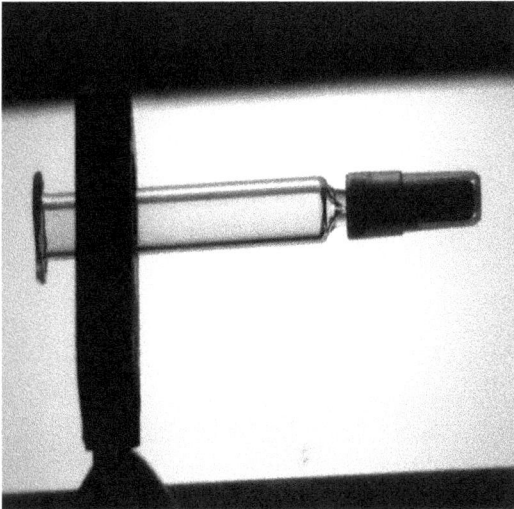

Figure 137 - 640 x 480 pixel image

The series of images in Figure 133 was actually only a small section of the image that was captured by the camera. 11,881 pixels is only a 109 x 109 square! This area is considered to be the "region of interest" for a particular inspection.

The actual image shown in Figure 137 was captured with a 640 x 480 CCD. It is from an inspection of glass syringes with a plastic cap on it, the purpose of the application is to determine if the cap is on tightly.

Components of the System

Lenses and Exposure

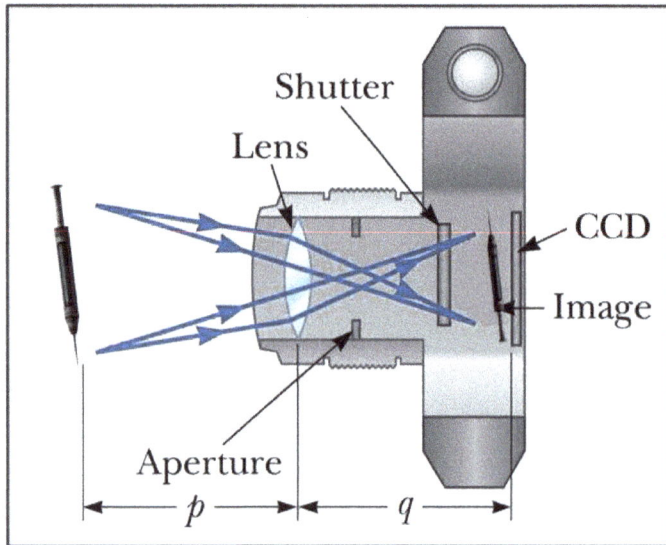

Figure 138 - Camera and Lens

In order to capture a good image, there need to be good sharp edges and lots of contrast in the captured image. Because the camera usually needs to be fairly close to the target, there is a relatively small range within which an object will be in focus. The closer the object is to the lens, the smaller the range. In an automated system, the camera is usually a fixed distance from the target, so the lens can be chosen based on the required **Field of View** (which is based on the size of the target).

A Field of View (FOV) chart is shown in Appendix C at the back of this book for reference.

Figure 139 shows the narrowing range of FOVs as lens sizes become smaller, a 4mm lens has a wider angle of view than a 12mm lens.

Of course, lighting is also an important part of getting a good image, but there are two other things that affect how much light exposure an image gets. The shutter or **exposure time** determines how long the image is imposed onto the

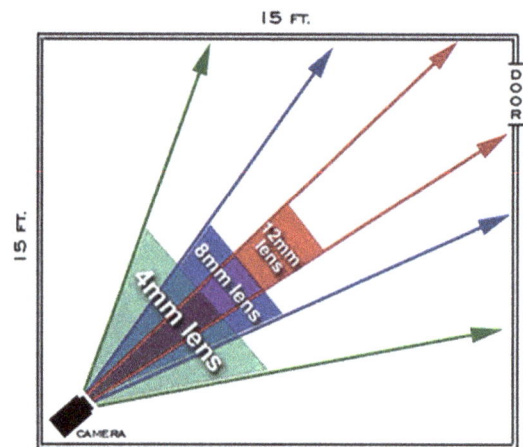

Figure 139 - Relative Fields of View

array of pixels. In machine vision, the shutter is electronically controlled, usually on the order of milliseconds.

The lens has an adjustment called the **aperture** that is opened or closed to let more or less light enter the lens. In combination with the exposure time, this determines the amount of light on the pixel array before lighting is even applied. With electronic shuttering, the exposure time is controlled electronically.

Depth of Field (DOF) is the range in distance of the object that is in acceptably sharp focus. If the object moves with respect to the camera, the depth of field influences the range of acceptable motion. DOF is determined by three factors – aperture size, distance from the lens, and the focal length of the lens. A lens with a large aperture (or small F number) gives a shallower DOF compared to a small aperture.

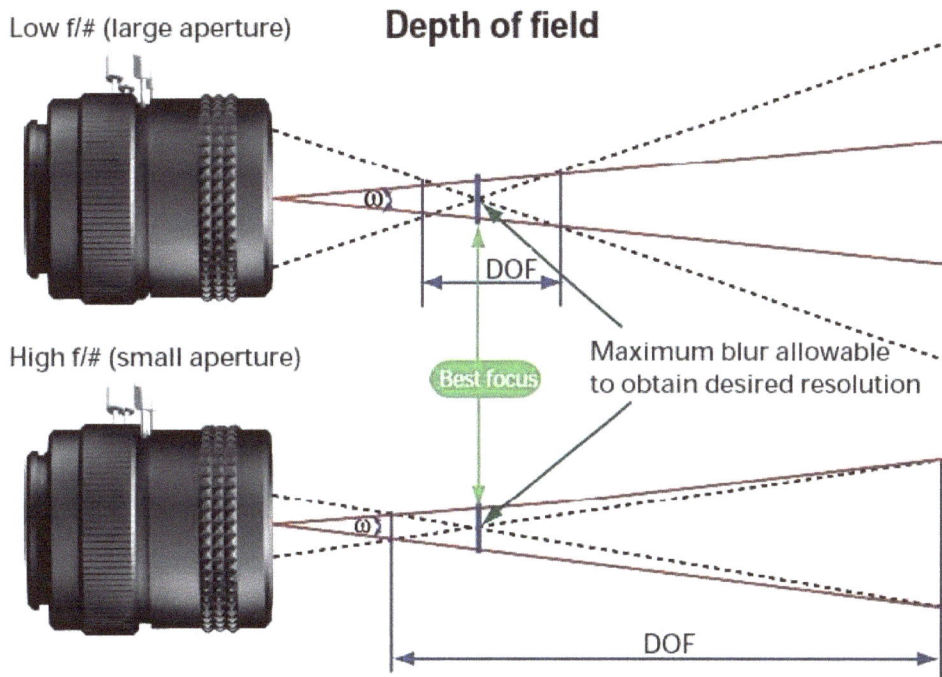

Figure 140 - Depth of Field

Resolution and Gain

The image of the syringe that we looked at before was from a 640x480 pixel CCD. This is known as the **resolution** of the camera or image. A larger array of pixels provides more detail and information about the image. Pixel arrays can be as large as 5000 x 5000 pixels for some machine vision applications!

Applying more power to the pixel sensors is known as adjusting the **gain**. This can allow the lighting or exposure time to be reduced but applying too much gain can reduce the image quality. Part of a good machine vision setup is a good balance of lighting and gain.

$$\% \text{ distortion} = \frac{(AD-PD) \times 100}{PD}$$

- AD = Actual distance of the image points from the center of the field
- PD = Predicted distance that the real world points would be from the center of the field if distortion was not present

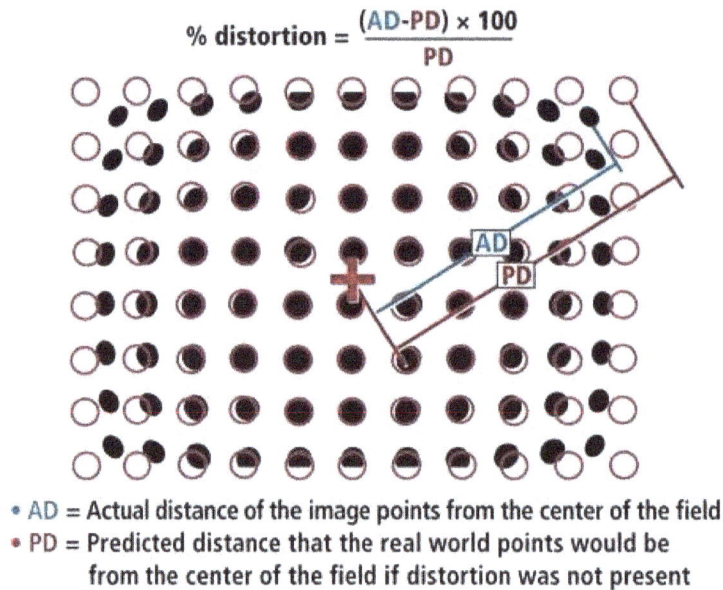

Figure 141 - Perspective Distortion or Parallax (Cognex)

Perspective distortion (parallax) is the illusion that the further an object is from the camera, the smaller it appears through a lens, and is caused by the lens having an angular field of view. Perspective distortion can be minimized by keeping the camera perpendicular to the field of view and further minimized by using a **telecentric** lens. These lenses maintain magnification over the depth of field, thus eliminating perspective distortion.

Machine vision systems can correct for image distortion mathematically. An image with dots is placed in front of the camera and true distances are entered into the software. This is especially important when motion control and robotics are involved, or if accurate measurements of large objects are needed.

Figure 142 shows an object (top) captured using a fixed focal length lens (bottom left) and a telecentric lens (bottom right). Using a conventional fixed focal length lens, the two parts appear to be different heights even though they are exactly the same. In the image on the lower right, the telecentric lens has corrected for perspective distortion and the objects can be measured accurately.

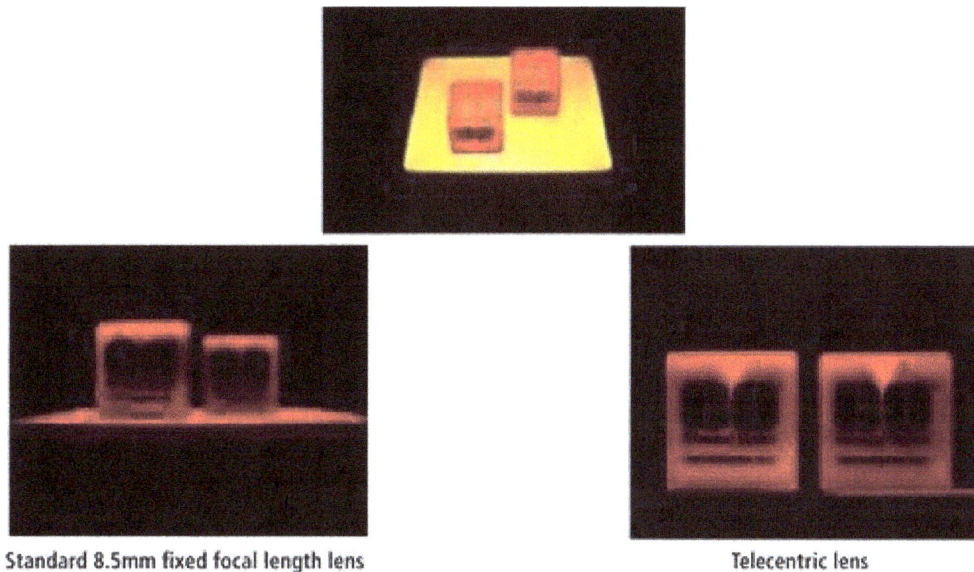

Standard 8.5mm fixed focal length lens   Telecentric lens

Figure 142 - Perspective Size Distortion

To avoid perspective distortion, keep the camera perpendicular to the FOV!

Lighting

There are many different methods of applying lighting to the image. The direction the light comes from, its brightness, and its color or wavelength compared to the color of the target are all important elements to consider when designing a machine vision environment.

Lighting may be the most important contributor to a successful machine vision application. There are many different types of lighting available; ring lights, LED lights in various colors and strobes are just a few.

Figure 143 - Various Lighting Formats

Lights can be used to either evenly illuminate the target and reduce glare or shadows, or lighting can be applied at an angle in order to _create_ shadows to highlight features. It is important to use DC lighting rather than AC to avoid the flickering that can be seen with 50 or 60Hz lighting.

Images are usually captured quickly, and the amount of time the electronic shutter of the camera is exposed relates to how much light is needed. Shorter image capture times require brighter light and moving objects will blur if the capture time is too long.

The color of the light is also important. For instance, a red object will appear very bright with a red light but will appear dark or black with a green light. White light contains various colors or wavelengths of light and is common when even illumination of multicolored objects is needed.

Illustrations on the following page show just some of the possible lighting configurations used in machine vision. These can be used in combination to create the desired effect.

Direct Ring Light: Ring lights provide lighting almost directly on the same axis as the camera's CCD or pixel array. Many cameras come with a built-in ring light around the lens. This reduces shadows, but can create glare.

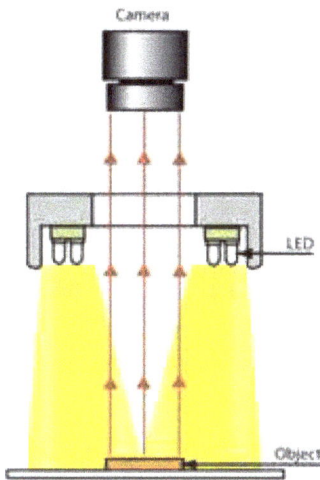

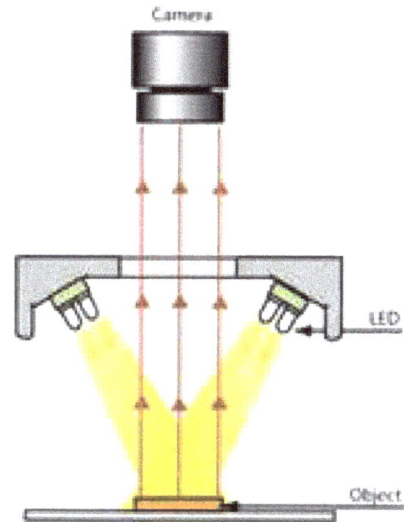

Figure 144 - Direct Ring Light

Angled Ring Light: Applying the light at an angle to the target can help create shadows (if desired) and reduce glare.

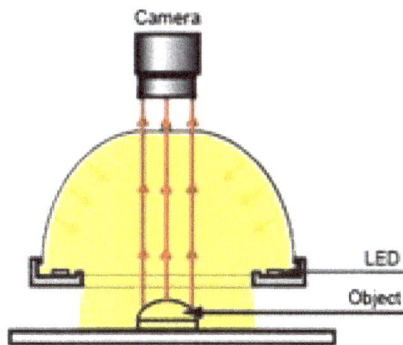

Figure 145 - Angled Ring Light

Dome/Tent Indirect Lighting: By reflecting the light from a white dome or tent, light can be placed diffusely across the target. This can reduce shadows and glare.

Figure 146 - Indirect Dome Light

Direct On-Axis Light (DOAL): By placing the camera on the backside of a one-way mirror, the light can be reflected directly in line with the pixel array. This reduces shadows as much as possible on perpendicular surfaces.

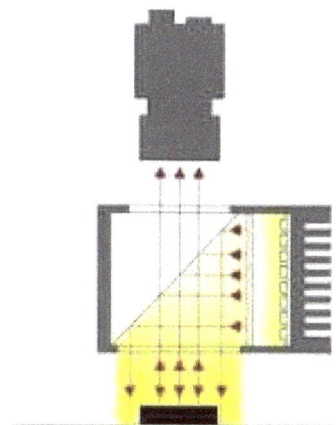

Figure 147 - Direct On-Axis Lighting

Machine Vision Configurations

Machine vision manufacturers provide their products in different form factors. Some machine vision cameras are completely self-contained "smart cameras", others may have a dedicated controller that attaches to multiple camera heads, while others use a frame grabber board in a computer, using the PC to run software that processes images.

Smart Cameras

Figure 148 - Cognex Insight 5000 "Smart Cameras"

"Smart Cameras" have all of the processing, I/O and communications built into the camera. Figure 148 shows a couple of cameras that even have the lighting built in!

Communications and 24 vdc digital signals are used to interface with the control system. The advantage of these types of cameras are that they are self-contained and don't require the use of a connected computer, other than for programming and setup. The disadvantage is that memory is limited, so an external computer needs to be added to store images. The processing power and speed is also often greater in larger systems.

Dedicated Processors

Figure 149 - Keyence CV Series

Systems with a dedicated processor can host 2, 4, or even more cameras in the same system. The digital I/O can be shared among the cameras, and they generally have more processing power than a "smart camera". That said, they are generally more expensive.

They may have the ability to store images in removeable memory cards, but generally communicate with a computer for this. The cameras themselves can be smaller and come in various resolutions. Programming and setup can be done with a computer or a simple monitor and mouse.

Computer-Based Systems

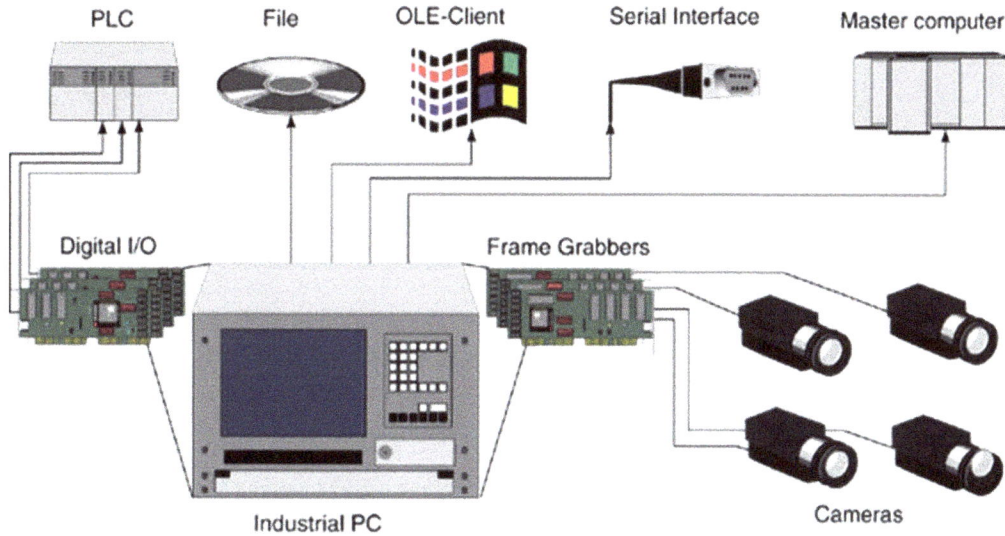

Figure 150 – Distributed computer-based machine vision system

Using a computer for processing images allows for a lot of flexibility but is generally the most complex. Figure 150 shows some of the elements that might be part of such a system.

The individual elements of this type of system can make it less expensive than dedicated processor types for a multi camera system, but there are so many choices and combinations of components that usually only system integrators and OEMs use them. There is a huge range of vendors for both hardware and software, and the machine vision trade shows and publications often focus on these computer-based products.

Software and Tools

So now we have captured an image, what can we do with it? There are many software tools that can be used to process the image. Some are applied before analysis (preprocessing) while others are used to determine the properties of the object being examined.

Preprocessing

Before examining an image for information, it is often important to "clean it up" or preprocess it. This makes the image easier to analyze.

Preprocessing can include things like image resizing, but usually involves geometric or color transformation. Color images may be converted to grayscale, or contrast between lighter and darker pixels could be increased.

Data can be augmented also; objects can be scaled or edge-enhanced, or holes can be eroded to make them appear larger.

Basic Software Tools

Machine vision tools operate by detecting differences in pixel color or value. For instance, a sudden change from dark to light or vice versa represents the edge of an object.

The typical method of creating simple tools is to draw the tool directly on the image or around the "area of interest".

Edge Detection: Detecting the edge of an object can be done by drawing a line across the area where the edge is expected, or by drawing a rectangular shape across it.

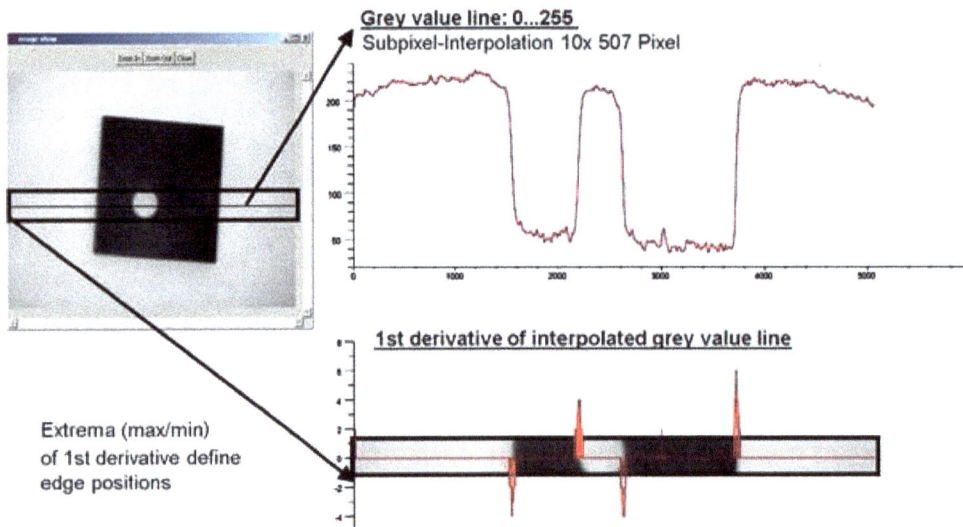

Figure 151 - Edge Detection

Figure 151 shows an edge detection tool drawn across a dark object with a hole in it. Notice the grayscale graph to the right of the object representing the pixel values along the line.

The 1st derivative graph below the grayscale graph is the contrast along the line, showing the transition from light, to dark, to light, to dark, and to light again. Note the spikes indicating where the edges of the image are.

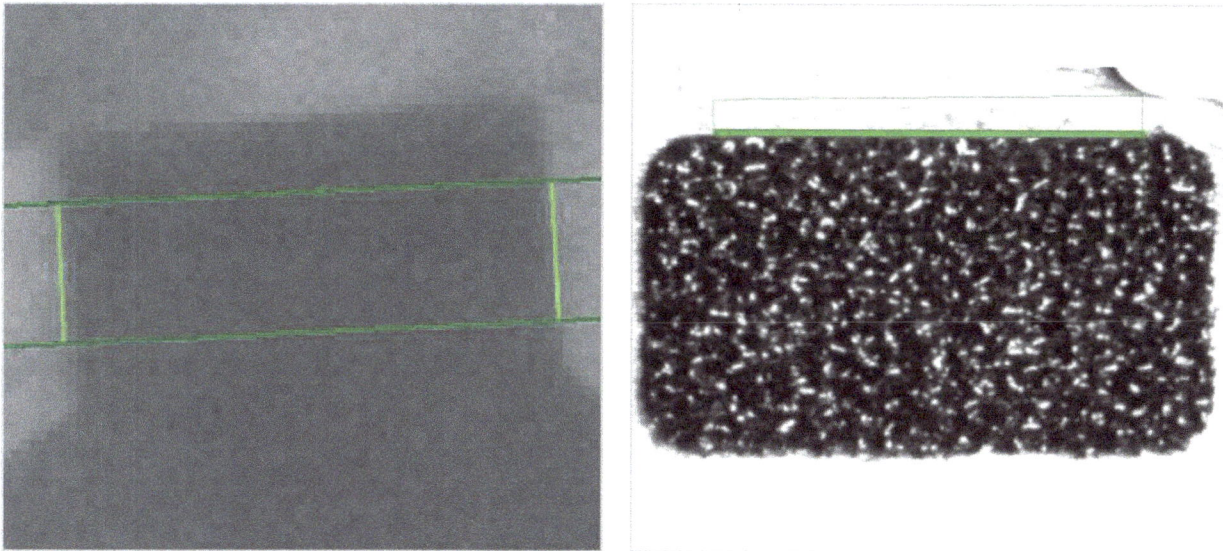

Figure 152 - Edge Finders - Rectangular Tool (Cognex)

By drawing an edge rectangle over an area, the tool will highlight where it finds contrast and can be made to fit the edge into a straight line as shown in Figure 152. These examples are particularly difficult: the left image has very low contrast, and the image on the right has an irregular edge.

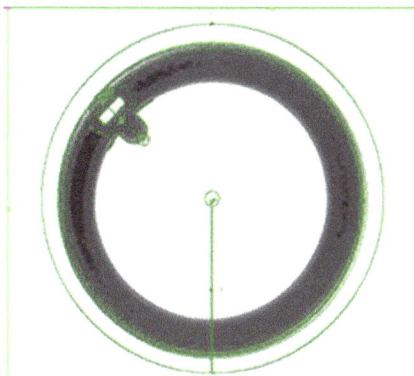

By drawing the edge tool as a circle, the edges of round objects can be found and evaluated for quality and size. Mathematical algorithms can be used to give the "roundness" a percentage value. The center of round objects can be calculated from where the edges are detected and established as a reference point.

Figure 153 - Circular edge tool

These tools can be drawn from inside of the object to the outside or vice versa and can be set to find light to dark or dark to light edges.

Measurements: Once edges or other object features are located, they can be evaluated for size or the distance between different objects.

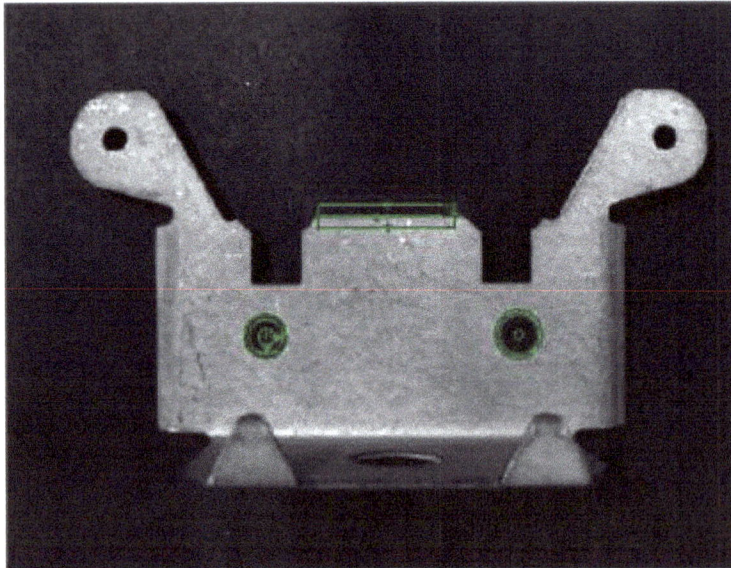

Figure 154 - Features on Object

Figure 154 shows two circles or holes and an edge. Each of these features have different measurements or positions that are associated with them. Pixels are counted in an X-Y direction, so the centers of the holes can be expressed as XY coordinates. The diameter of the circles can also be calculated as a number of pixels but can also be easily scaled into engineering units such as inches or millimeters. The same is true of the distance between the centers of the holes or the distance between the edge line and the y coordinate of the centers.

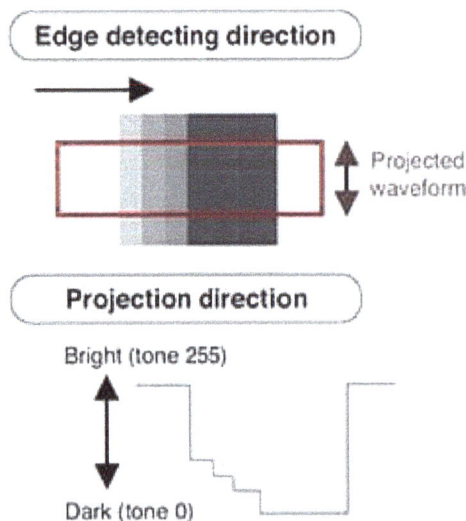

Figure 155 - Sub Pixel evaluation (Keyence)

Measurements can also be made to a **sub-pixel** level as shown in Figure 155. By examining the transition from light to dark pixels, up to 1/10 pixel accuracy can be obtained in a measurement.

Additional measurements that can be extracted from the data include angles and radius. Pixel measurements can be calibrated to real-world units of measurement.

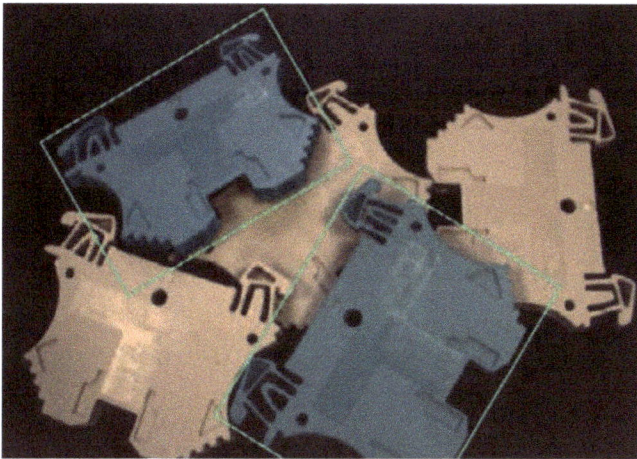

Color: As shown in Figure 156 colors can be extracted from the image. This can be used for color matching, sorting applications, and help with pattern matching.

Subtle color differences can be found by using additional software tools. Partial images can also be used with pattern recognition and color to accurately count objects.

Figure 156 - Color Sorting

Pattern Matching: A simple tool that can refine information about an object is known as a "**blob tool**". Blobs are self-contained shapes that can be sorted by basic parameters such as size or color, whether they have internal spaces or are eccentrically shaped.

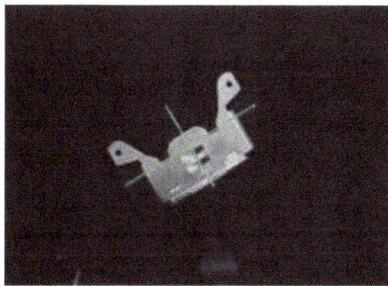

Figure 157 - Blob Tools and Pattern Recognition

More complex pattern recognition tools can "learn" a shape so that it can be recognized even if it is obscured by another object as shown in Figure 157.

Optical Character Recognition (OCR): Patterns can also be analyzed as alphanumeric characters and read as text. This allows for products to be categorized or information to be recorded. Figure 158 shows an example of information being extracted from an alphanumeric label.

Figure 158 - Optical Character Recognition (OCR)

1D and 2D Barcodes: Like OCR, machine vision systems can read both one and two-dimensional bar code information.

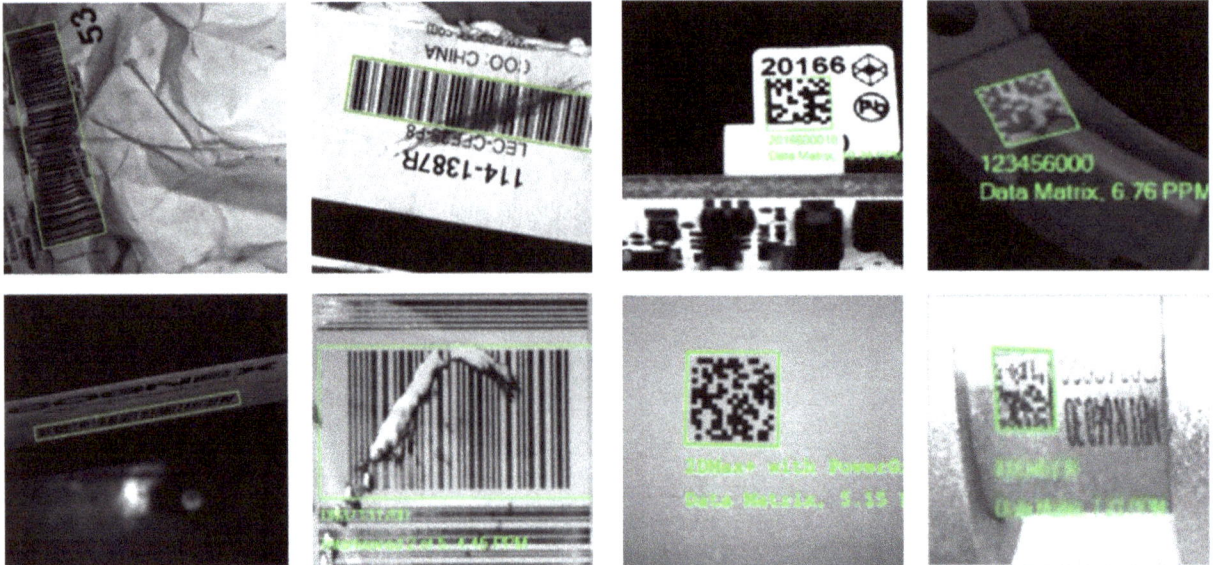

Figure 159 - One and Two Dimensional (DataMatrix) Barcodes

With both OCR and bar code data, the decoded information is usually saved to some type of spreadsheet or data registers in a PLC. This requires communication beyond simple digital signals. Most modern machine vision systems use ethernet for communication and have drivers for different common PLC protocols.

Image Location and Translation: Some of the previously listed tools such as edge, circle, and blob tools can be used to shift the positions of other tools. This is useful when the target can't be fixtured in the image area. If the area of interest is in view, the software sensors can move with the target image.

Troubleshooting and Maintenance

Most problems with machine vision systems are related to a change in the image. This may be related to lighting, lens focus, or position of the target or camera. In order to view the image, knowledge of how to operate the software is also necessary, though some applications will have a utility for this purpose. Of course, the system also should be designed properly so that inspections are stable and repeatable. Controlling the position of the inspection target is an important element.

Exercise 20

1.   What are the individual light sensors called in a CCD? _____

2. How many levels of light sensitivity are there in an image? _____

3. List some of the components of a vision system:

   _____
   _____
   _____
   _____
   _____
   _____

4. What are the three types of physical configuration for vision systems?

   _____
   _____
   _____

5. A gauging station is used to inspect and measure a 4" metal plate. Using Appendix D in the back of the book, choose an appropriate lens if the face of the camera is 12" from the target: _____

6. List some of the software tools used in vision systems:

   _____
   _____
   _____
   _____
   _____
   _____
   _____

7. Where should a troubleshooter start when troubleshooting a machine vision system? _____
   _____

Programmable Controllers (PLCs)

Programmable Logic Controllers or PLCs are the main method of controlling discrete machinery in manufacturing. The following information is extracted from *Advanced PLC Hardware and Programming, Frank Lamb, Automation Consulting LLC, April, 2019*:

What is a PLC, or Programmable Logic Controller?

A **PLC**, Programmable Logic Controller or Programmable Controller, is a digital computer used to control electromechanical processes, usually in an industrial environment. It performs both discrete and continuous control functions and differs from a typical computer in several important ways:

1. It has **Physical I/O**; electrical inputs and outputs bring real world information into the system and control real world devices based on that information.

2. It is **Deterministic**; it processes information and reacts to it within defined time limits.

3. It is often **Modular**; it can have I/O modules, communication modules or other special purpose modules added to it for expansion.

4. It is programmed using several defined **Languages**. Some languages allow the program to be changed while the machine or system being controlled is still running.

5. Software and Hardware are **Platform Specific**; components and programming software usually can't be used between different manufacturers.

6. It is **Rugged** and designed for use in industrial environments.

Unlike computers, PLCs are made to run 24 hours a day, 7 days a week and can resist harsh physical and electrical environments.

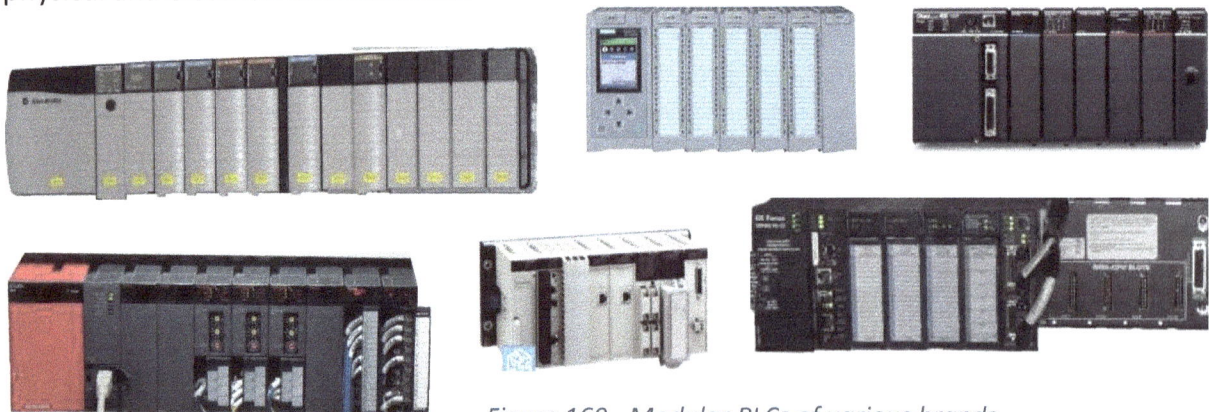

Figure 160 - Modular PLCs of various brands

Physical Layout of a PLC

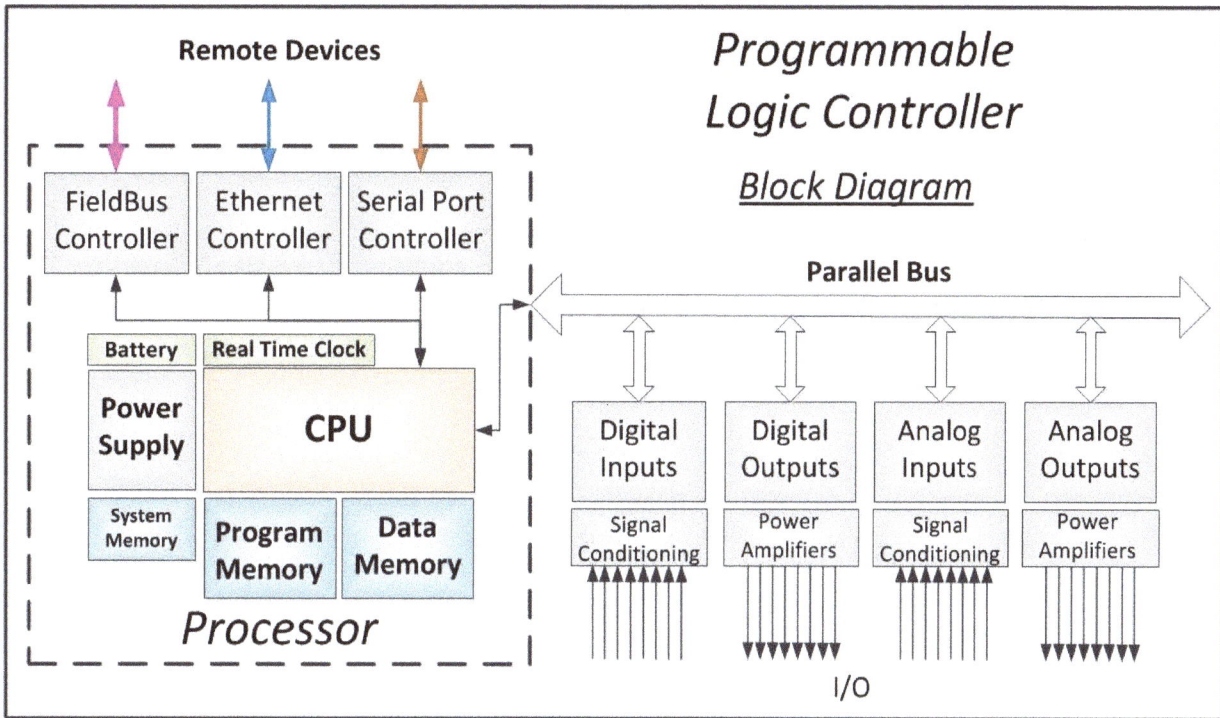

Figure 161 - PLC Layout

A block diagram of the physical arrangement of a PLC is shown in Figure 161. Not all the items shown in this diagram are present on every PLC, but this will give you an idea of a typical configuration.

Processor or CPU

The **CPU** processes all of the logic loaded into the controller and also contains the operating system. It usually has a real-time clock built in which is used for various functions. The system memory is also associated closely with the CPU.

Memory

PLC Memory consists of the operating system and firmware of the processor (sometimes called **System Memory**), the module firmware (if any), and the program and data that is used by the programmer. There are volatile and non-volatile areas of memory; the volatile part of memory needs a battery, "super-capacitor", or other rechargeable energy storage module to hold its program and/or data.

Though the program can be saved on Flash or SD RAM cards without a battery, the data exchange rate is too slow to use this for the actual interfacing of the program with its data. When the PLC is powered on, the program is loaded from non-volatile RAM cards into the user memory of the controller. Not all PLC platforms back up the user memory with a battery or other energy storage device, data memory may be lost when a processor loses power. Some platforms, however, ensure that the data is kept intact even when power is lost by use of battery backed RAM. This means that the values in data registers will be retained, and the program will start in its last state.

Other PLC platforms assign some parts of RAM to be "**retentive**" and other parts **non-retentive**. Omron separates its retentive bits into "holding" relays and non-retentive "CIO", and its data into the retentive DM Area and non-retentive "Work Area". Siemens allows its general "marker" memory to be assigned as retentive or non-retentive and defaults to only 16 bytes of retentive marker memory, but it can be changed. Siemens data blocks, however, are retentive unless defined not to be. Allen-Bradley's memory is all retentive.

The operating system itself on a processor is held in non-volatile System memory, called "firmware". To change the firmware on a PLC a "Flash" program or tool needs to be used to download it. This is usually included with the programming software.

I/O, communications and other modules also often have firmware built in. The firmware update tools can also update these modules and the firmware is usually available from the manufacturer's website. It is necessary to have software that is at least as up-to-date as the firmware being installed.

The RAM part of memory in a PLC can be separated into two general areas: **Program Memory** and **Data Memory**.

Program memory consists of the lists of Instructions and Program Code. This is what is sent to the processor. The act of sending the program instructions to the PLC is called "downloading" on most brands of PLC, however this may differ on some platforms.

Data memory includes the Input and Output Image Tables as well as Numerical and Boolean Data. You will find that most of the data used in the PLC program is internal memory and not directly related to I/O!

As the program executes, it keeps track of whether bits (BOOLS) are on or off and the values of numbers in Data Memory. Different platforms have different ways of organizing this data.

I/O (Inputs and Outputs)

Physical I/O can be **_Discrete_**, single signal bits that are either on or off, or **_Analog_**, signals that change amplitude in either voltage or current.

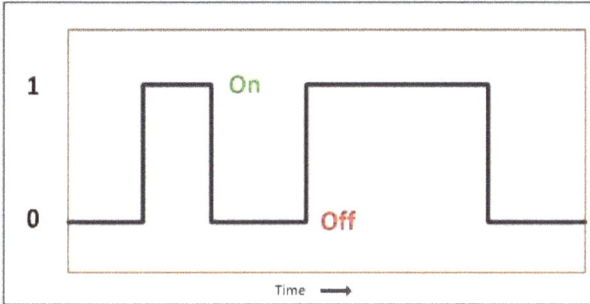

Figure 162 shows a **Digital** or discrete signal. Typical signal levels for discrete inputs and outputs are 24V DC and 120V AC, but other levels may be present depending on the type of device or input card. In addition to the designation one (1) and zero (0) or on and off, discrete signals may be described as true or false.

Figure 162 - Discrete Signal

Examples of digital input devices are shown below:

Pushbutton   Photoeye   Proximity Switch

Here are some digital output devices:

Pilot Light   Motor Starter   Solenoid Valve

And here is a device that is both a digital input and a digital output!

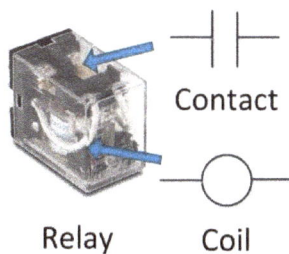

Contact

Relay   Coil

Figure 163 - Discrete I/O devices

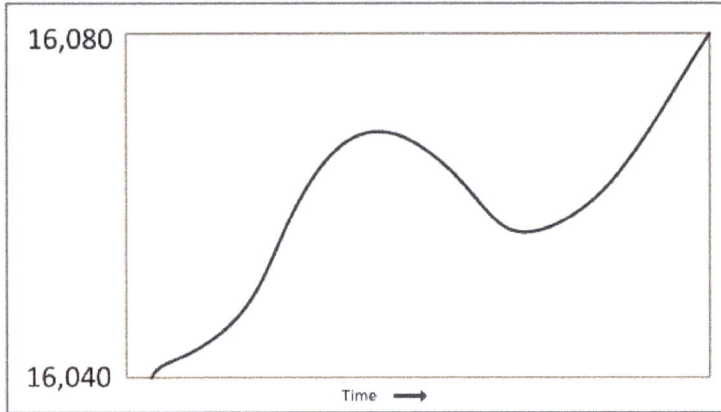

Analog signals vary in either voltage or current. Ranges are typically 0-10V DC or -10 to +10V DC (voltage type) or 0-20mA or 4-20mA (current type). The electrical signal is then converted into a number for use in the PLC program.

Figure 164 - Analog Signal

Here are some examples of analog input devices:

| Potentiometer | Pressure Transducer | Platinum RTD |

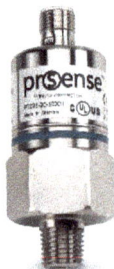

And these devices use analog outputs:

| Proportional Valve | Linear Actuator | VFD Speed |

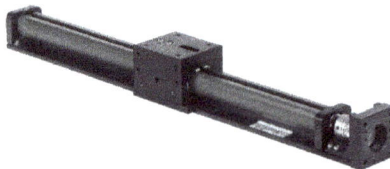

Figure 165 - Analog I/O Devices

Analog signals are converted by Analog to Digital Converters (ADCs) for inputs, or Digital to Analog Converters (DACs) for outputs.

Communications

As mentioned earlier in this book, there are a wide variety of communications methods used to communicate with PLCs and their devices. Serial communications, ethernet and various fieldbus protocols are just some of these.

Platform	Serial	Fieldbus	Ethernet
Allen-Bradley 500	DF1 (RS232)	DeviceNet, DH485	TCP/IP
Allen-Bradley 5000	DF1 (USB)	ControlNet	Ethernet/IP
Siemens Step 7	MPI, PtP	Profibus, Profinet	Profinet, TCP/IP
Siemens TIA Portal	MPI, PtP	Profibus, Profinet	Profinet, TCP/IP
Omron	Host Link	RS485/Modbus-RTU	Ethernet/IP
Beckhoff	TwinCAT	Modbus-RTU/ASCII	EtherCAT

Note that some of these protocols are the same across different brands, even then, they may not communicate properly without modification of parameters.

Serial and Ethernet are the two most common methods of connecting with a PLC, and Fieldbus protocols are usually used to communicate with remote or distributed I/O devices. Ethernet is used both to program PLCs and for distributed devices, but it is important that I/O communication is deterministic, as described in the previous PLC definition.

PLC Data

Depending on the platform, PLCs treat data in different ways. Older PLCs were usually either Byte based (Siemens, 8 bit) or Word/Integer based (Allen-Bradley, 16 bit). This had a major effect on how memory was stored and used. Some PLCs have registers assigned to specific data types, that is, Bit, Integer or Real, (Allen-Bradley SLC and Micro), whereas other brands may separate data by whether it is retentive (Holding Relays, Omron) or place all data together ("V Memory", Koyo/Automation Direct).

It is important when learning a new PLC's programming platform to first understand how its memory is organized. In the older **GE** PLCs for instance, data memory and I/O share the same space. It could be quite embarrassing if you were to cause actuators to move when intending to simply save an integer to a register!

A-B SLC	
O	Outputs
I	Inputs
S	System
B	Bits
T	Timers
C	Counters
R	Control
N	Integers
F	REALs

Siemens S7	
I	Digital In
Q	Digital Out
PIW	Analog In
PQW	Analog Out
M	Memory (M)
DB	Memory Blocks

Omron	
CIO	Basic I/O (Discrete)
CIO	Special I/O (Analog)
CIO	CPU Bus I/O
W	Work Area
H	Holding Area
A	Aux Relay Area
TR	Temp. Relay Area
D	Data Memory Area
E	Ext. Memory Area
T	Timers
C	Counters
TK	Task Flags
IR	Index Registers
DR	Data Registers

Koyo		
T	Timer Curr. Val	V0-377
	Data Words	V400-777
CT	Counter Curr. Val	V1000-1377
	Data Words	V1400-7377
	System	V7400-7777
	Data Words	V10000-35777
	System	V36000-37777
GX	Remote Inputs	V40000-40177
GY	Remote Outputs	V40200-40377
X	Disc. Inputs	V40400-40477
Y	Disc. Outputs	V40500-40577
C	Ctrl. Relays	V40600-40777
S	Stages	V41000-41077
T	Timer Status	V41100-41117
CT	Counter Status	V41140-41157
SP	Special Relays	V41200-V41237

Figure 166 - Various PLC Data Structures

Figure 166 shows the layout of several PLCs' memory areas. The first list, **Allen-Bradley's SLC** and **MicroLogix** family, shows that data is segregated into numbered files, O0, I1, S2…F8. Each data file is expandable up to 255 words but after that, new file numbers must be added, for instance N9, B10, and so on.

The next table shows **Siemens' Step 7**. I/O is assigned during hardware configuration rather than by slot number as with Allen-Bradley. The general memory area "M" is of a fixed size, whereas Memory Blocks or Data Blocks (DBs) contain a mix of data types and can be up to 64KB in size!

Omron's memory sizes are fixed in dimension for each data type; memory is not allocated dynamically as in the previous two examples. It is unique in that it separates retentive memory (Holding Area) from non-retentive (Work Area).

Koyo (Automation Direct) uses a large data area much as the GE system described earlier; each type of data is fixed in size and can't be expanded. All data can be accessed by directly addressing the "V" addresses.

I/O addressing varies from brand to brand. Inputs may be addressed as I or X, outputs as O, Q or Y, and analog I/O designations may use a completely different format than digital ones.

Some brands, such as Allen-Bradley, designate I/O based on the slot number where the card is assigned while configuring hardware; this can't be changed. Other platforms, like Siemens, have a default location where I/O is assigned during configuration, but this can be overridden by the programmer. Addressing may also be Octal, Decimal or even Hexadecimal!

	Allen-Bradley	Allen-Bradley	Siemens	GE	Omron	Mitsubishi	Codesys
	SLC-500	ControlLogix	S7	311	CP1E	FX2N	
Digital Input	I:1/3	Local:1:I.Data.3	I0.3	I0103	I0.03	X003	%IX4000.3
Digital Output	I:2/5	Local:2:O.Data.5	Q3.5	Q0205	Q1.05	Y025	%QX4002.5
Analog Input	I:3.2	Local:3:I.Ch1Data	PIW272	AI06	CIO200	D302	%IW2022
Analog Output	O:4.3	Local:4:O.Ch2Data	PQW800	AQ007	CIO210	D403	%QW2036
Address Base	10	10	8	10	10	8	

Figure 167 - PLC I/O Addressing

Figure 167 shows some examples of I/O addressing from several platforms.

Viewing Data Types: PLC software allows the user to view a number in many of the formats listed here, removing the need for a calculator. It is not always clear in what format the data is being viewed, but there are often designators. In Siemens, a signed integer (decimal or Base 10) will have no designator; however a Hex number will have a W#16 prefix, indicating that it is base 16. A REAL will have a decimal point or be expressed with an exponent, while a binary representation may have a prefix or appear as a string of ones and zeros.

Dot Fields and Separators: If a single bit of an integer is designated, it may be shown with a separator such as a slash or a dot; for example, N7:5/3 (Allen-Bradley, the fourth bit of the sixth word; numbering starts at zero) or Q3.2 (Siemens, the third bit of the fourth output byte).

Dot fields are also often used to designate an element of a complex data type, such as a timer. For example, Timer1.ACC designates the accumulated value (integer or double integer) of Timer 1. It is important to understand how memory is addressed for your particular PLC before beginning a program.

Tags: Many modern PLC platforms don't use numerical data registers at all. Instead, they allow users to create memory objects as required in the form of text strings. Allen-Bradley's ControlLogix and Siemens' TIA Portal platforms are examples of this. Most major PLC manufacturers make a PLC with tag-based data. Tags are also called Symbols on some platforms, but a symbol is not necessarily a tag; it may simply be a mnemonic address or shortcut to a register address. Tagnames are downloaded into the PLC and used instead of an address.

Tags are usually created in a data table as required. Instead of numeric addresses such as "B3:6/4" or "DB2.DBW14", symbolic names such as "InfeedConv_Start_PB" or "Drive1402.ActualSpeed" are created as memory locations. As tags are created, details such as the data type (BOOL, Timer, REAL) and the display style (Hex, Decimal) need to be selected.

Tags have the advantage of being more descriptive than numerical register numbers. In addition, descriptions and symbols from register addresses were only present in the computer and were not downloaded into the PLC. With tags, since the address is the actual register location, a tag-based program can usually be uploaded straight from the PLC.

Also, the same tags from the PLC program can be used in the HMI or SCADA program directly. This saves time rather than having to map PLC addresses to HMI tags.

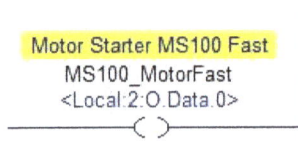

Motor Starter MS100 Fast
MS100_MotorFast
<Local:2:O.Data.0>

Of course, I/O addresses are still created from the hardware configuration of the PLC, but manufacturers have created various ways to connect I/O addresses with tags. One of the most useful of these is Allen-Bradley's ControlLogix platform, where any tag or address can be "**aliased**" to any other and both show up in the ladder logic, as shown in the figure to the left. Modicon's newer

Figure 168 - Aliasing

platforms also allow "Symbols" or tags to be connected to I/O points in a similar way.

Data Groups: In addition to the "atomic" or elemental data types such as BOOL, Byte, Integer, or REAL as described back in the Data section of this book, individual elements can be grouped.

Array: An Array is a group of _similar_ data types. For instance, an array can be defined that contains ten integers, or 50 REALs, or 32 BOOLs. Data types can't be mixed in an array.

Complex data types such as Timers, Counters, or User Defined Types (UDTs) can also be placed into an array. Typically, an array will be shown with square brackets, such as Delay_Tmr[6]. This designates element number 7 of the array if the array begins at 0.

Some platforms allow multidimensional arrays to be defined, such as Integer [2,4,5]. This means Integer number five of the fourth group of the second group.

Elements that are composed of more than one data type are known as **Structures**. A structure may be defined by the programming software, as with instructions, or by the programmer.

User Defined Type (UDT): A UDT is a group of _different types_ of data, or a structure. Later in this course, data types that consist of more than one type of data will be discussed; for instance, Timers and Counters consist of two integers or double integers and several bits, all combined into a structured data type called "Timer" or "Counter".

A UDT can only be used with Symbols or Tags; this is because a UDT is not data. Instead, it is a _definition_ of data. After defining a UDT, a tag or symbol must be created using the new data type.

A common reason to build a UDT is to describe an object more complex than a simple element of data. As an example, a Variable Frequency Drive has many pieces of data that may be associated with it. For instance, a motor needs to start and stop. It has various numerical parameters to describe its movement such as commanded speed, actual speed, acceleration, and deceleration. We may also want to know its status: whether it has faulted and what kind of fault has occurred.

UDT Name:	"Drive"	
Name	Type	Description
Run	BOOL	Run Command
Stop	BOOL	Stop Command
Alarm	BOOL	Drive in Alarm
Running	BOOL	Run Status
CMD_Speed	REAL	Commanded Speed %
Act_Speed	REAL	Actual Speed %
Accel	INT	Acceleration ms
Decel	INT	Deceleration ms
AlarmStatus	SINT	Alarm Number

Name	DataType	Status	Description
Drive_5207	"Drive"		Spindle Drive VFD 5207
Drive_5207.Run	BOOL	0	Run Command
Drive_5207.Stop	BOOL	1	Stop Command
Drive_5207.Alarm	BOOL	1	Drive in Alarm
Drive_5207.Running	BOOL	0	Run Status
Drive_5207.CMD_Speed	REAL	27.34	Commanded Speed %
Drive_5207.Act_Speed	REAL	0.00	Actual Speed %
Drive_5207.Accel	INT	40	Acceleration ms
Drive_5207.Decel	INT	50	Deceleration ms
Drive_5207.AlarmStatus	SINT	4	Alarm Number

Figure 169 - UDT (Left) and Tag made from the UDT (Right)

On the left is a UDT named "Drive" defined in the software, and on the right is a tag made from the UDT. The definition does not get downloaded to the processor; it can only be modified on the programming device.

The sub-elements of the tag are an example of the dot fields described earlier. By making a UDT, many drives can be added to a program without a lot of extra typing. UDTs are in important element in Rapid Code Development.

TIP: On non-tag-based systems, UDTs can cause problems if the commented program is not available. Remember: descriptions and non-tag symbols are not kept in the processor. This is why it is difficult to reconstruct a Siemens S7 program if you don't have the original code; the downloaded data blocks do not contain the names of the elements.

Hardware Configuration

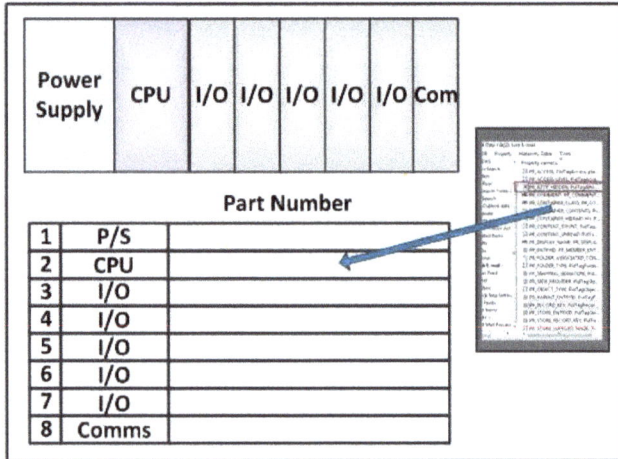

		Power Supply	CPU	I/O	I/O	I/O	I/O	I/O	Com

Part Number

1	P/S	
2	CPU	
3	I/O	
4	I/O	
5	I/O	
6	I/O	
7	I/O	
8	Comms	

Figure 170 - Hardware Configuration

The first step in beginning a new PLC program is configuring the hardware. This is because different processors have different amounts of memory, and because the addresses for I/O are determined by the configuration. As cards are added, new addresses or tags are generated and available for selection in the program.

You can't write a program until you know the I/O addresses and memory configuration!

Some platforms assign I/O addresses by location in the rack (Slot Number) while others allow the programmer to assign addresses. Usually there will be a default address that can be modified in the properties of the card. As explained later in this course, some brand's memory allocation can overlap the I/O area, so careful configuration and planning is important. Selecting hardware will often also include entering a hardware or firmware number for each card. If a rack is used, selecting the rack size will be necessary before inserting cards.

Depending on the platform, the next step in setting up a new program is to configure the data areas. Newer tag-based systems allow tags to be added one at a time either locally or globally, but many of the older platforms require data tables to be sized offline.

Exercise 21

1. What is a tag? Is it the same thing as a symbol?

    _____

    _____

    _____

2. Where does a PLC begin its scan (in which routine)? _____

3. Before writing code in a PLC, the _____ must be configured.

Programming

In order to make a PLC do what you want it to, it must be programmed. Programming can be done offline, after which it is downloaded to the PLC. Before the download process, the software checks the program for errors and compiles it into machine code.

Programs can often be edited online also. When downloading a program, the processor must be stopped and all control ceases; editing online allows changes to be made while the program is running, without interruption.

Program Organization

Organization of a program is very important both from a functional standpoint and so that operations can be found easily. Different platforms provide the ability to create Tasks, Programs, Routines, and "Sections".

Routines: All PLC platforms have a routine that is designated to run first. It is important when learning a new platform to identify which routine this is. This is sometimes known as the "Main Routine".

As with data memory, the program itself can be organized in different ways. Major PLC platforms all have some form of subroutine, though they may be called by different names. Allen-Bradley's PLC5 and SLC500 platforms organize their subroutines by file number, where file 2 or "Ladder 2" is the routine that runs first and calls the other routines. As new routines are created, they are called Ladder 3, Ladder 4, etc.

Siemens' routines can take different forms; OBs or Organization Blocks each have a special purpose based on their number. For instance, OB1 is the continuous routine that runs first, OB86 runs if there is a network fault, and OB35 runs on a periodic basis set by the programmer, such as every 100 milliseconds. Functions, or FCs, are much like standard subroutines; however, they have local temporary memory. Function Blocks, or FBs, are like FCs but have retentive memory in the form of the data blocks mentioned earlier.

Koyo or Automation Direct's PLCs have subroutines, but they also have special routines called "stages" that automatically deactivate the stage from which it was called. Omron has subroutines called "sections" and also periodic routines called "tasks".

```
Controller Organizer
  ⊟ 🗃 Controller Sample
       🗋 Controller Tags
       🗀 Controller Fault Handler
       🗀 Power-Up Handler
  ⊟ 🗃 Tasks
     ⊟ 🗃 MainTask
        ⊞ 🗐 MainProgram
        ⊟ 🗐 Cell_1_Robot
             🗋 Program Tags
             🗐 a_System
             🗐 b_Inputs
             🗐 c_AutoSequence
             🗐 d_Outputs
             🗐 e_Faults
             🗐 f_DataAcq
        ⊟ 🗐 Cell_2_Dial
             🗋 Program Tags
             🗐 a_System
             🗐 b_Inputs
             🗐 c_AutoSequence
             🗐 d_Outputs
             🗐 e_Faults
             🗐 f_DataAcq
          🗀 Unscheduled Programs / Phases
     ⊟ 🗃 Motion Groups
          🗀 Ungrouped Axes
        🗀 Add-On Instructions
     ⊟ 🗃 Data Types
```

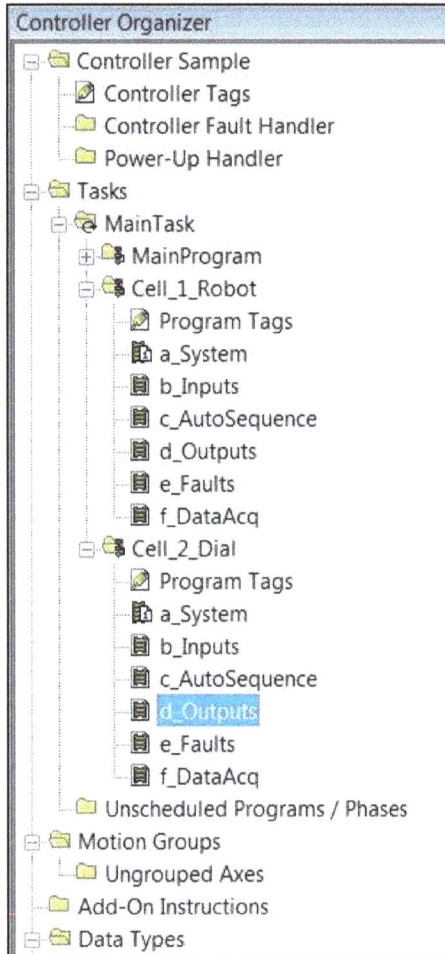

Figure 171 - Program Organization

TIP: It is very important to consider whether memory will be **Global** (available to all programs and routines) or **Local** (only available to part of the program) before beginning a program. Think about whether you will have multiple instances of the same code.

More powerful PLCs may also allow multiple programs to be placed into a task, as shown in this Allen-Bradley ControlLogix Controller Organizer figure. While the programs are still scanned one at a time, this allows data tables or tag lists to be assigned to one program, rather than being global. Programs are then scheduled to run in a specific order under the task.

This also allows programs to be duplicated under different names, but with the same tag names. This allows for rapid code development, since a program can be written and tested then copied, addresses and all.

Tasks can also be assigned to run on a periodic basis as with the OB35 block mentioned earlier in the Siemens description. Analog I/O processing is often done this way to accommodate PID instruction cycle frequency.

Languages (IEC 61131-3)

PLCs have evolved in different ways depending on the manufacturer. Programming software and methods of handling data can differ immensely from platform to platform. Because of this, the International Electrotechnical Commission (IEC) created an open standard in 1982 that defines what equipment, software, communications, safety, and other aspects of programmable controllers should look like. After the national committees had reviewed the first draft, they decided it was too complex to treat as a single document. They originally split it into five sections as follows:

Part 1 - General Information

Part 2 – Equipment and Testing Requirements

Part 3 – Programming Languages

Part 4 – User Guidelines

Part 5 - Communications

Currently, the standard is divided into 9 different parts, and a tenth is being worked on.

The third part, **IEC 61131-3**, defines the languages that are used in programming. It describes two graphical languages and two text languages, along with another graphical method of organizing programs for sequential or parallel processing. It also describes many of the data types described previously in this document.

The first two graphical languages described are Ladder Logic (LAD) and Function Block Diagram (FBD). The text languages are Structured Text (ST) and Instruction List (IL), while the organizational method described above is Sequential Function Charts (SFC), which is also graphical. An additional extension language is Continuous Function Charts, (CFC), which allows graphic elements to be positioned freely; it can be considered as an extension of SFC.

The following examples illustrate the five IEC programming languages; the addresses used are generic and the logic shows selection of Auto and Manual modes, along with a timer enabling "Cycle". These examples do not come from an actual programming language or brand but are meant to illustrate uses of the languages.

The following examples each perform the same function written in all 5 IEC languages:

Ladder Logic (LAD)

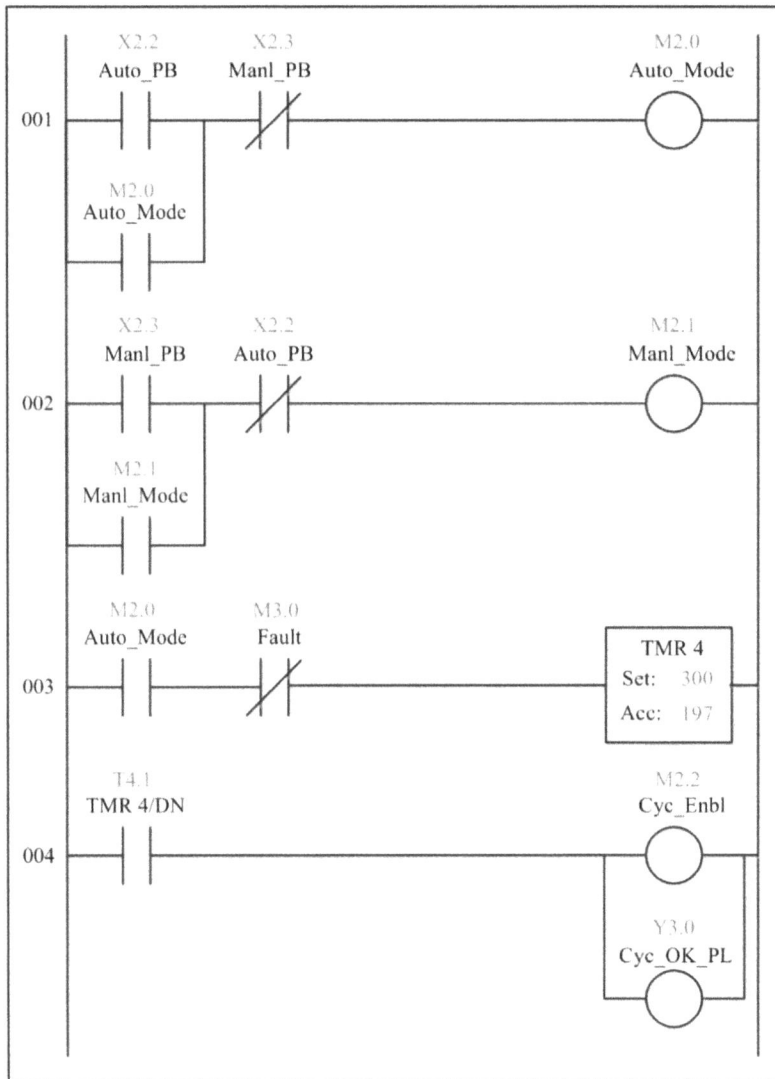

Figure 172 - Ladder Logic (LAD)

Ladder Logic evolved from electrical circuit drawings, which resemble the shape of a ladder when drawn. As a graphical language, the instructions represent electrical contacts and coils; the vertical sides of the ladder diagram are known as "rails" and the horizontal circuits are often called "rungs". In Siemens software, the rungs are known as "networks".

The "X" addresses represent physical inputs, while the "Y" address is a physical output. The "M"s are internal memory bits.

Because of the variety of addressing schemes as described previously, register representations can mean different things on different platforms.

When monitoring Ladder Logic in real time, usually contacts and coils change color to indicate their state in the logic. If a path of continuity exists from the left rail to the coil, the address will be said to be "On" or "True".

The timer shown in the diagram may also show a time base. If the above timer's preset is three seconds, the time base would then be 10 milliseconds.

More on ladder logic will be discussed in a later section.

Function Block Diagram (FBD):

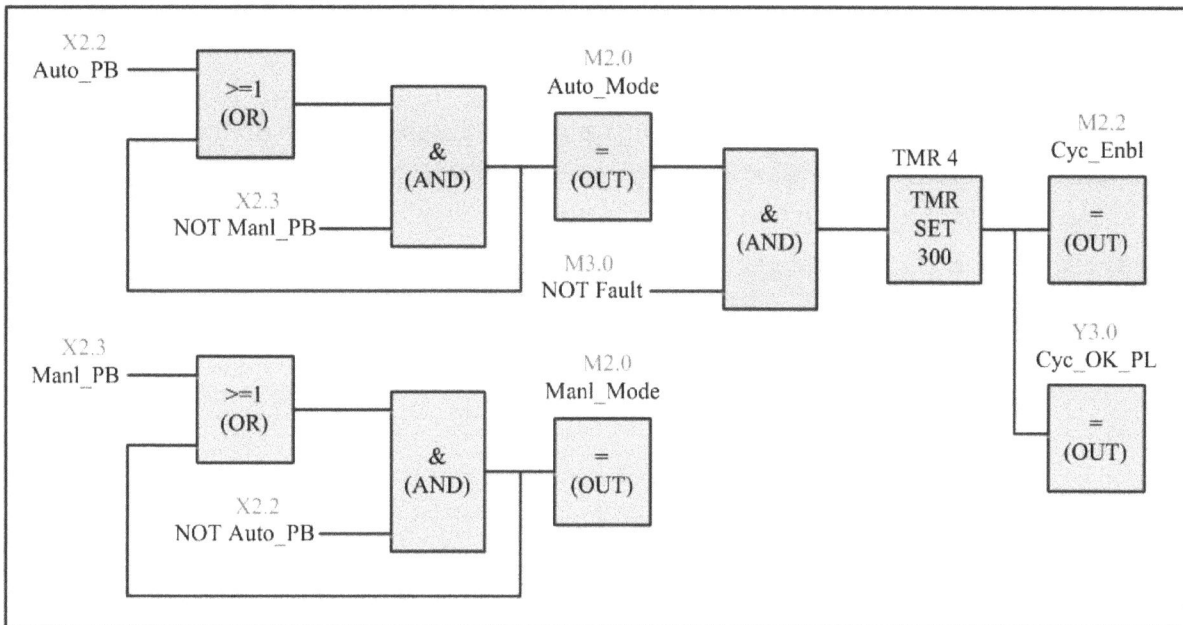

Figure 173 - Function Block Diagram (FBD)

Figure 173 shows the same functionality as that in the Ladder Logic diagram. The function blocks evolved from Boolean algebra, the AND and OR representing basic logic. More complex blocks are used for math, loading, comparing and transferring data, timing, and counting. As with the previous example, this does not represent any particular brand of PLC.

There are some functions, such as XOR (Exclusive OR), that cannot be easily represented in Ladder Logic. Also, because of the complex nature of some FBD drawings, logic can often extend across many pages. Off-page connector symbols are used to show these connections.

```
LD X2.2 Auto_PB
O M2.0 Auto_Mode
AN X2.3 Manl_PB
= M2.0 Auto_Mode
LD X2.3 Manl_PB
O M2.1 Manl_Mode
AN X2.2 Auto_PB
= M2.1 Manl_Mode
LD M2.0 Auto_Mode
AN M3.0 Fault
= TMR 4 Set 300
LD T4.1 TMR 4/DN
= M2.2 Cyc_Enbl
= Y3.0 Cyc_OK_PL
```

Figure 174 - Instruction List (IL)

Instruction List (IL)

Graphical languages are usually converted into a text language called Instruction List before being compiled into another low-level code called Machine Language. Before the advent of personal computers, handheld programmers were used to type instructions into the PLC before compilation. These devices often had pictures of Ladder Logic contacts on the keys.

Some platforms, such as Siemens, make extensive use of IL in programming; Siemens' version of Instruction List is called **Statement List**, or **STL**. Statement List has many commands that are not possible in Ladder Logic or FBD. Other platforms simply use Instruction List as a "steppingstone" to machine language; Allen-Bradley is one of these.

Because everything can be converted to Instruction List/STL, yet the opposite case is not true, it is considered more efficient than FBD or Ladder.

The table below shows an example of **Assembly Language**, a close relation of machine code. Note the list of addresses and Hexadecimal equivalents to the assembly instructions; this is very similar to the conversion of Instruction List to machine language.

Address	Assembly	Hex	Comment
6050	SEI	78	Set Interrupt Disable Bit
6051	LDA #$80	A9 80	Load Accumulator HEX 80 (128 Decimal)
6053	STA $0315	8D 15 03	Store Accumulator to Address 03 15
6056	LDA #$2D	A92D	Load Accumulator HEX 2D (45 Decimal)
6058	STA $0314	8D 14 03	Store Accumulator to Address 03 14
605B	CLI	58	Clear Interrupt Disable Bit
605C	RTS	60	Return from Subroutine
605D	INC $D020	EE 20 D0	Increment Memory Address D0 20
6060	JMP $EA31	4C 31 EA	Jump to Memory Address EA 31

Figure 175 - Assembly Language

TIP: Since IL is text based, it is easy to manipulate in third party text or spreadsheet editors such as Microsoft Excel. Instruction List can usually be imported or exported to and from PLC software in the form of .csv (comma-delimited) files or XML (eXtensible Markup Language). This makes it easy to create tables of addresses or tags with a common structure and then convert it into many repetitive rungs or blocks with different addresses. When writing large amounts of repetitive code, this can be a big time-saver!

Structured Text (ST)

```
// PLC Configuration

CONFIGURATION DefaultCfg

VAR_GLOBAL
        Auto_PB      :IN @ %X2.2        // Auto Pushbutton
        Manl_PB      :IN @ %X2.3        // Manual Pushbutton
        Cyc_OK_PL :OUT @ %Y3.0          // Cycle OK Pilot Light
        Auto_Mode  :BOOL @ M2.0         // Automatic Mode
        Manl_Mode  :BOOL @ M2.1         // Manual Mode
        Cyc_Enbl     :BOOL @ M2.2       // Cycle Enable
        Fault          :BOOL @ M3.0     // Machine Fault
        TMR 4        :TIMER @ T4        // 10ms Base Timer
END_VAR

END_CONFIGURATION

PROGRAM Main

STRT  IF (Auto_PB=1 OR Auto_Mode=1) AND Manl_PB=0 THEN Auto_Mode=1
        ELSE IF (Manl_PB=1 OR Manl_Mode=1) AND Auto_PB=0 THEN Manl_Mode=1
        End IF

        IF Auto_Mode=1 AND Fault=0  THEN
        START TMR 4
        END IF

        IF TMR 4.ACC GEQ 300 THEN
        Cyc_Enbl=1
        Cyc_OK_PL=1
        END IF

        JMP STRT

END_PROGRAM
```

Figure 176 - Structured Text (ST)

Structured Text resembles high-level programming languages such as Pascal or C. Variables are declared as a data type at the beginning of routines as well as configuration of other parameters. Comments are shown in this program as starting with "//"; this may differ depending on the brand.

Linear programming languages such as Structured Text use constructs like "If-Then-Else", "Do-While", and "Jump" to control program flow. In these languages, syntax is very important, and it can be difficult to find errors in programming. Debug tools allowing for partial execution of the code one section at a time are common.

While writing PLC code in Structured Text can be difficult, it is also a much more powerful language than Ladder Logic or Function Block. Libraries can be developed to perform complex tasks such as searching for data using SQL or building complicated mathematical algorithms. At the same time, since the program proceeds step by step, it is more difficult to respond to multiple inputs at the same time; program control can be complex with many loops.

Different PLC platforms may use a different designation for these IEC languages. Siemens' version of Structured Text is called Structured Control Language (SCL).

Sequential Function Charts (SFC)

SFC makes use of blocks containing code that typically activates outputs or performs specific functions. In many platforms, the blocks or "steps" can contain code written in other IEC programming languages such as Ladder or FBD. The program moves from block to block by means of "transitions", which often take the form of inputs.

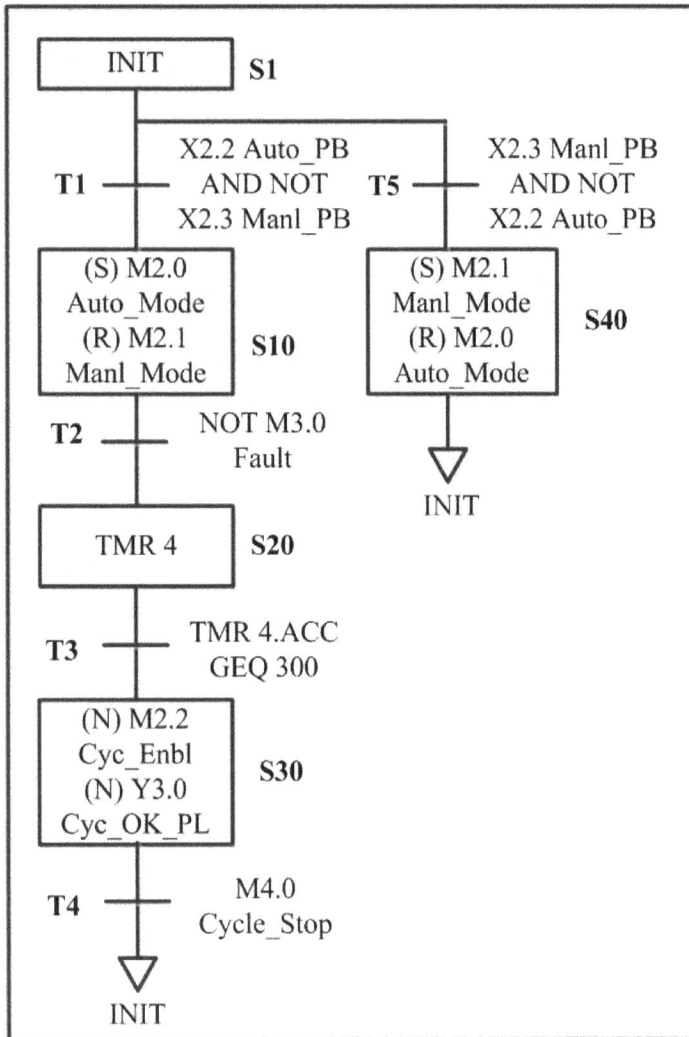

Figure 177 - Sequential Function Chart

SFC is based on Grafcet, a model for sequential control developed by researchers in France in 1975. Much of Grafcet is, in turn, based on binary Petri Nets, also called Place/Transition nets. Petri Nets were developed in 1939 to describe chemical processes.

Steps in an SFC diagram can be active or inactive, and actions are only executed for active steps. Steps can be active for one of two reasons; either it is defined as an initial step, or in was activated during a scan cycle and not deactivated since. When a transition is activated, it activates the step(s) immediately after it and deactivates the preceding step.

Actions associated with steps can be of various types, the most common being Set (S), Reset (R) and Continuous (N). N actions are active for as long as the step is, while Set and Reset operate as in the other PLC languages.

Actions within the steps and the logic transitions between them can be written in other PLC languages. Structured Text is common in the action blocks, while ladder is often used for transitions. Steps and transitions are labeled as S# and T#. The top of the program will always contain an initial step; the program starts here, and this is also where it returns after completion. A program will scan the logic in a step continuously until its associated transition logic becomes true; after this the step is deactivated and the next step is activated.

How Does it Work?

Scanning

The PLC processor or CPU controls the operating cycle of the program. The operating cycle, or **Scan**, consists of a sequence of operations performed sequentially and continuously.

There are four parts to the scan cycle as listed below:

1. Read physical inputs to the Input Image Table.

2. Scan the logic sequentially, reading from and writing to the memory and I/O tables.

3. Write the resulting Output Image Table to the physical outputs.

4. Perform various "housekeeping" functions such as checking the system for faults, servicing communications, and updating internal timer and counter values.

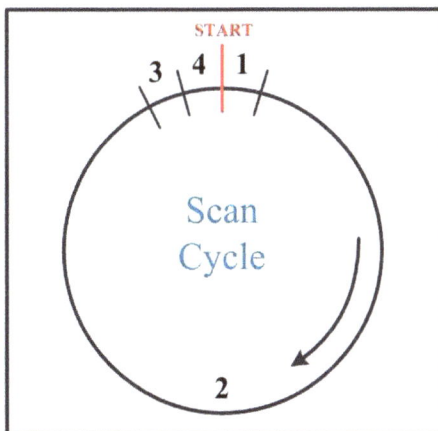

Figure 178 - PLC Scan Cycle

The scan cycle can be visualized as shown in the diagram on the left. When the processor is placed into "run" mode, the state of the analog and digital inputs is captured and saved into memory registers dedicated to the configured inputs. In the second part of the scan, the logic is evaluated sequentially. If the program is written in ladder logic, the rungs are evaluated one rung at a time, left rail to right, top to bottom. As the logic is processed, output and memory coils are energized or de-energized and their status is saved into their respective registers. In the case of the outputs, they are saved into the Output Image Table, which is generated during hardware configuration.

The time that it takes to execute a scan depends on the number and types of instructions in the program. Scans may be as short as 3-5 milliseconds (a very short program or a very fast processor) or as long as 60-70 milliseconds (a longer program). If the scan time exceeds this period by very much, physical reactions of actuators start to become noticeable; it may be time to evaluate a change to a more powerful PLC processor or even use multiple processors.

If a scan takes too long to complete as defined by a system-configured "**watchdog timer**", the PLC processor will fault.

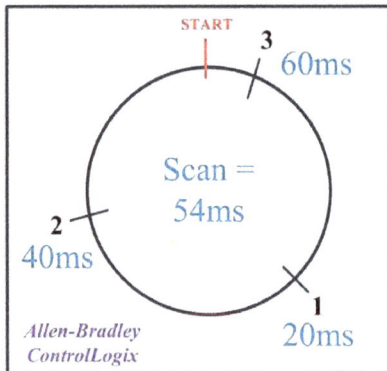

Figure 179 - ControlLogix Scan

An exception to the scanning method described previously is that of Allen-Bradley's ControlLogix platform. Instead of accessing the Input and Output Image Tables at the beginning and end of the scan, each I/O card is configured with a "Requested Packet Interval", or RPI. I/O tables are updated at this rate, as shown in Figure 179.

Subroutines

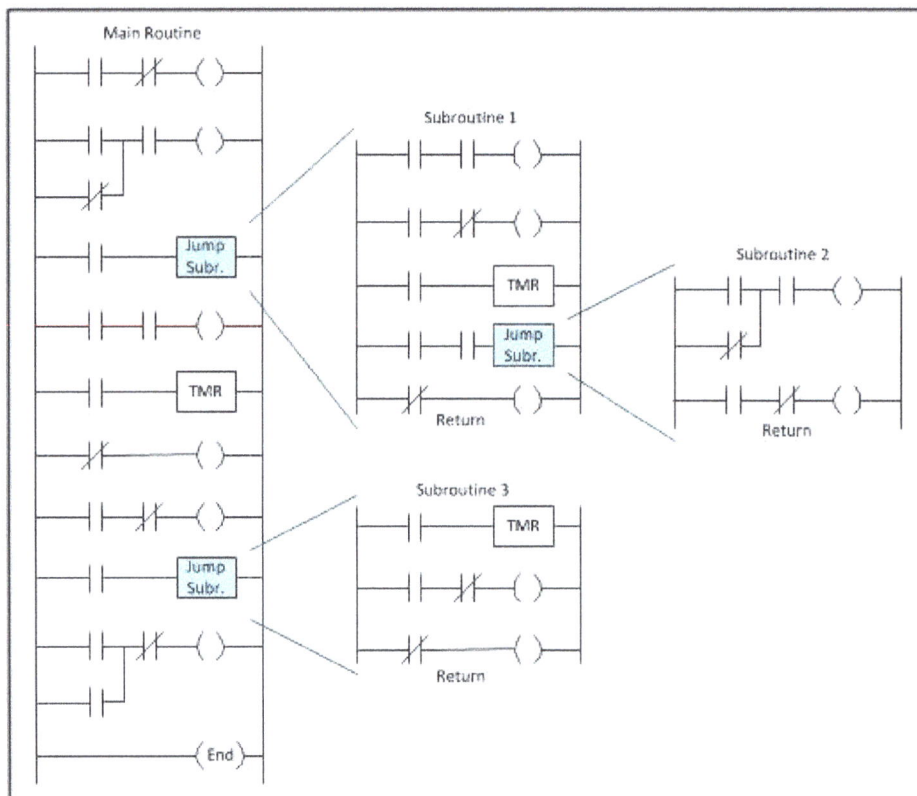

Figure 180 - Subroutine Scans

Scanning begins with the first rung of the routine designated as the main routine; all PLCs have a routine with this designation. As logic is processed, calls to other routines will occur as shown. After a subroutine is scanned, the scan returns to where the jump or call was made. Eventually, the scan always ends at the end of the main routine.

In Figure 180, the scan begins at the top of the main routine and is evaluated left to right, branch by branch, top to bottom. When the scan reaches a Jump Subroutine (also known as a "call"), the scan continues in the new routine, in this case Subroutine 1. Subroutine 1 then calls Subroutine 2; when the end of Subroutine 2 is reached, the scan continues in Subroutine 1. When the end of Subroutine 1 is reached, the scan continues back in the main routine.

Subroutine 3 is called, is scanned to the end, and returns to the Main Routine. When the end of the Main Routine is reached, the resulting Output Image Table is written to the physical outputs and the housekeeping functions (Step 4) is completed. The scan then starts over at the top of the Main Routine.

Scan times will vary from cycle to cycle, depending on the number of instructions and routines active at any given time, the load on the processor from "housekeeping" activities, and communications' connections. Execution times for each type of instruction can be found in each manufacturer's documentation.

Hardware also affects scan times; newer models of processor are much faster than older ones due to advances in technology. For example, Allen-Bradley's new ControlLogix processors are nearly 150 times as fast as their older PLC5, which was the most powerful platform they made in the 1990s!

PLC Modes

PLCs can be placed into several operational states. When the processor is executing its program and scanning in the normal way, it is said to be in "Run" mode. When initially downloading a program, the CPU is in "Stop" or Program mode; it is not executing the program and I/O is not changing state.

The mode of a PLC can often be changed by means of a switch on the front of the processor; for security, this may be a key. There is often a third position on the key switch labeled "Remote"; this allows a computer to be used to change the state or mode. This adds an additional layer of safety; when the switch is in Program or Run the state cannot be changed from the computer.

While the computer is connected, the program can sometimes be changed while the PLC is scanning. Not every PLC allows this, and it is not always done in the same way. Some platforms such as Allen-Bradley allow each line or rung of code to be changed while online followed by accepting, testing, and assembling (compiling) the program; others such as Siemens allow each block to be compiled and downloaded without interrupting program execution.

Additionally, many platforms allow the processor to be placed in "test" or "debug" mode. This allows breakpoints to be placed in the code to stop execution at that point. This can be useful if monitoring "looped" code.

I/O can also be "forced" on many platforms. While this is not a mode as such, it does change the operation of the program. There is more information on forcing in the Maintenance and Troubleshooting section of this book.

Exercise 22

1. List the five languages defined in IEC 61131-3

   _____

   _____

   _____

   _____

   _____

2. List the four parts of the scan cycle in a PLC

   _____

   _____

   _____

   _____

3. Can a PLC program be changed while it is running? _____

4. If a PLC program takes too long to complete its scan, what will happen?

   _____

Ladder Logic

The most common and first PLC programming language is Ladder. This is because it is based on physical wiring diagrams, which most electrical maintenance personnel are familiar with. This makes this graphical type of programming easy to read and troubleshoot.

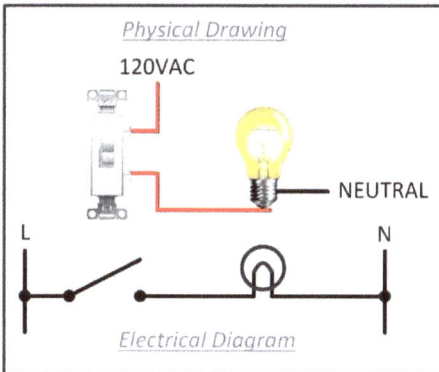

Figure 181 - Physical Circuit

The simplest circuit to understand in an electrical system is a switch turning on a light. The top picture shows a physical drawing of a light switch passing 120 Volts AC to an incandescent light bulb; the bulb also needs a neutral wire to complete the circuit.

The diagram on the bottom shows the electrical equivalent of the physical drawing. The L symbolizes **Line** voltage and the N signifies **Neutral**.

There are many symbols used in schematic diagrams to signify the type of device used in an electrical circuit. The device on the left is generally known as a *Switching Device*, while the device on the right is called the *Load*.

Discrete Logic

Just like electrical circuit drawings, there are symbols or figures signifying the type of device used in the circuit; however, they are classified by their function rather than by the type of device (i.e. a pushbutton, switch or sensor uses the same contact icon).

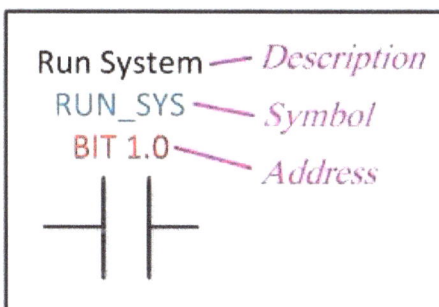

Figure 182 – Component Information

For the logic diagrams in this section, the conventions to the left are used. The **Address** is the register that is updated as the logic is processed. This applies to non-tag-based systems; even they use addresses for I/O.

The **Symbol** is used as a short cut for the address. In most PLC platforms, typing the symbol will automatically call up the address. In the case of a tag-based platform, there may be no address; the symbol is the only address that is used. For tag-based systems, the symbol is actually downloaded into the PLC as part of the program. For address-based systems, it usually is not; it is a part of the program saved on the computer but not present in the CPU. Symbols or tags are usually limited in the number and choice of characters allowed. For example, underscores (_) may have to be substituted for spaces.

A **Description** is purely for the convenience of the programmer. The description is only present in the programming computer and is not downloaded to the PLC. Typically a description can be several lines in length, unlike a symbol. It can usually contain any text character and is used to fully describe aspects of the device or contact, such as its physical location, numerical designation, or purpose.

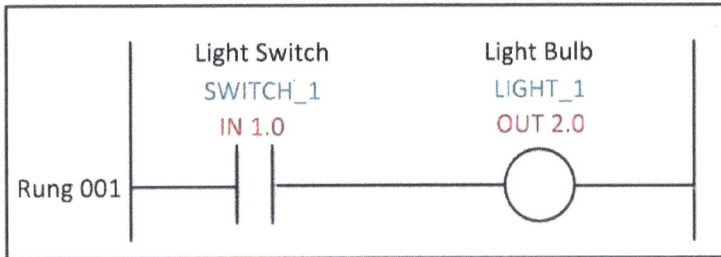

Figure 183 - Contact and Coil

This diagram shows the same circuit as in the electrical diagram on the previous page but in Ladder Logic. Like the electrical diagram, if the switch contacts on the left are closed, the light will be energized; however, the only information you can get from the contact while

offline is that it is a *Normally Open* contact.

The device on the left is called a **contact**, and the device on the right is a **coil**. These names come from parts of a relay, which is the basis of Relay Ladder Logic. This type of contact is known as Normally Open, or NO; it is assumed to be not energized or activated in its normal condition. There are also contacts that are in the energized or "on" condition when in its normal condition; these are known as Normally Closed, or NC contacts.

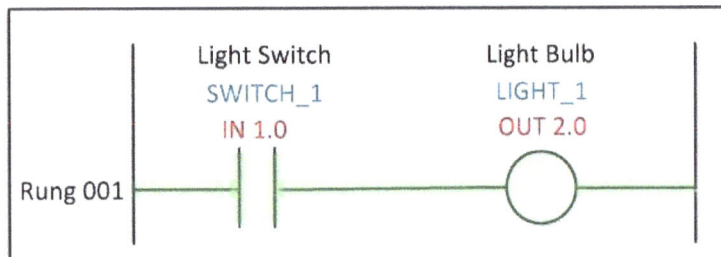

Figure 184 - Energized Circuit Logic

When a contact is viewed in its online condition, it is often highlighted in some way to indicate continuity. In the case of the light switch logic shown before, this indicates that while the switch is Normally Open, something has activated it and the contacts are closed. The green highlight around the coil indicates that it is energized.

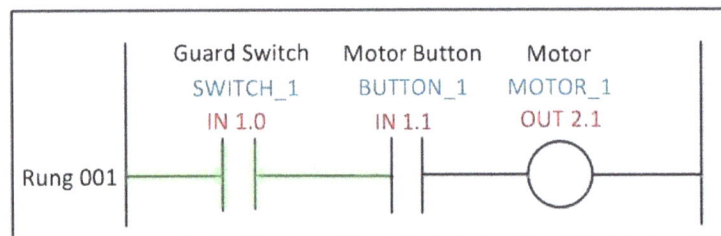

Figure 185 - AND Logic

Figure 185 shows a normally open guard switch in series with a normally open motor pushbutton. If the guard switch is a switch on a motor cover, it would be assumed that if the cover is closed, the switch would allow the button to jog the motor. The switch is wired to be closed with the cover on; if the cover is removed, the switch opens, turning off the motor even if the button is being pushed. In other words, if the Guard Switch is closed **AND** the button is pressed, the motor will run. This is a classic example of an AND circuit, two contacts in series. In this case, if the circuit was being monitored through the software, the cover is on the motor (Guard Switch) but the button is not being pressed, so the motor is not running.

Figure 186 shows a Normally Closed contact from a conveyor in parallel with a Normally Open button. If the conveyor is on, (that is, if the output for running the conveyor is turned on

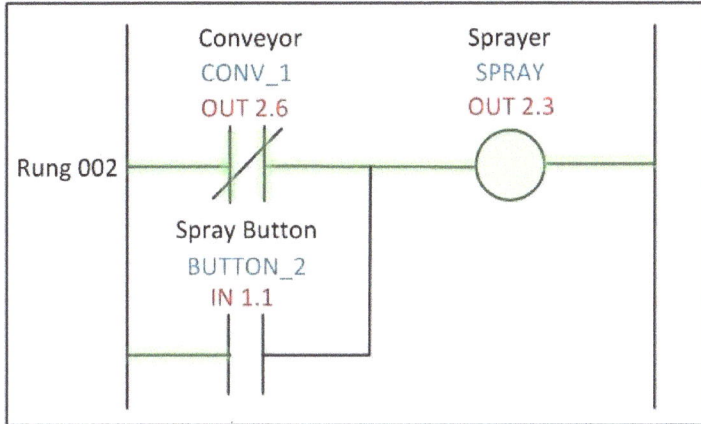

Figure 186 - OR Logic

elsewhere in the program), the sprayer will not be energized UNLESS the button is pushed. This illustrates a new concept, the **OR** circuit. If the conveyor is off OR the button is pushed, the sprayer output will activate. In this case, since the coil is activated and the highlighting extends through the conveyor contacts, the conveyor is not running! Note also that it is possible to use output addresses as contacts.

In the previous examples, the coil has indicated the status of the preceding logic. In other words, if there is a continuous path from the left rail to the coil, the coil is on. If not, the coil is off. It is also possible however to **SET** or **RESET** the coils address, making the logic Retentive. This is also sometimes called "Latch" and "Unlatch".

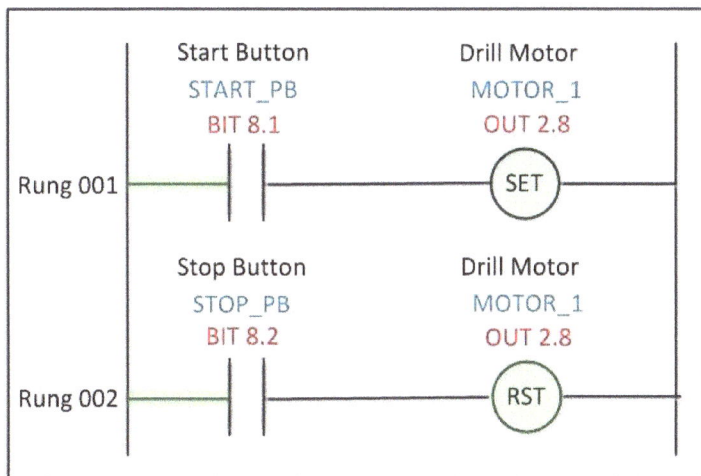

Figure 187 - Set Reset or Latch Unlatch pair

Figure 187 shows an output that energizes a motor. It can be "latched" on or off by pressing the start or stop button. Notice that the coil is energized, but there is no path or continuity from the contacts to the coils. This means that the output will be maintained until the Stop Button is pressed.

Also notice that in this case a "bit" address is used to control the motor. Since these aren't input register addresses as in the previous examples,

that means that they must come from somewhere <u>other</u> than physical inputs. It is hard to tell without doing a cross reference to see where the bit's coil might be, but if a coil for that address can't be found in the program, it likely comes from an HMI or operator interface.

Contrary to the drawing, it is not advisable to directly latch output addresses. Instead, a memory bit should be latched and unlatched that controls the output.

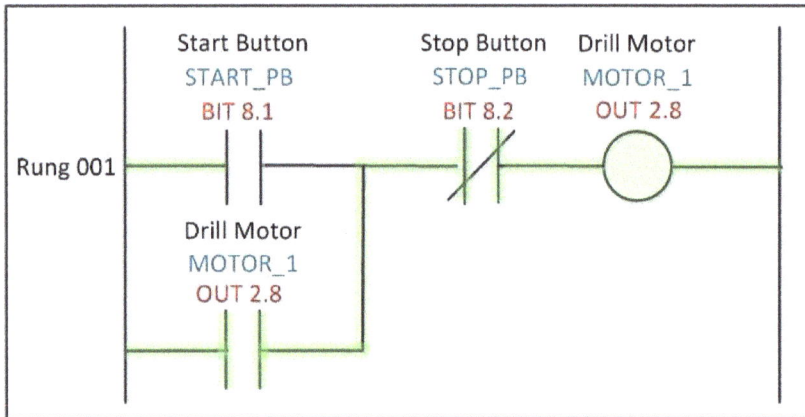

Figure 188 - Hold-In Logic

This logic performs exactly the same function as the latching circuit but in a different way. Sometimes called a "hold-in" or "seal-in" circuit, if the start button is pressed and the stop button is not being pressed, the motor coil will energize. Even if the start button is released, the "hold-in" contact of the motor coil will ensure that the motor stays on until the stop button is pressed. Notice that in this case, the stop button needs to be a normally closed contact, even though both buttons are <u>physically wired</u> normally open in both diagrams.

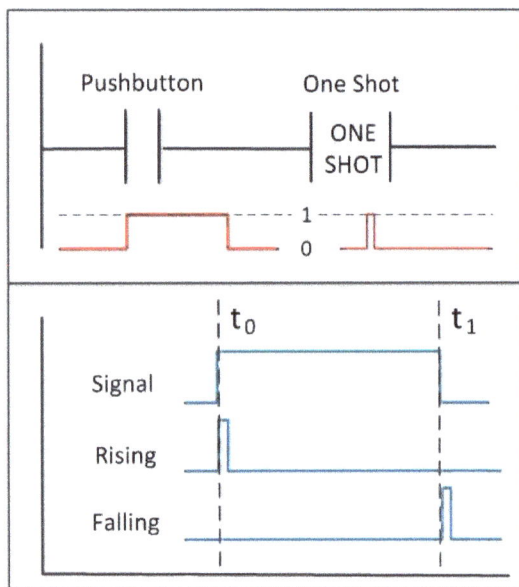

Figure 189 - One Shot

A **One-Shot,** single shot, or differential pulse is used for a variety of reasons in ladder logic. This diagram shows the result of using a one-shot; when a signal changes state, the one-shot creates a pulse exactly one scan in length. For example, when the Pushbutton is pressed, no matter how long it is held down, the logic will generate one single-scan pulse.

One-shot pulses can be generated from the rising or falling edge of a signal. They are known by different names on different platforms: Allen-Bradley uses OSR and OSF contacts called One-Shot Rising and One-Shot Falling; ONS one-shots are also sometimes used on the rising edge. Siemens calls its one-shots Positive and Negative differentials, they look like –(P)- and –(N)- in ladder logic.

Omron's one-shots are called DIFU (Differential Up) and DIFD (Differential Down) for the rising and falling edge versions. They look similar to this: -|DIFU|- and -|DIFD|-

Mitsubishi calls them PLS or Pulse: -|PLS|-

Some types of one-shot are placed at the end of the rung as a "box" type instruction. These have two addresses: one for storage and one to be used as the output. These output contacts can be used at multiple places in a program. In-line one-shots must each have their own memory bit address; *do not use the same address for multiple in-line one-shots!*

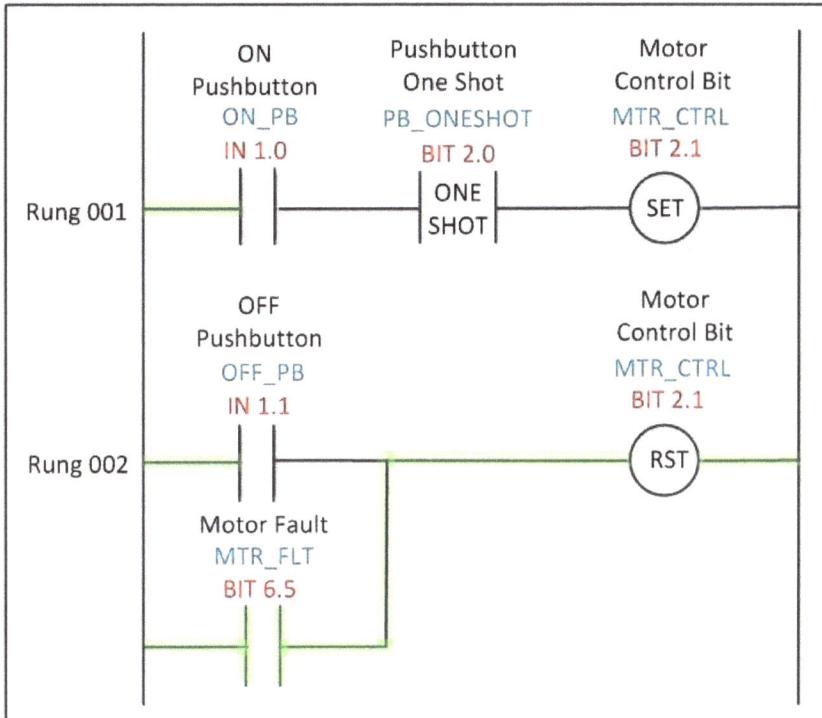

One- shots are often used with latching circuits to ensure that logic defaults to the off state as shown. If the one shot is of the "in-line" type described previously, each must have its own address.

If the Motor Fault is present, the motor control bit will be held in the off state. To turn the bit back on, the fault will have to be cleared and the ON pushbutton will need to be pressed again.

Figure 190 - Set Reset with One Shot

Notice also that even though the rung is highlighted all the way to the RST coil in Figure 190, the coil itself is not highlighted, indicating that the address is off. Contrast this with the latched coil picture in Figure 188; this is typical when monitoring PLC logic with the software. The state of the <u>address</u> is what is being indicated, not the state of the coil.

Exercise 23

1. Are descriptions usually downloaded to the PLC with the program? _____

2. Complete the following using Series and Parallel:

   Contacts placed in _____ form an AND function, while contacts placed in _____ form an OR.

3. A one-shot creates a pulse one _____ in length.

4. Does a one-shot use one or two addresses? _____

5. A hold-in circuit performs the same function as a _____

   Programming Exercises:

6. Draw ladder logic to place a machine into either Auto or Manual mode by pressing physical pushbuttons. Ensure that the machine is placed into Manual if a Fault occurs.

7. Draw ladder logic to accomplish the following:
   a. Turn on a motor using a start button and a stop button using a hold-in contact. Both buttons should be <u>wired</u> Normally Open (NO). Include a NC Fault contact that will stop the motor.
   b. Latch the Fault bit if the motor's overload trips or the guard door is opened while the motor is running.
   c. Turn on a Red Light if there is a fault.
   d. Turn on a Green Light if the motor is running.
   e. Reset the Fault if there is no fault and a Reset Button is pushed.

   Assign addresses to all of the contacts and coils using the generic method shown in this document; i.e. BIT, IN, OUT

Timers

The purpose of a timer in ladder logic is to delay the on or off state of a signal. Timers can be used to track the accumulation of time in a process, create a pulse of a fixed length, or determine whether a fault has occurred.

Before looking at how a timer operates, it is important to view its data structure.

Preset	
Accumulated	
Status Bits	

Figure 191 - Timer Data Type

The Preset and Accumulated values are usually either an integer or double integer value; however, Siemens uses a BCD value that incorporates the time base as part of a word (16 bits). The status bits always include a "**Done**" bit but may also include bits for Timer Enabled or Timer Timing. If the status bits don't include these, they can easily be generated using logic structures.

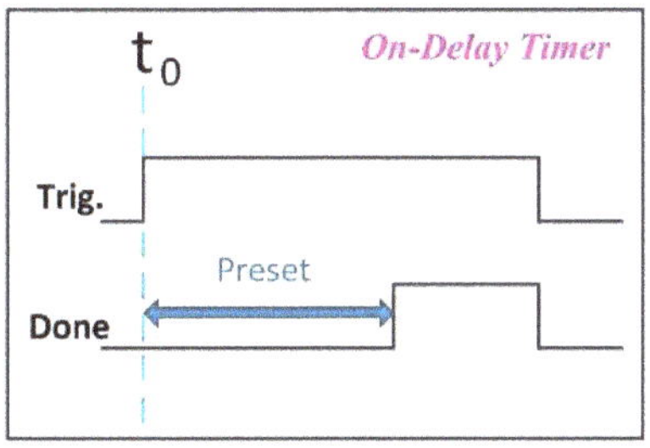

Figure 192 - On-Delay Timing Diagram

On-Delay Timer

An On-Delay Timer is used to delay the ON state of its done bit as shown in this timing diagram.

When the button is pushed, the timer begins timing. After the preset time (Set) has expired, the done bit will energize and remain on until the pushbutton has been released. If the Pushbutton is released before the Accumulated time (Acc) reaches the preset, the timer will reset to zero. When the done bit comes on, it will energize the output, Light 1.

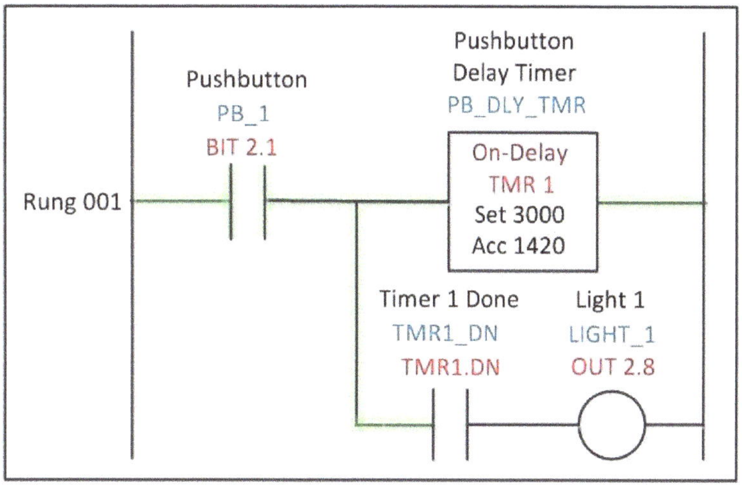

Figure 193 - Delayed Light

TIP: Notice that the preset is in thousands. According to IEC 61131-3, a timer counts time in milliseconds, so 3000 is three seconds. Also, the timer will count <u>up</u> as it times. Not all platforms' timers perform this way; some time down, some have a time base that allow them to count in 10ms, 100ms, 1 second, or even 10 second increments. Most major manufacturers now have a timer available that meets the IEC definition.

Off-Delay Timer

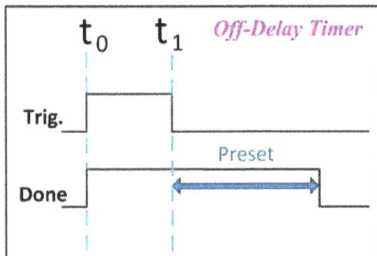

Figure 194 - Off-Delay Timing Diagram

An Off-Delay Timer delays the OFF state of its done bit. Unlike the On-Delay, the done bit turns on immediately. It does not begin timing until the timer is deenergized.

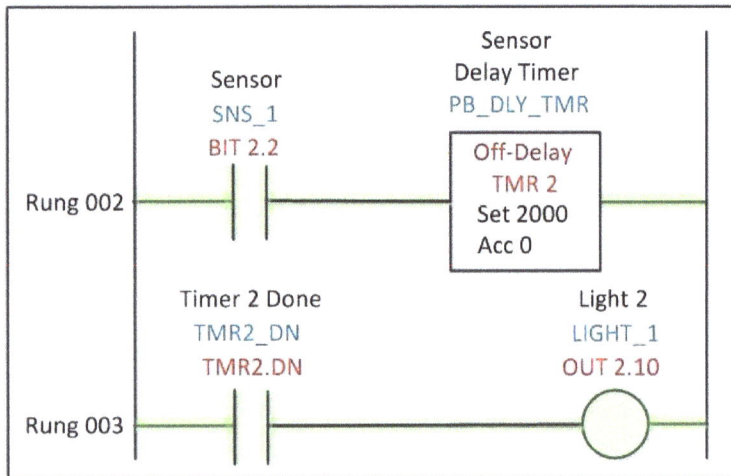

Figure 195 - Delayed Off (Time Stretched" Light

When the sensor is activated, the done bit comes on immediately; therefore, the light comes on. When the sensor turns off, the timer begins timing and runs until the preset (Set) value is reached, then the done bit turns off. During this time, the accumulated value is counting up in milliseconds.

Notice that with this type of timer, the done contact can't be placed in a branch after the trigger contact as with the on-delay timer. If it was, turning off the sensor would also not allow the light to come on, though in some platforms it will. The off-delay timer can be thought of as a "pulse stretcher".

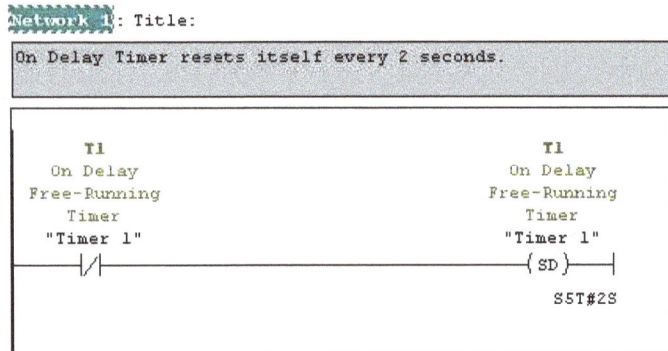

Figure 196 – "Free Running" Timer

TIP: With many PLCs, the done bit is shown having the same address as the timer. Figure 196 shows a rung or "Network" from a Siemens Step 7 program. Note that the Symbol T1 is on both the SD (On-Delay) coil and the NC contact; the contact is the "done" bit. This rung creates a one scan-length pulse every 2 seconds.

Another aspect of a timer that is not part of its data type is the **RESET** coil. This coil is used to set the accumulated value to zero. All timers have this capability, though the reset is most commonly used in retentive timers.

Retentive On-Delay Timer

A retentive timer keeps its accumulated value even when the energizing contact is not made. This is useful for accumulating run-time on a device or product. This means it must be reset.

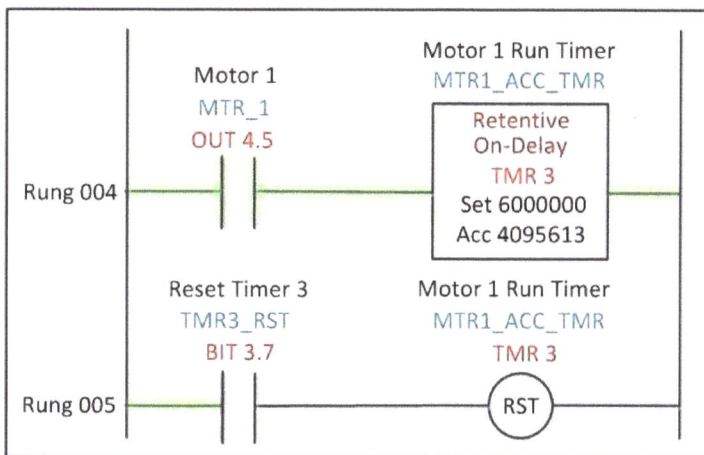

Figure 197 - Retentive On-Delay Timer

When the motor runs, the timer accumulates time. When it stops, the timer retains its accumulated value and does not reset. When the accumulator reaches the preset value, the timer stops timing, the done bit is energized, signifying that it is time to service the motor.

Usually, rather than placing a high number in the preset, a number such as 60000 (60 seconds) is placed there. The done bit is then used to increment a "Minutes" counter, which resets the timer. When the Minutes counter reaches 60, it increments an "Hours" counter; service on devices is often specified in hours. After the maintenance has been performed, the timer and counters can be reset. *A Siemens Retentive On-Delay Timer acts differently: when the trigger signal is removed, the timer continues timing until done. Siemens timers also time down from their preset.*

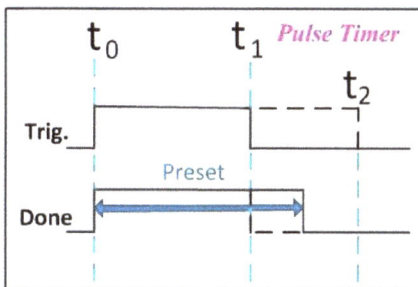

Figure 198 - Pulse Timer Timing Diagram

Pulse Timer

Pulse Timers create a pulse of a fixed length when energized; the done bit energizes immediately and stays on until the preset is reached. Some pulse timers' done bits will de-energize if the trigger is removed from the timer early.

If the trigger signal stays on either shorter (t1) or longer (t2) than the Preset, the pulse output ("done" bit) stays on for the preset time. (Siemens S_PEXT). If the trigger signal is released before the preset time, some pulse timers will turn off the done bit as shown at t1 (Siemens SE). For PLC brands without a Pulse Timer, the pulse can be created using two On-Delay timers.

Counters

A **Counter** is used to add or subtract counts in a register. As with the Timer, a counter also has a data type associated with it.

Preset	
Accumulated	
Status Bits	...

Figure 199 - Counter Data Type

The Preset is the value at which the done bit will come on in most counters, again with the exception of Siemens. The Accumulated value is the current number of counts in register. Unlike a timer, which stops accumulating when the done bit activates, a counter will continue incrementing. Preset and accumulated values may be an integer or double integer depending on the platform.

Other Status Bits that MAY be present besides the "Done" bit:

a. Count Up (CU): active while up count trigger is active

b. Count Down (CD): active while down count trigger is active

c. Overflow (OV): active when Accumulated value exceeds maximum value for an Integer or Double Integer, depending on brand/platform

d. Underflow (UN): active when Accumulated value exceeds minimum value for an Integer or Double Integer, depending on brand/platform

TIP: The IEC61131-3 definition of a counter states that a counter's done bit will change state when the accumulator reaches the preset. However, Siemens counters, once again, act differently; the done bit is on if the accumulator contains a value higher than zero. Because of this, Siemens programmers often use a different method to count, Accumulators and Decumulators.

There are three types of counters: Up Counters, Down Counters, and Up-Down Counters. While every brand needs an Up and a Down counter, the Up-Down type is not available on all PLCs.

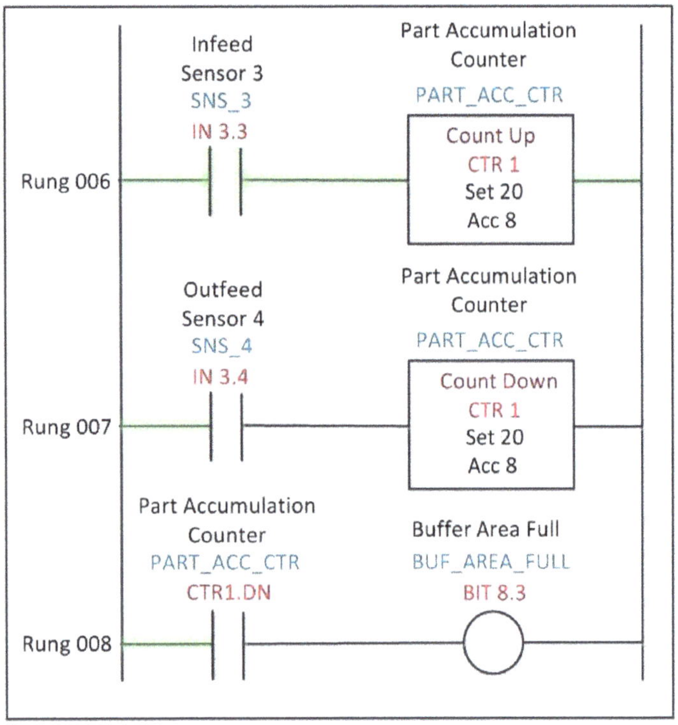

Figure 200 - Up and Down Counters

Separate up and down counters can be used with the same effect. As with retentive timers, counters need a Reset bit. Some counters will also use a "Set" bit to place the preset value into the accumulator.

This logic tracks parts into and out of a "Buffer" area on a conveyor system. A sensor is located at the entry and exit of the area, so parts entering make the counter count up, while parts exiting make it count down. When the "Done" bit is active, a gate could be closed preventing new parts from entering the buffer area until parts have exited.

Notice that the same address was used for both counters. This is important to ensure that the same address is counting up and down.

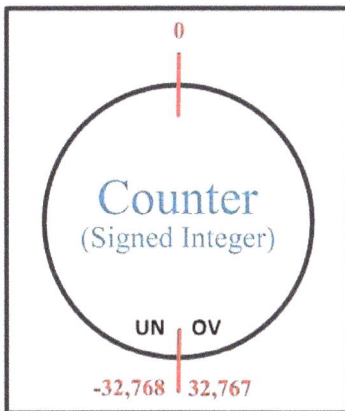

So, what happens when a counter reaches the overflow or underflow value? It rolls over to the negative range if the positive value is exceeded and to the positive side if counting down in the negative direction. This is the purpose of the overflow and underflow status bits; they indicate that this boundary has been crossed. They can be reset (unlatched) after determining what you wish to do about the over or underflow.

If the counter is a Double Integer based device, the limits are

-2,147,483,648 to +2,147,483,647.

Figure 201 - Counter Values

The reset for the counter works the same as the one shown in the previous timer diagrams.

In addition to the Reset bit, a Set coil may be used to load the preset value into the accumulator. This is used for down counters; as mentioned previously, Siemens counters operated differently in the older Step 5 and Step 7 platforms. The Done bit was on when the accumulated value was NOT zero.

Figure 202 - Reset Bit

Exercise 24

<u>Programming Exercises:</u>

1. Draw ladder logic that counts parts into a box. When the box is full (10 parts), latch a memory bit that controls a light, the light tells an operator to remove the box. Use a sensor to detect that the box has been removed. When the box has been removed, reset the counter and the memory bit that controls the light.

2. Draw ladder logic that uses a Retentive On-Delay timer to accumulate run time on a motor. When the timer reaches one minute, use the done bit to increment a "Minutes" counter and reset the timer. When the counter reaches 60 minutes, use its done bit to increment an "Hours" counter and reset the Minute counter. When the Hour counter reaches 10,000, it is time to perform maintenance on the motor.

   Do not reset the Hours counter automatically; use a pushbutton that is pressed by the maintenance technician, confirming that maintenance has been done. How can you incorporate logic to ensure that the technician really did do the maintenance?

   Don't forget to reset the timer and other counters when the operator presses the button!

Data Manipulation and File Movement

An important part of programming is the manipulation, modification, and movement of data. The simplest method of doing this is to simply move a number from one location or register to another.

Move

Despite the common term "Move" that is used for this operation, it is actually a copy since the value also remains where it was moved from.

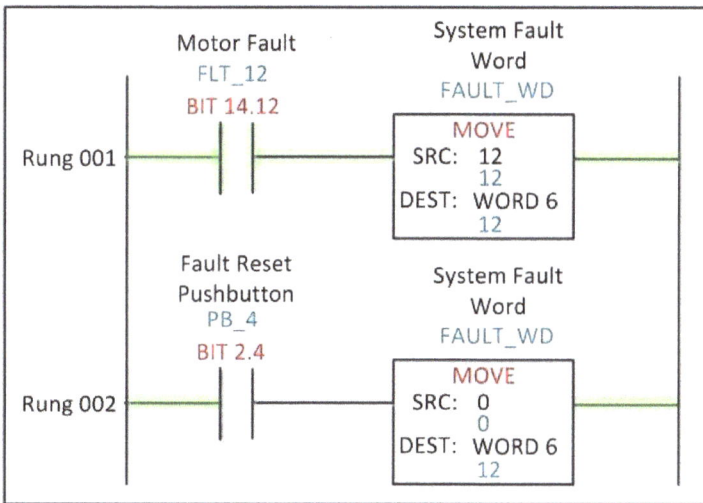

Figure 203 - Integer Moves

This logic shows an integer being moved into a register based on a numbered fault. The number can be used to display a message on a touchscreen or in a comparison to set the fault status by latching a bit. The numbers in blue show the actual number in the register (Word 6) if the logic was being monitored online.

In this example, constants are being moved, but numbers can also be moved from one register to another.

This logic shows a speed being moved into a VFD speed command. The numbers are floating point or REALs, so they require a double word, or 32 bits, for the register size.

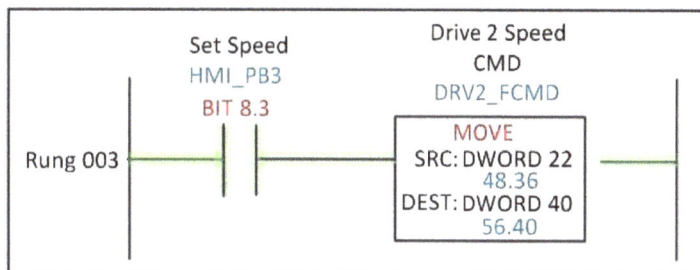

Figure 204 - Setting VFD Speed

For larger data structures, a different instruction is usually used. For Allen-Bradley, this instruction is a COP, or "Copy". This instruction can also move multiple files in an array. In Siemens, a pointer must be used in Statement List (STL), Siemens' version of Instruction List.

Masked Move and Shift

Parts of numbers can also be moved. If it is necessary to extract a specific part of a 16 or 32-bit number, a **Mask** can be used to move only that part, as shown in the diagram.

Source	1 1 0 1	0 1 1 1	0 0 1 1	1 0 0 1	1 0 0 0	1 0 1 0	0 0 1 1	1 0 0 1
Mask	0 0 0 0	0 0 0 0	1 1 1 1	1 1 1 1	0 0 0 0	0 0 0 0	0 0 0 0	0 0 0 0
Dest.	0 0 0 0	0 0 0 0	0 0 1 1	1 0 0 1	0 0 0 0	0 0 0 0	0 0 0 0	0 0 0 0

Figure 205 - Masked Move

The Source is a 32-bit number, which is four bytes. If one wanted to move only the second byte into a register, the mask would contain ones wherever the data to be moved and zeroes elsewhere. Masks are usually entered as a Hexadecimal number, which in this case would be 00FF0000, or just FF0000.

Of course, then you have the problem that the byte in the destination is in the wrong location within the double integer; this is easily remedied by using a **Shift** instruction. The result is shown below. In this case the shift is to the right sixteen spaces.

Source	0 0 0 0	0 0 0 0	0 0 1 1	1 0 0 1	0 0 0 0	0 0 0 0	0 0 0 0	0 0 0 0
Dest.	0 0 0 0	0 0 0 0	0 0 0 0	0 0 0 0	0 0 0 0	0 0 0 0	0 0 1 1	1 0 0 1

Figure 206 - Bit Shift Right

An interesting result of shifting bits is that shifting to the left **multiplies** the contents by two for each shift, while shifting right **divides** the number contained by two for each shift operation.

Shifting is also used to track parts on a high-speed bottling or canning line. A sensor is placed at the point where products enter, and the register is shifted when the line is indexed.

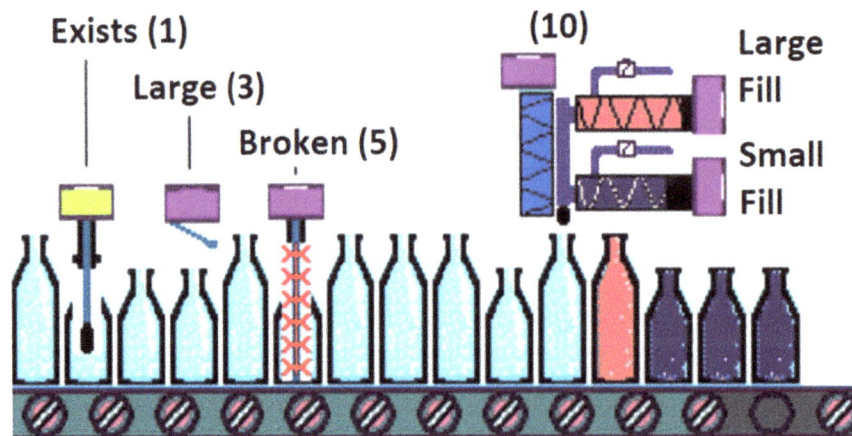

Figure 207 - Bit Shifts on Bottle Line

Figure 207 shows a bottling line with various stations that use the positions of different shift registers to determine whether an operation should be performed.

A register assigned to track all bottles would set a bit at position 1, another register operating in parallel would set a bit at position 3 if the bottle is large. Another register would set bits at position 5 if a bottle was broken for later rejection. If the bottle is present, large, and unbroken it would fill from the Large Filler at position 10; if present, unbroken and NOT large, it would fill at position 10 from the small filler. Bits are all shifted together when the line is indexed.

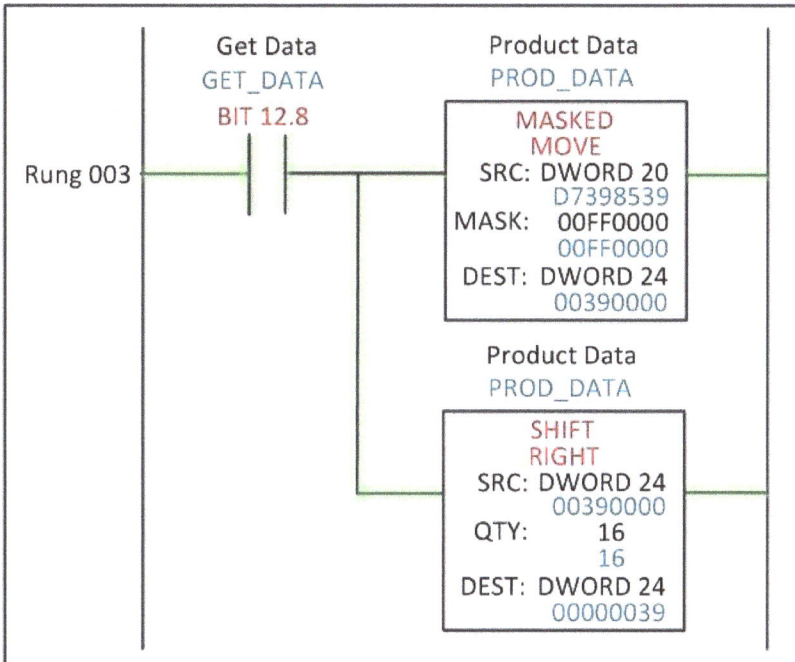

Rung 003

Get Data
GET_DATA
BIT 12.8

Product Data
PROD_DATA

MASKED
MOVE
SRC: DWORD 20
D7398539
MASK: 00FF0000
00FF0000
DEST: DWORD 24
00390000

Product Data
PROD_DATA

SHIFT
RIGHT
SRC: DWORD 24
00390000
QTY: 16
16
DEST: DWORD 24
00000039

The ladder logic for a Masked Move and a Shift are shown in Figure 208. The shift instruction may not have a different destination as shown in the diagram; however it is shown here to indicate the number before and after the shift. In other words, the Source and Destination registers can be the same.

Shift Left instructions can also be used.

Figure 208 - Masked Move with Right Shift

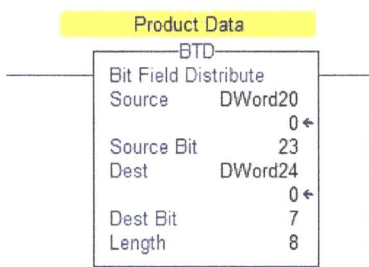

Product Data
BTD
Bit Field Distribute
Source   DWord20
                    0 ←
Source Bit   23
Dest   DWord24
                    0 ←
Dest Bit   7
Length   8

TIP: The Allen-Bradley ControlLogix 5000 platform has an instruction that does both of these functions at once (Move and Shift), called "Bit Field Distribute" or BTD (Figure 209)

Figure 209 - BTD Instruction

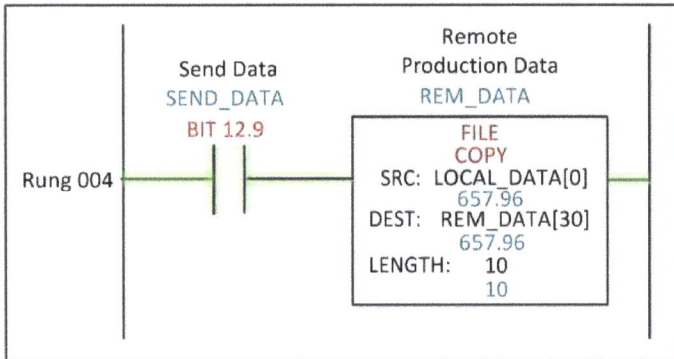

Figure 210 - File Copy Instruction

File Copy

File instructions are used to move data structures that are larger than a single element or 32 bits. If a file is a single structure, such as a UDT or an individual data type such as an Array, a simple "Copy" type command can be used. If it is necessary to move specific overlapping sections, more complex pointer-based commands may be used.

TIP: This logic illustrates the use of pointers in Siemens S7 platform. SFC20 is a Block Move command that allows the use of pointers to designate the type of data (Local and Marker Memory Bytes), the size of the data (10 Bytes) and the location within the structure (0.0 to 150.0). This command allows movements to be specified to the bit level.

The RET_VAL output is present on many Siemens instructions to allow for error return values.

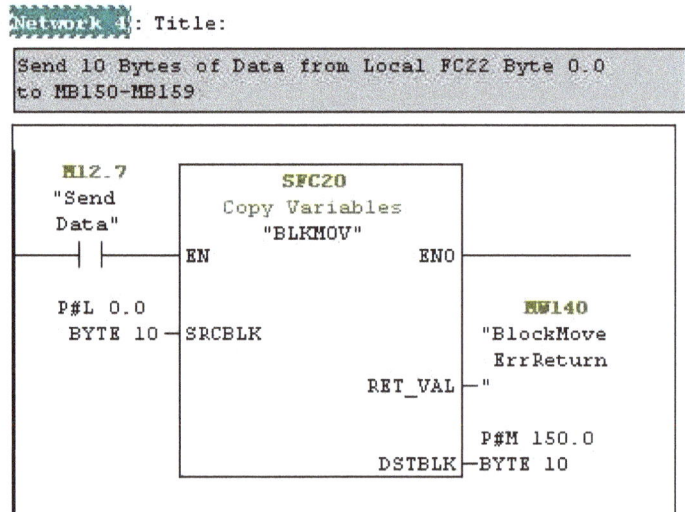

Figure 211 - Siemens Block Move

Comparisons

Data comparison is an important element of PLC programming. Though comparisons deal with numerical values, they are input type instructions. That is, they evaluate to either true or false.

Standard comparison instructions that are found in any PLC include Equal (=), Not Equal (<>), Greater Than (>), Less Than (<), Greater Than or Equal (>=), and Less Than or Equal (<=).

Some PLC platforms require that the data type being compared is the same, while others allow comparisons between different types. Where data types must be the same, a conversion instruction may be required.

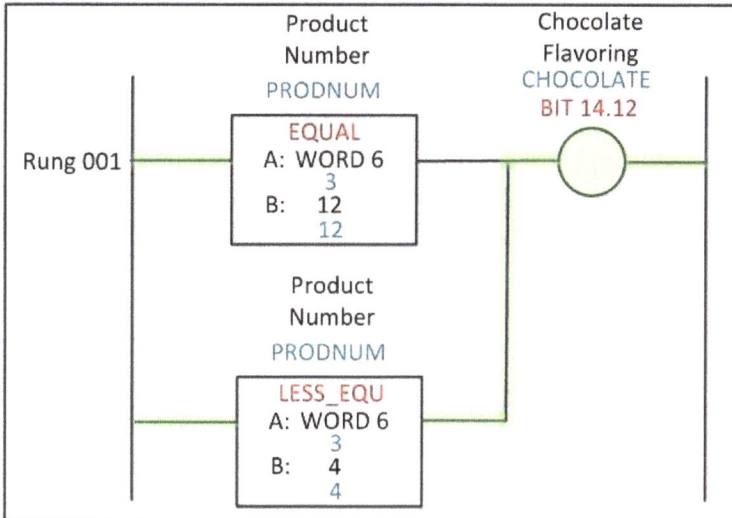

Figure 212 - Comparisons used to Activate Coil

Figure 212 illustrates the use of two comparison instructions used to activate a bit that is used to add chocolate flavoring to a recipe. If the selected product number is between 0-4 (inclusive) or equal to 12, the bit will be true. In this case the selected product number, in Word 6, is 3.

Comparisons are also often used in automatic sequences to move from one step to another or to determine when an output will be activated.

This logic increments a sequence to

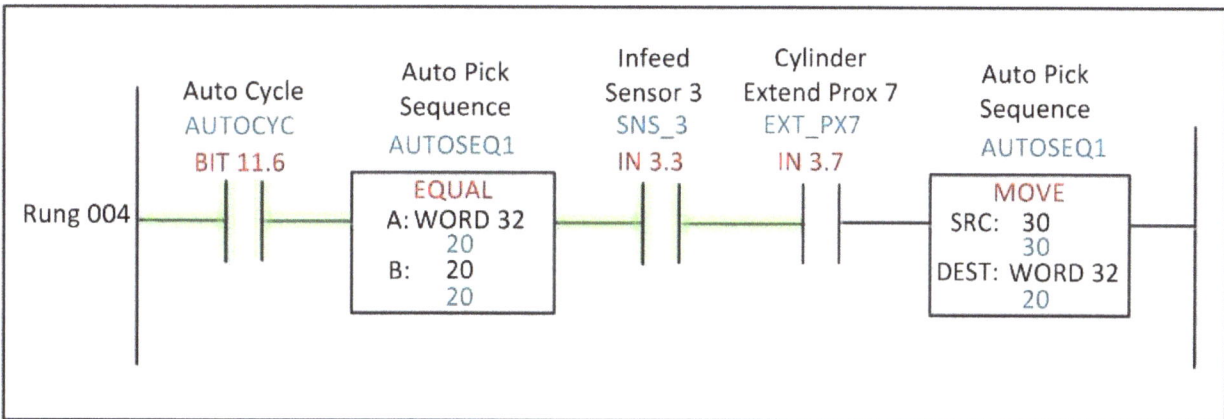

Figure 213 - Sequence Increment Logic

the next step if the machine is in Auto Cycle mode and a couple of sensors are activated. This only happens if the sequence is in Step 20.

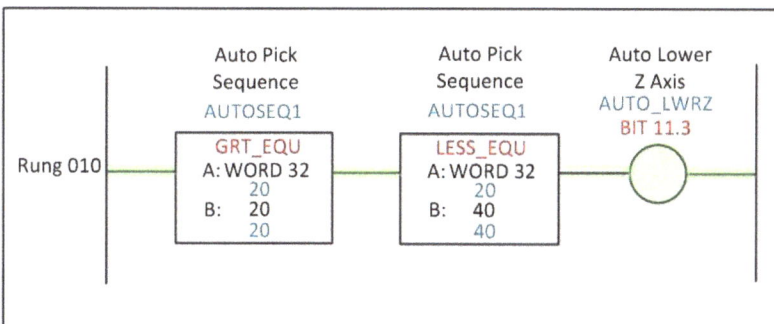

Figure 214 - Comparisons to Activate Output Command

Comparison instructions can be placed in series to form a "window" during which a statement is true. In this case, the command to lower the Z axis of a pick-and-place mechanism will be on if the sequence is in steps 20-40, inclusive.

Auto Sequence State
```
            ----LIM----
  Limit Test (CIRC)
  Low Limit                20

  Test  PNPAutoSequence_State
                    <MemoryWord[2]>
                              0 ←
  High Limit               50
```

Some PLC platforms have instructions that form this range window in one instruction.

TIP: The Allen-Bradley "Limit" instruction can be used to form a window where a sequence state is tested between a low and high value. If the Low Limit value is greater than the High Limit value, the instruction operates in reverse, where if the tested value is outside of the range, the instruction is true! Siemens TIA software uses "IN_RANGE" and "OUT_RANGE" instructions for this.

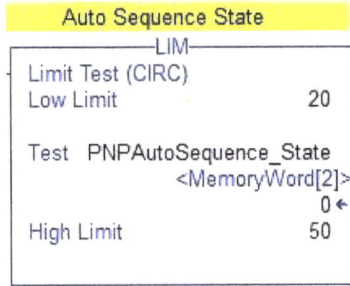

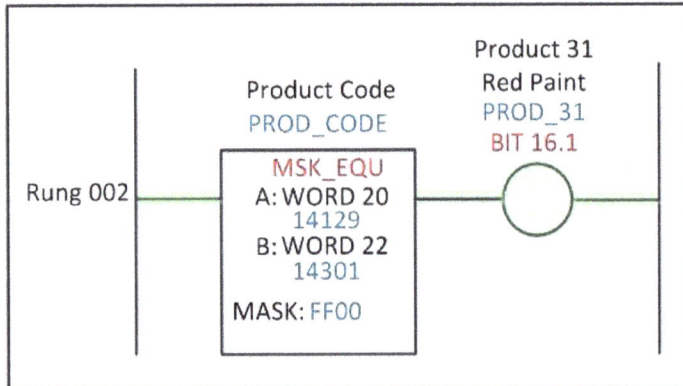

Figure 215 - Limit or "Range" Instruction

```
                    Product Code      Product 31
                                      Red Paint
                     PROD_CODE        PROD_31
                                      BIT 16.1
                     ----MSK_EQU----
  Rung 002           A: WORD 20           ( )
                        14129
                     B: WORD 22
                        14301

                     MASK: FF00
```

Figure 216 - Masked Equal Instruction

A Mask can also be used with an equal instruction on some platforms. As with the Masked Move instruction, wherever there are ones in the mask, the values will be compared.

In this logic, it appears that the two values in Word 20 and Word 22 are different, yet the coil is energized. This is because only the 8 most significant bits are being looked at, while the lowest 8 bits are ignored.

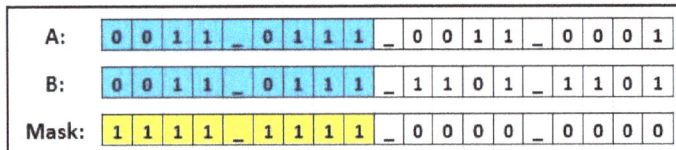

The numbers are equal!

This is what the numbers 14,129 and 14,301 look like in Binary.

A:	0 0 1 1 _ 0 1 1 1 _ 0 0 1 1 _ 0 0 0 1
B:	0 0 1 1 _ 0 1 1 1 _ 1 1 0 1 _ 1 1 0 1
Mask:	1 1 1 1 _ 1 1 1 1 _ 0 0 0 0 _ 0 0 0 0

Figure 217 - Words 20 and 22, Mask

Exercise 25

1. If the Mask in a Masked Equal Instruction is 00FF, are 31,290 and 4410 Equal? _____

2. Write Ladder Logic that increments an Auto Sequence through three different steps based on I/O and/or internal bits. On the last step, reset the sequence to zero.

3. Why do you think the Auto Sequence examples in this chapter increment by 10, rather than 1? _____

   _____

4. What is the difference between a numerical operation and a file operation?

   _____

   _____

Math Instructions

Processing of data in a PLC often involves performing mathematical operations on data. Not all processors allow math to be performed between different data types, so it may be necessary to convert one data type to another.

Conversion

Common conversions include the following:

1. **Integer to Double Integer, Double Integer to Integer**. The inherent problem in this type of conversion is that the number in the DINT won't fit into the INT. Remember that these are usually signed values, so the largest value you can put into an integer is -32,768 to 32,767.

2. **Double Integer to Real, Integer to Real**. Some platforms will have DINT to REAL but not INT to REAL. In this case the integer will have to be converted to a double integer first.

3. **Real to Double Integer, Real to Integer**. In this case you will lose the value after the decimal point. Again, not all platforms have Real to Integer conversion.

4. **Integer to BCD, BCD to Integer**. Remember that your BCD value will contain more bits than your integer.

Platforms in which data conversions are necessary include Siemens and Koyo. BCD data conversions are possible on most platforms.

Addition and Subtraction

Addition and subtraction are commonly used for all data types. A typical use might be to increment or decrement a register by some amount:

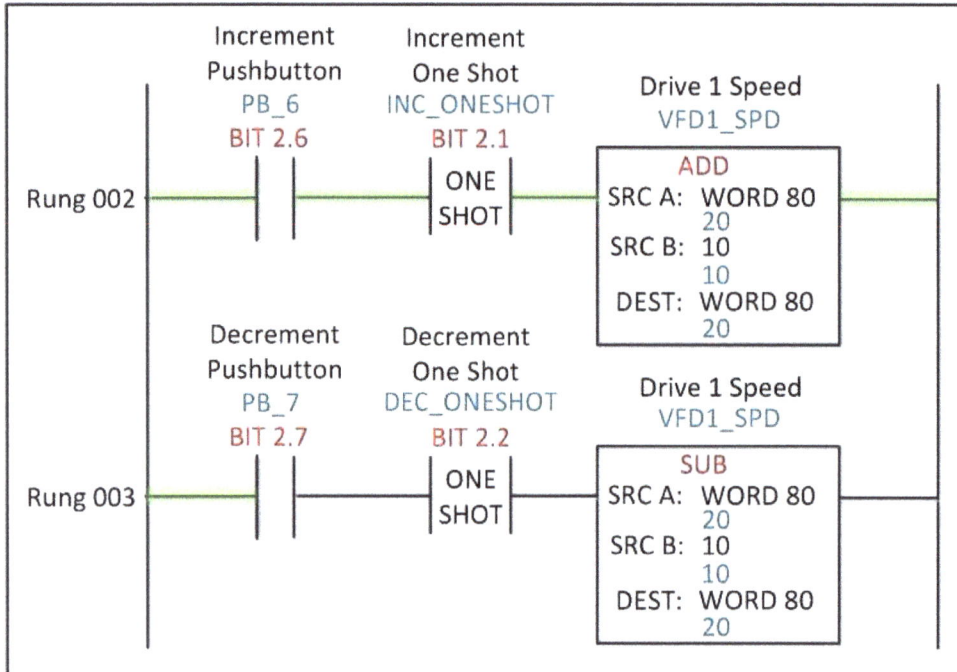

```
                Increment      Increment
                Pushbutton     One Shot
                  PB_6         INC_ONESHOT     Drive 1 Speed
                  BIT 2.6      BIT 2.1           VFD1_SPD
                                ┌─────┐        ┌─────────────────┐
                                │ ONE │        │      ADD        │
 Rung 002 ──────────┤├──────────┤     ├────────┤ SRC A: WORD 80  ├──
                                │SHOT │        │          20     │
                                └─────┘        │ SRC B: 10       │
                                               │        10       │
                                               │ DEST: WORD 80   │
                                               │          20     │
                                               └─────────────────┘

                Decrement      Decrement
                Pushbutton     One Shot
                  PB_7         DEC_ONESHOT     Drive 1 Speed
                  BIT 2.7      BIT 2.2           VFD1_SPD
                                ┌─────┐        ┌─────────────────┐
                                │ ONE │        │      SUB        │
 Rung 003 ──────────┤├──────────┤     ├────────┤ SRC A: WORD 80  ├──
                                │SHOT │        │          20     │
                                └─────┘        │ SRC B: 10       │
                                               │        10       │
                                               │ DEST: WORD 80   │
                                               │          20     │
                                               └─────────────────┘
```

Figure 218 - Increment or Decrement a VFD Speed (Add/Subtract)

In this logic, pushbuttons are used to either add or subtract 10 from a register each time the button is pushed. Notice that the number is subtracted from a register and then placed back into the same register; this is the normal way to accomplish this. Also notice that a one-shot is used on the pushbuttons. If the one shot is not used, the instruction will not add or subtract every time the button is pushed, instead it will add or subtract *every scan* while the button is being pressed! You could end up with a very large number in Word 80 if you aren't careful!

A common name for Rung 002 is an **Accumulator**, while the following rung is sometimes known as a **Decumulator**.

TIP: A Siemens counter has quite a few limitations. The "Done" bit is on if the counter is not at zero, and the preset and accumulated values are signed BCD numbers; they only count from -999 to +999.

Because of this, Siemens programmers will often use Accumulation/Decumulation logic to count. In this case, the "Done" bit will be created by using a Greater Than or Equal instruction, the Reset command moves zero into the accumulator value, and the registers would increment and decrement by 1.

Multiplication and Division

As with addition and subtraction, multiplication and division can be done with any data type, however, greater care must be taken for several reasons.

If you multiply two integers, it is important to ensure that the result will fit into the destination address. For instance, if 20,000 is multiplied by 20,000, the answer of 400,000,000 will not fit into an integer register or tag.

When dividing, if the denominator is too small (or zero!) the same thing will happen.

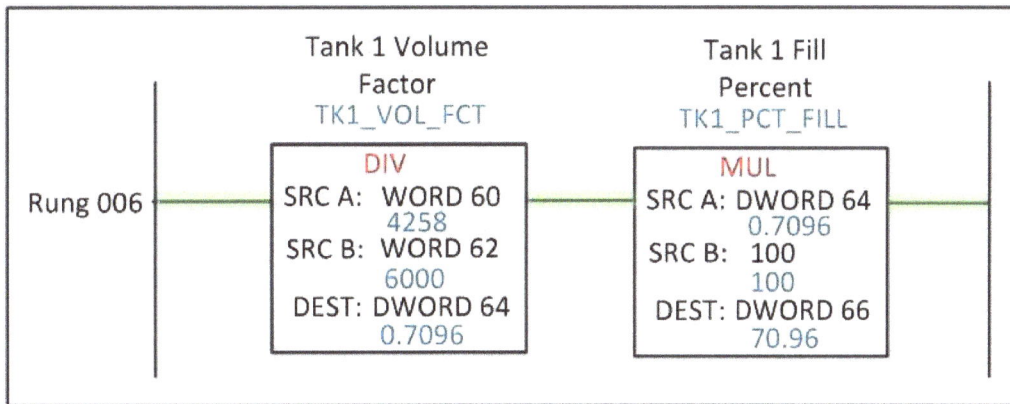

Figure 219 - Calculating Percent (Divide/Multiply)

This logic calculates the percentage of fill of a 6,000-gallon tank. The measured volume (WORD 60) is divided by the total volume (Word 62), and then multiplied by 100. In this example, data types are mixed; an integer is divided by an integer and the result is placed in a REAL. The REAL is then multiplied by an integer and the result placed in another REAL. If the PLC being used did not support this functionality, additional data conversions from INT to REAL would be necessary.

On some brands of PLC, the data types must be the same to perform math. In this case, the numbers must be converted as shown in the Siemens S7 example below.

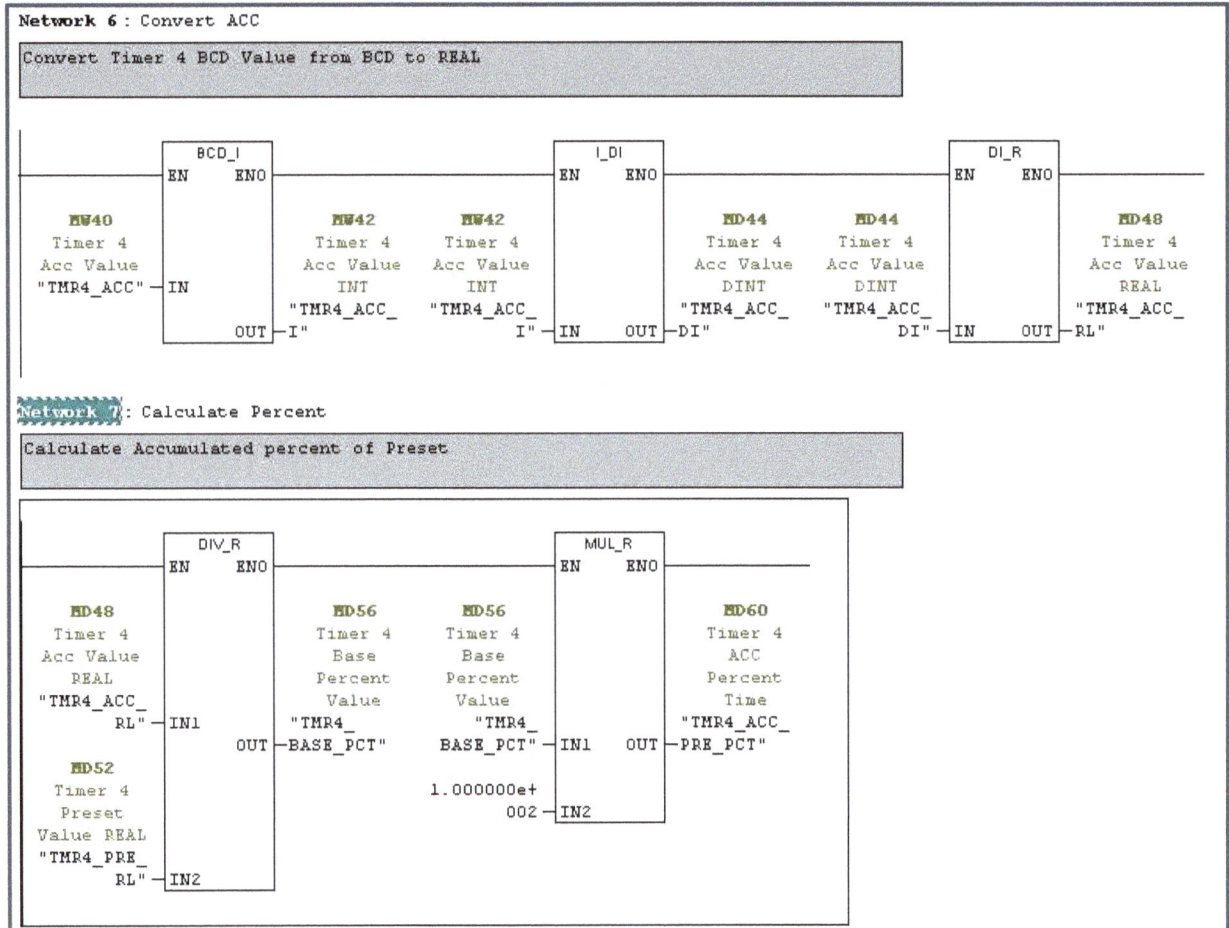

Figure 220 - Percent Calculation With Conversions (Siemens)

As in the previous example, this logic calculates a percentage, which in this instance is a timer's percent of completion. Siemens' timer accumulated values are in BCD, so the number is converted from BCD to Integer to Double Integer to REAL. Siemens' timers also time from the setpoint down to zero, so the final value would need to be subtracted from 100% to present the remaining time to completion (100.0 – TMR4_ACC_PRE_PCT, or 100 - MD60).

Exercise 26

1. A Variable Frequency Drive (VFD) is used to control a conveyor. Its maximum speed is 1750 RPM, but the number sent from the drive is in integer form. At full speed, the integer reads 31,760, while it reads zero when stopped. Write ladder logic to calculate the percentage of a drive's actual speed to its total speed, also providing a REAL speed in RPM.

2. Production and reject data from a manufacturing line is entered manually at the end of each shift. After completion of the third shift, there are three registers named Shift1_Prd, Shift2_Prd and Shift3_Prd containing the total parts made that shift, and 3 more registers named Shift1_Rej, Shift2_Rej and Shift3_Rej that contain the number of failed parts. Write ladder logic to calculate Total Parts, Total Rejects and Total Good Parts for the day.

Scaling

An important mathematical function is that of converting raw analog values into usable units of measurement or converting one unit into another. This is known as **Scaling**, and it follows a standard formula, $y = mx + b$. **Y** is the units of the Y axis, and **X** is the units of the X axis. **B** is known as the **Offset** while **M** is the **Scalar**, determined by dividing the "rise", or increase of the Y axis, by the "run", or increase of the X axis.

As an example, let's look at the graph below:

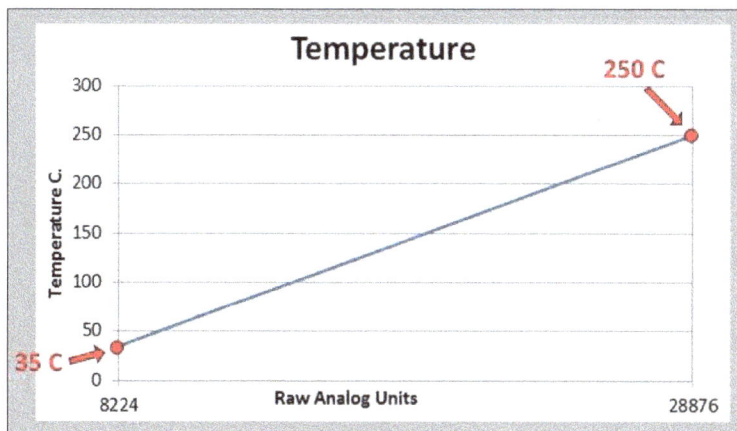

A temperature sensor produces a 0-10v signal. This is wired into an analog card which produces a signal that ranges from 0-32,767, a signed integer. The process operates at about 150 degrees Centigrade.

Figure 221 - Temperature Scaling

A thermometer is used to measure the actual temperature at two different points and the raw value is recorded for each measurement; the first point P1 is recorded as 8,224 on the analog card at 35 degrees C, while the second (P2) is recorded as 28,876 at 250 degrees C.

The first step in scaling the raw measurements into degrees is to calculate M, the Scalar. The "rise" or difference in Y values is Y2-Y1, 250-35, or 215. The "run" or difference in X values is X2-X1, 28,876-8224, or 20,652. Dividing the rise by the run produces a Scalar M of 0.01041061.

The next step is to calculate the offset B. Since y=mx+b, the B factor can be calculated as B=Y-MX. Substituting the values for P1, which were Y1 and X1, the calculation becomes B = (35-(0.01041061*8,224)), which yields an offset of -50.6168894. These two constants, M and B can be used to calculate Y for any inserted value of X.

As an example of how to use this formula in calculating a temperature, assume that the temperature sensor reads a value of 10,512 into the analog card. If the formula y=mx+b is used, the temperature is (0.01041061*16,512) – 50.6168894, or 121.28 degrees C.

This formula can also be used to calculate all of these variables in one group of calculations. Allen-Bradley's Scale with Parameters (SCP) instruction allows the programmer to enter Raw High and Low and Engineering High and Low units into the instruction along with the measured value from the card. It then outputs the scaled value to another variable.

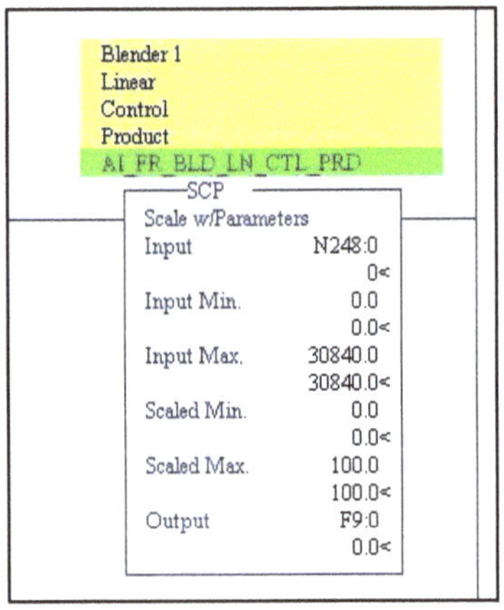

The input value is N248:0, a signed integer value from an analog card. The Minimum and Maximum Input values are taken from observed values coming from the card, while the Minimum and Maximum Scaled values of 0.0 and 100.0 represent 0 to 100 percent of the Input values. The result is moved into F9:0, a Floating Point or REAL register.

Figure 222 - Scale with Parameters (SCP) Allen-Bradley RSLogix 500

Unfortunately, many PLCs do not have this instruction, including A-B's ControlLogix, but the math can easily be reproduced as shown in the following logic:

This logic can also be packaged inside of a subroutine, function, or "Add-On Instruction" (AOI) where parameters can be passed into it and the results passed out.

The first two rungs use High and Low variables passed "by reference", (as shown in the SCP instruction), to calculate the internal (local) variables M and B. These are then used in the third rung to scale the Analog Sensor Input to Engineering Units Out.

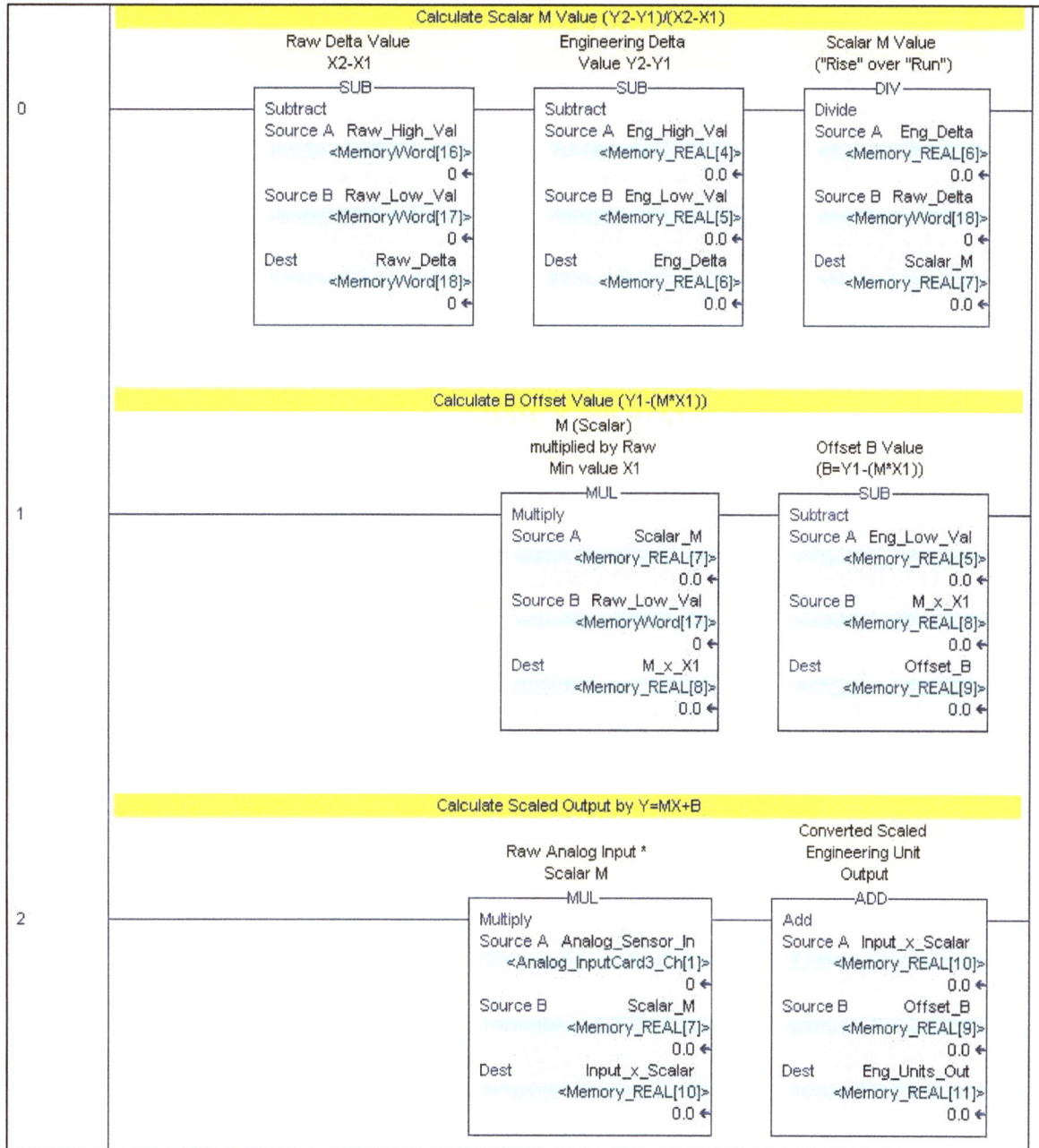

Figure 223 - Use of Formula Y=MX+B for Scaling

Advanced Math

In addition to the addition, subtraction, multiplication, and division instructions mentioned previously, here is a list of more advanced math instructions along with their purpose:

Exponent – An exponent signifies the number of times that a number is to be multiplied by itself. The exponent is usually indicated by using the "^" sign, so 3^4 is 3 x 3 x 3 x 3.

Logarithm, Natural Log (LOG, LN) – A logarithm (LOG) is the inverse of an exponent. For instance, if 2^3 is 2 x 2 x 2 = 8 where 3 is the exponent, then the LOG of 8 in Base 2 is 3. Where a LOG is typically base 2, a Natural Log (LN) has a base of 2.718. Natural Logs are often used in math and physics (calculating decibels and pH), while LOG is often used for computer calculations.

Sine, Cosine, Tangent (SIN, COS, TAN) – Also known as Trigonometric functions, these are used to calculate geometrical coordinates. These functions -- along with their reciprocals Cosecant, Secant and Cotangent and their inverses Arcsine, Arccosine and Arctangent -- are often used in motion control applications.

Modulo (MOD) – This function calculates the remainder after a division operation.

Absolute Value (ABS) – Returns the positive version of a number even if it is negative. The Absolute Value of both -15 and 15 is 15.

Other Advanced Instructions

There are a wide variety of other instructions available on different PLC platforms in addition to those listed previously. These are just a few that are common to some of the major brands of PLC.

String Operations

As mentioned in the data section of this manual, strings are arrays of SINTs, or Single Integers (Bytes). The array elements contain ASCII characters, which can be thought of as printable characters with a few non-printable commands included. Values contained in strings can be displayed as decimal or hexadecimal numbers, or as text characters. If in text, they are often displayed with a "$" sign before the character, such as Text = $T, $e, $x, $t characters. These equate to the decimal numbers 84, 101, 120, 116 or the hex numbers 54, 65, 78, 74. These can be found in a standard ASCII table; there is one in the appendix of this manual.

Strings may also contain a length (LEN) field that contains the number of characters that exist in the string. For instance, if a string has space for 80 characters, but is filled with the characters "Today is Tuesday, September 13" then LEN = 30.

Concatenate (CONCAT) – Connect two strings together, one after another.

Middle (MID) – Copies a specified String into the middle of another String at a specified location.

Find (FND) – Locate the starting position of a specified String within another String. Usually returns the position of the found String.

Delete (SDEL) – Removes characters from a String at a specified position.

Insert (INS) – Adds characters to a String at a specified position.

Length (LEN) – Finds the number of characters in a String if length is not part of the string definition.

PID Instructions

PID, or Proportional-Integral-Derivative instructions, control a process variable such as flow, pressure, temperature, or level.

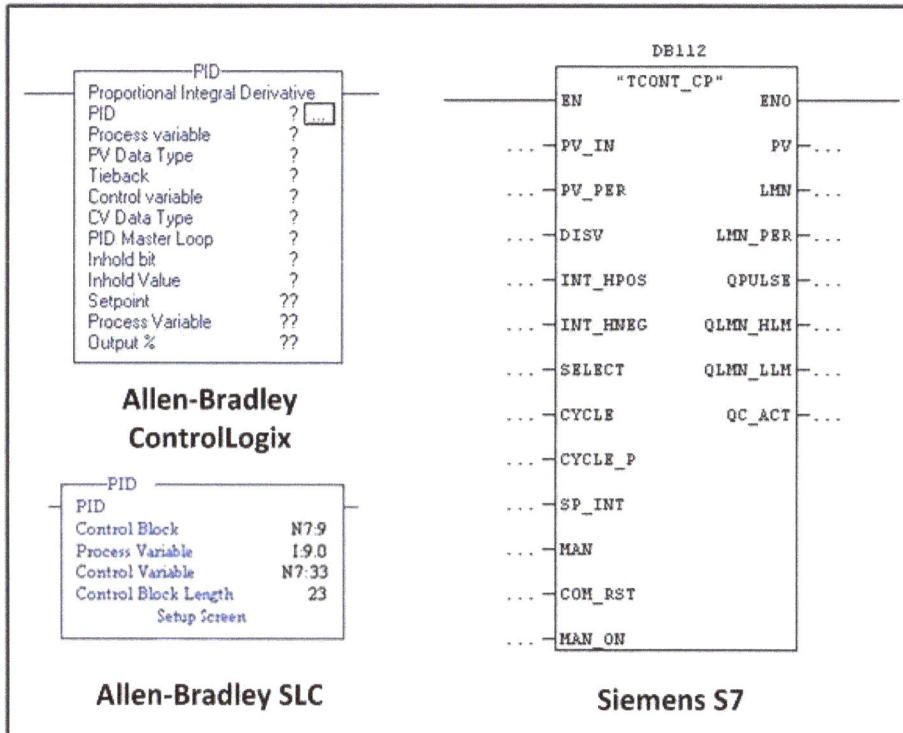

There are a wide variety of parameters that can be set for control. These may be passed in as variables as shown in the ControlLogix and S7 diagrams or set up on a special setup screen as shown for the SLC.

Figure 224 - PID Instructions, Different Platforms

Motion Control Instructions

Newer PLC platforms may use multi-axis controller cards to coordinate movement. There are many possible commands associated with controlling axes: start, stop, jog, and direction are just a few. Numerical values such as speed, acceleration, deceleration, and "go to position" commands are also common.

In addition to these individual axis commands, many coordinated movement commands are also included.

Allen-Bradley's RSLogix5000 software (ControlLogix and CompactLogix) includes 6 folders with 43 different instructions relating to motion control. These are available in Ladder and sometimes Structured Text or Function Block:

Motion Configuration – 4 Instructions. Related to tuning and parameter assignment for axes.

Motion Event – 6 Instructions. Related to arming and disarming events in the motion control card.

Motion Group – 4 Instructions. Related to issuing simultaneous command to multiple axes.

Motion Move – 12 Instructions. Related to individual issuing of motion commands to an axis.

Motion State – 8 Instructions. Related to directly controlling the operating state of an individual axis.

Multi-Axis Coordinated Motion – 9 Instructions. Related to controlling multiple axes in coordinate movement as with robotics. XYZ, XY, Articulated and SCARA configurations are all addressed.

Communications Instructions

Communications instructions are used to access a port in order to send or receive data.

Siemens Step7 software contains a range of System Functions and System Function Blocks that are used to send and receive data in a variety of ways. These are chosen based on the type of data, type of communications, and, in some cases, the protocol used.

Many of these system blocks are available in the CPU; they just need to be called as a routine. Different parameters need to be filled in on the block. Most of these system blocks are categorized as COM_FUNC (Comm. Function), DP (Profibus), PROFIne2 (Profinet), and occasionally TEC_FUNC for ptp, a Siemens point to point or RS422 protocol.

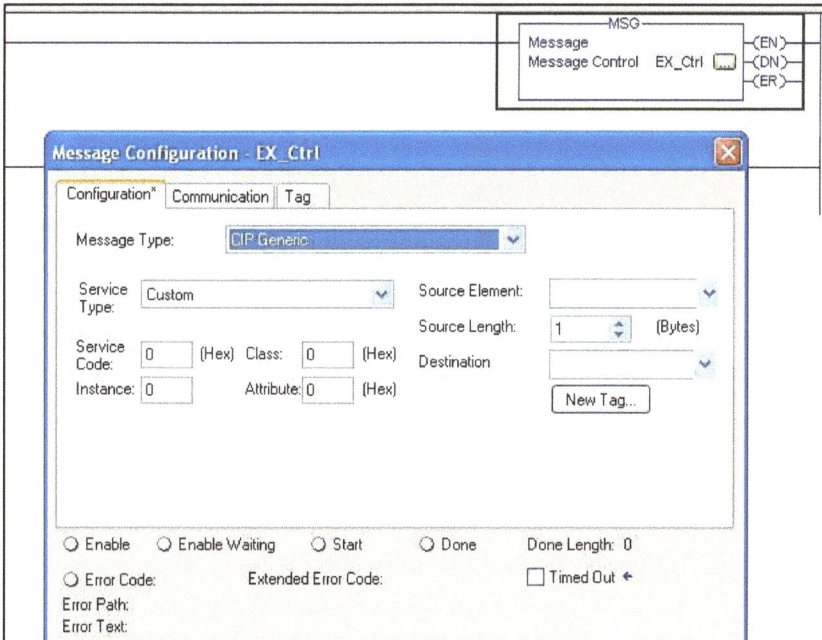

Figure 225 - Allen-Bradley MSG Instruction

Allen-Bradley's messaging is generally handled by the MSG instruction:

This instruction requires a message control tag specified at the controller level, shown as EX_Ctrl in this image.

After defining the control tag, a message configuration screen is accessed and the type of communications and, ideally, the path to the remote device are specified.

Protocols for Ethernet, DH485, DH+, and Serial DF1 along with SERCOS to motion control devices are available. The target node can be specified as PLC2, PLC3, PLC5, SLC, and Generic CIP devices.

There are also several ASCII serial port instructions available for reading, writing, handshaking, and buffer control on a basic level.

Allen-Bradley's ControlLogix platform also has a method of directly linking tags in one controller with those in another controller. These are called "Produced-Consumed Tags".

Program Control Instructions

Program control includes jumping to or calling subroutines, disabling parts of a program by jumps or "MCR" commands, looping by jumping or by using "For/Next", or redirecting the program flow in other ways. Following are some of the more common program control instructions:

Jump Subroutine/CALL – These instructions redirect the program scan to the start of a subroutine or function. At the end of the called subroutine, the flow is redirected to just after the jump or call statement.

Jump/Label – These instructions redirect the scan to a labeled point in the same routine. If jumping forward, some code will not be executed. If jumping backward, code within the zone will be executed over and over (looping) until redirected by another jump, usually associated with a counter which is preset with the number of loops to be executed.

End/Temporary End - This ends the scan of the routine and does not scan code past that point. This is often conditional, controlled by a BOOL or other logic.

For/Next, Do While – Similar to a loop as described above, a For/Next instruction is usually set to operate a specific number of times. A Do/While statement executes at least once and remains active until a defined condition is met. Both of these instructions are seen most in Structured Text. If used in Ladder, care must be taken not to exceed the Watchdog Timer.

Master Control Relay (MCR) – This instruction is used in pairs. If the first instruction is true, the program proceeds normally. If false (not activated), the physical outputs within the MCR zone will be de-activated. This is not true for outputs that are latched on. ***The MCR instruction should not be used to replace hardware MCRs.***

Miscellaneous/Other Instructions

There are a wide variety of other instructions available, far too many to list here. Every manufacturer has its own instruction set and different names for the instructions.

Here a few general categories of instructions with their uses:

LIFO and FIFO Instructions – LIFO is an acronym for "Last In, First Out", while FIFO stands for "First In/First Out". These instructions operate on a "Stack", which can be configured two different ways.

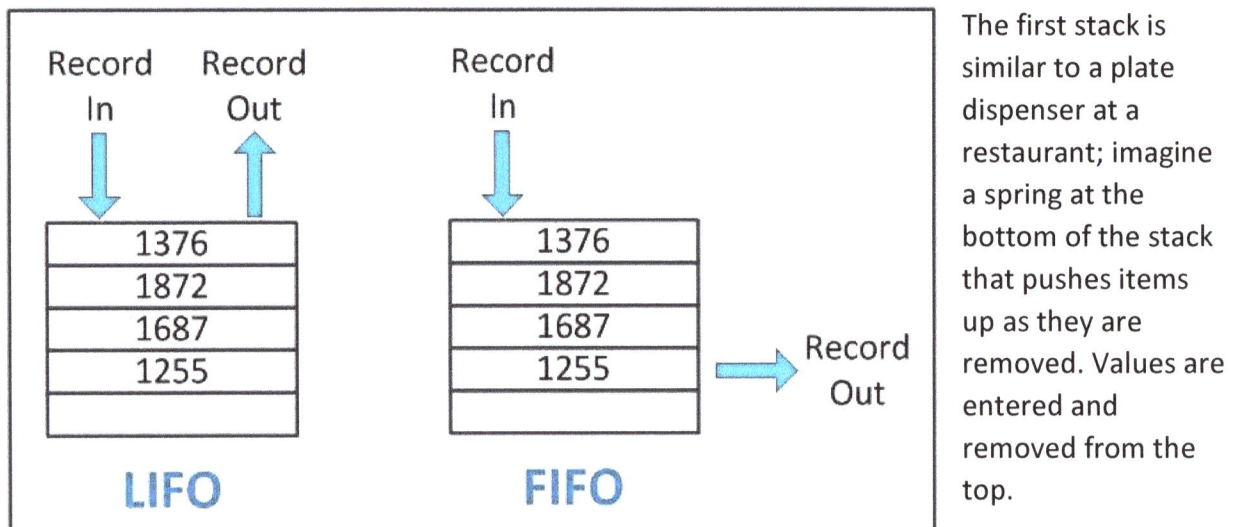

The first stack is similar to a plate dispenser at a restaurant; imagine a spring at the bottom of the stack that pushes items up as they are removed. Values are entered and removed from the top.

Figure 226 - LIFO and FIFO Concepts

The second stack or FIFO allows values to be entered from one end and removed from the bottom of the stack. Each of these stacks have multiple instructions that may be used to manipulate the values. The major instructions are **Load**, which places a new value or record on the stack, and **Unload**, which removes a record or value.

Sequencer Instructions – A sequencer, sometimes called a "drum sequencer", monitors and controls repeatable operations. These instructions also use a stack, but the numbers in the stack are treated as binary values that represent conditions or drive outputs.

A Sequencer Input (SQI) instruction is used to detect when conditions are correct to index the sequencer. If the bit pattern in the designated register matches that of the next position in the sequencer, the sequencer's position value register will increment by one. The bit pattern often represents physical input states.

The Sequencer Output (SQO) instruction is used to set output conditions. These are also represented by a bit pattern, often mapped to physical outputs. The SQI and SQO instructions are usually used in pairs with the SQI dictating the conditions that index the SQO.

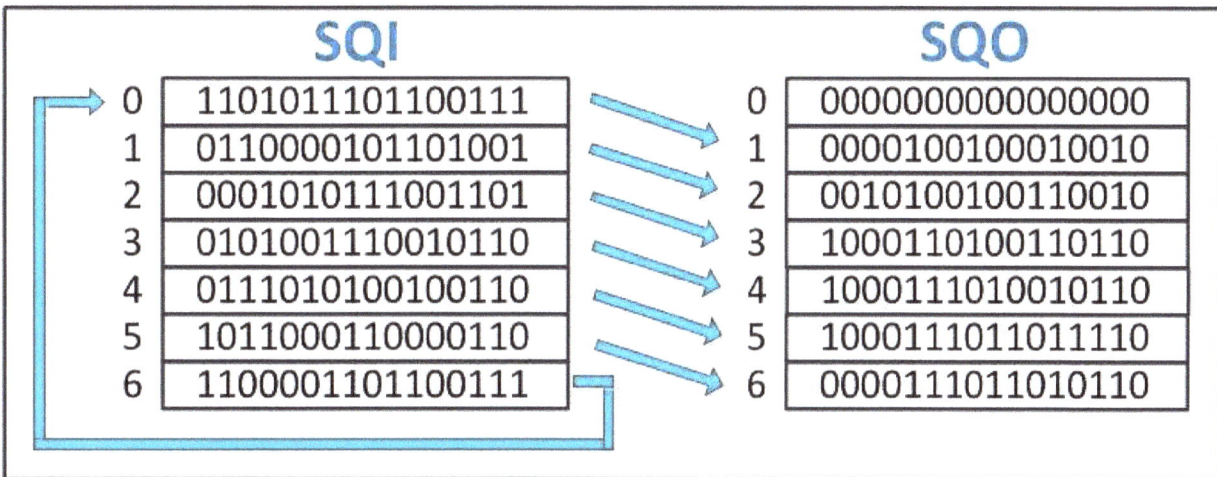

SQI		SQO	
0	1101011101100111	0	0000000000000000
1	0110000101101001	1	0000100100010010
2	0001010111001101	2	0010100100110010
3	0101001110010110	3	1000110100110110
4	0111010100100110	4	1000111010010110
5	1011000110000110	5	1000111011011110
6	1100001101100111	6	0000111011010110

Figure 227 - Allen-Bradley Sequencer Instructions

The Sequencer Load (SQL) instruction can be used to place values (bit patterns) into the sequencer's register. This is much like a "teach" function; the instruction looks at a register, often representing inputs. When the SQL is executed, the pattern is loaded into the next location in the stack.

Sequencer Compare (SQC) is another instruction that is sometimes used in order to index the position number of a sequence.

Statistical Instructions – There are various instructions available on some platforms to perform statistical math, such as standard deviation, moving averages, and finding minimum and maximum signals in a specified period of time.

Other Instructions are available for Safety Functions, Signal Filtering, VFD (Drive) Control, Equipment Phasing (State Programming), and many others. For full knowledge of a specific platform, it is a good idea to read the programming manual and help files for instructions. Don't forget that many of these require the use of languages other than ladder! It is also possible to build these instructions yourself by using Add-On Instructions or Functions.

Exercise 27

1. A tank used for blending juice holds approximately 8,000 gallons of liquid. There is a pressure transducer that produces a 4-20mA signal, it is wired into channel 1 of an analog card.

   The tank is filled with 6,000 gallons of juice; the reading of the analog card is recorded as 24,780. The tank is then drained completely and the value from the transducer is recorded as 96.

   Draw ladder logic that scales the raw reading from the transducer into gallons. There are 3.78541 liters in one gallon. Also calculate the number of liters.

2. What kind of applications are Trigonometric functions used for?

   _____

3. Decode the following hexadecimal ASCII characters using the table in Appendix E:

   47 _____ 6F _____ 6F _____ 64 _____ 20 _____ 4A _____ 6F _____ 62 _____ 21

   _____

4. Can a JUMP instruction be used to move backwards in a program?

   _____

5. What do the acronyms "FIFO" and "LIFO" stand for?

   _____

6. What is the purpose of a Sequencer instruction?

   _____

   _____

PLC Maintenance and Troubleshooting

There are a number of tools and techniques common to all PLC platforms that can aid a technician in isolating the causes of problems. An important thing to remember when using these tools is that *The PLC's program cannot change without someone changing it!* Programs can't change themselves; they either run or they don't.

Forcing

One method of determining whether an input or output is working properly is using a **Force**. In the case of an input, it is obviously not possible to physically force a point on a card. You would have to place a voltage on the point in order to energize it. So, what are you forcing when you force an input? *Only the input table*. This means that when you apply the force, the contacts or values related to that point will change only in the program, not on the card itself.

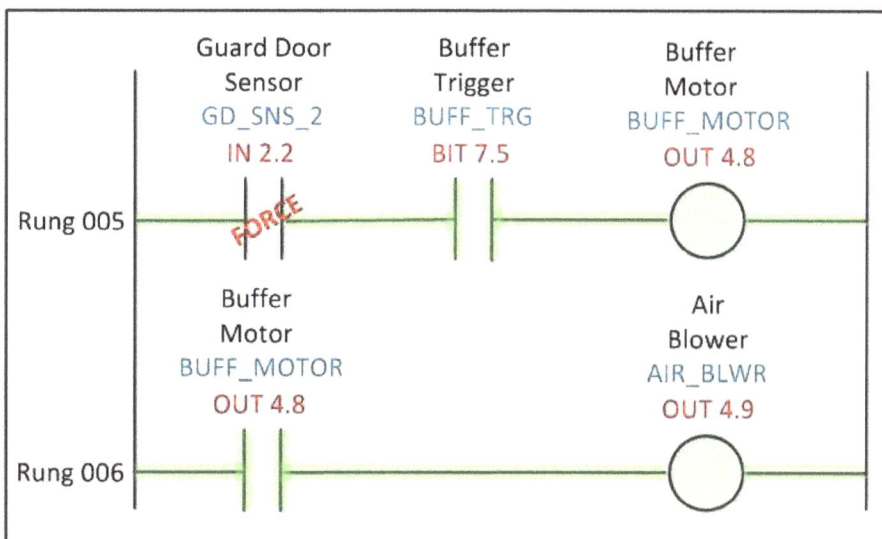

In this case, the Guard Door Sensor would not allow the Buffer Motor to run even though the trigger signal is shown to be active. A force is applied to input 2.2 and the motor runs. This is a verification that the electrical signal to the input needs to be checked; maybe it is a bad sensor, maybe the wire is disconnected, or

Figure 228 - Input Force

maybe the input point on the card itself is bad. In this case, the forcing of the input has helped to determine the problem. After the problem has been isolated, the force can be removed, and the problem fixed. *The force should not be used to hide the problem!*

On most PLC platforms, there will be some kind of indication on the contact showing that the input or output is forced. There is also usually a light on the processor itself indicating that a force is present.

An important note on forcing inputs: when the input is forced, every place the contact is used in the program will be at the state of the force; you are "lying" to the processor. If you simply want to affect the outputs on the rung in question, you can always place a bit in parallel with the input and use it to control the rung as shown in Figure 229.

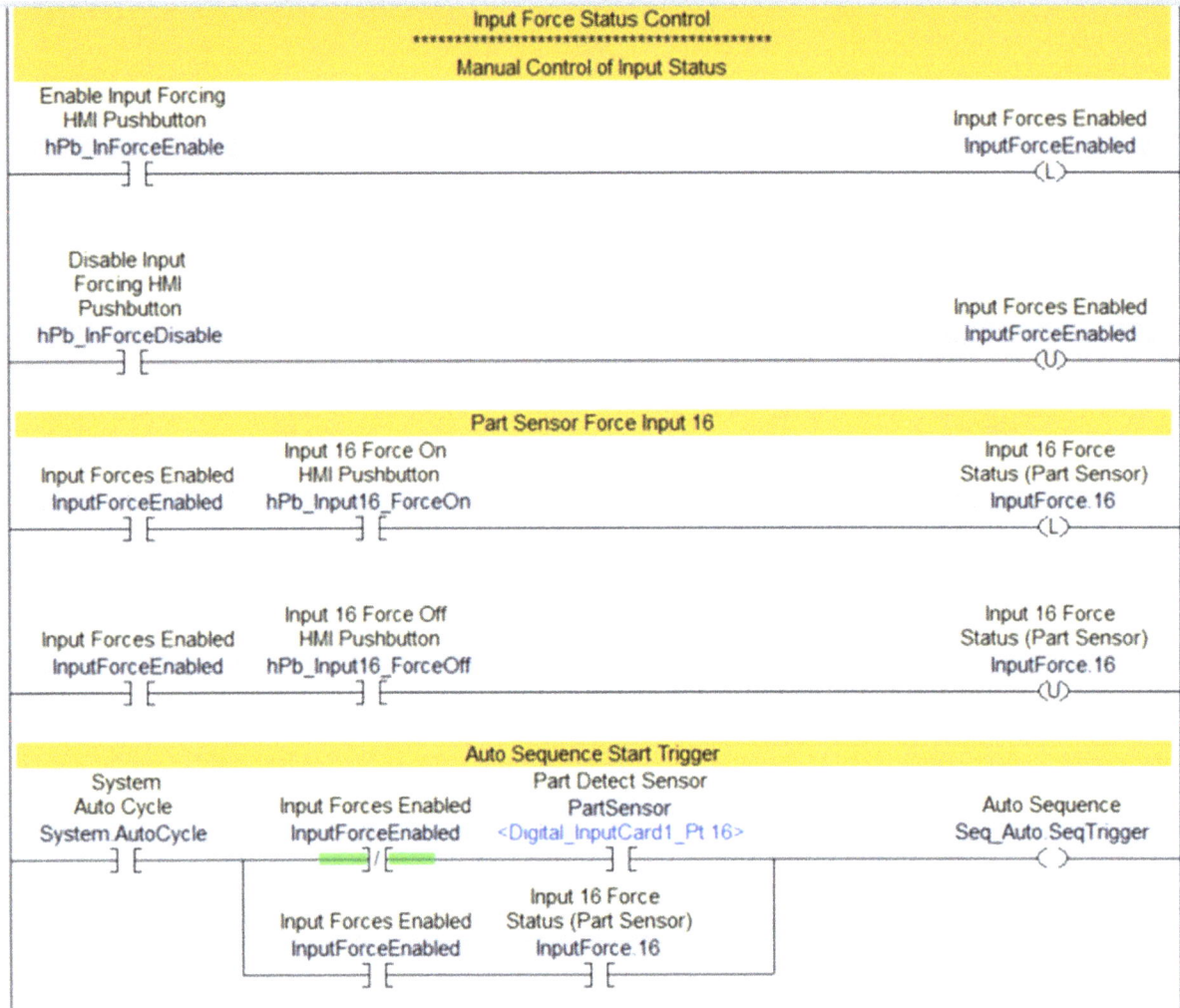

Figure 229 - Manual Control of Input Status Without Forcing

This method of input control takes more code but has the advantage of being selective when it comes to bypassing inputs. The Part Sensor InputForce.16 bit can be easily changed from a touchscreen rather than using programming software and can even be manipulated with the program logic.

Programs also often use mapping routines to connect a physical input to a status bit, which is then used in place of the input throughout the program. In this case the input status bit can simply be re-mapped to a different physical terminal. If so, it is important to document or "redline" the changes in the schematics.

Forcing an output is the _exact opposite_ of forcing an input. If a force is placed on the output point, the physical output will be energized but _the output image table will not be affected_.

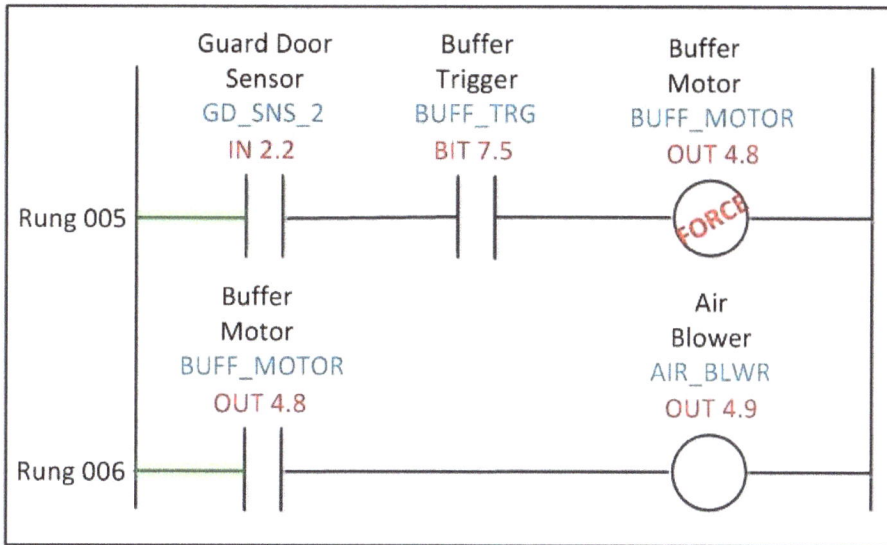

Figure 230 - Output Force

The logic energizing the Buffer Motor is not true, but a force is applied to the output. The physical output comes on, and the Buffer motor runs. Note, however, that the Air Blower that usually comes on whenever the motor runs is not energized. This is because _the image table is not affected by the force_.

The forced coil is actually transparent to the logic; if conditions are true up to the coil, the image table will be updated, and the Air Blower coil will be energized.

In most PLCs only inputs and outputs can be forced, but some platforms allow the forcing of memory also.

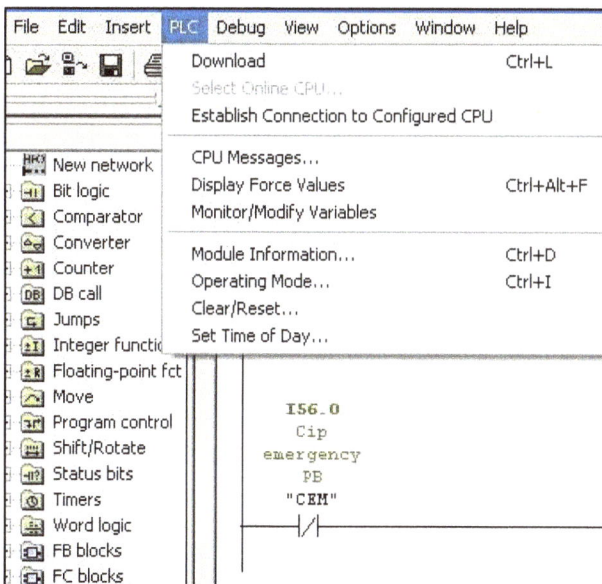

Figure 231 - Siemens Forcing

Installing and activating forces is generally a multi-step process. The force is installed in one step then activated as a separate action. This is because **forces can be very dangerous if implemented incorrectly**; you are telling the PLC to do something unnatural, outside of its coding. A force can, however, be helpful in the troubleshooting of a system.

The image to the left shows how forces are accessed in Siemens Step 7 software. From the editor, a Force table is opened from the PLC menu. Force addresses are entered into the table, and then activated.

When active, a red "F" appears by the address indicating that a force is being used.

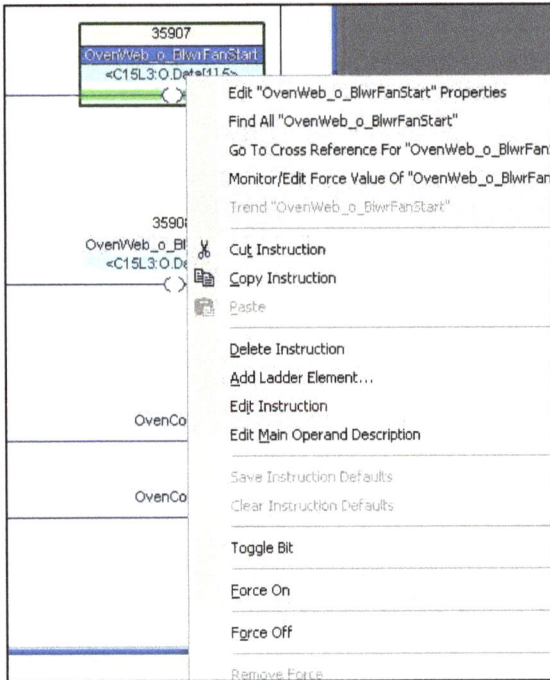

In Allen-Bradley's software, forces are installed by right clicking on an address in the program. After installing the force, forces must be enabled using the dialog shown below:

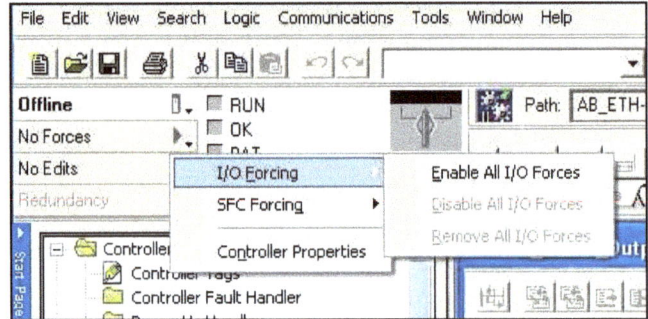

Figure 232 - Allen-Bradley Forcing

This two-step process ensures that the programmer truly intends to create a force.

Searching and Cross-Referencing

In order to diagnose problems in a machine, it may be necessary to trace logic through a program. There are a number of tools available to determine the location of addresses, check whether addresses have been used, and substitute one address for another. Searches can also locate words in the comments of a program.

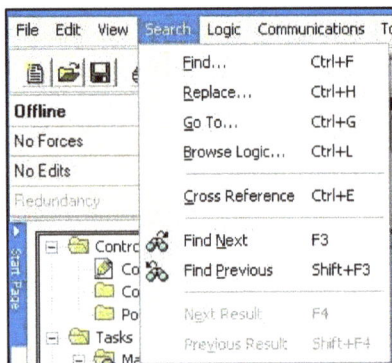

There is usually a tab in the software that will allow various "search" or "Go To" options as shown. Right-clicking an address in the program will also often bring up a selection allowing other instances of the address to be found.

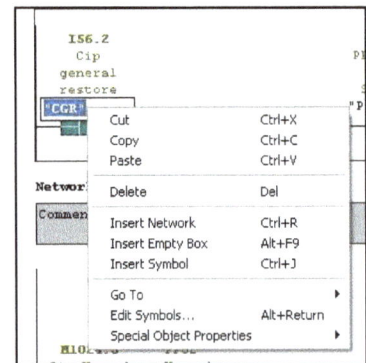

Figure 233 - Allen-Bradley Cross-Reference

Figure 234 - Siemens "Go To"

One of the most useful tools when trying to determine why an output is not being activated is the **Cross-Reference**. A cross-reference shows all of the places in a program where an address has been used.

Usually, troubleshooting starts with finding the coil of an address. Right clicking the address brings up a selection for cross-reference or "Find All"; this in turn brings up a list of all the places where the address is used in the program.

Selecting the location of the coil takes you to the rung or network where the coil (OTE) is activated. With proper programming technique, there should only be one place where a coil will be located for any address!

Addresses can then be traced from rung to rung until finally the cause of the problem can be identified.

Figure 235 - Cross-Reference Table

The following rungs illustrate tracing the cause of an output failing to come on:

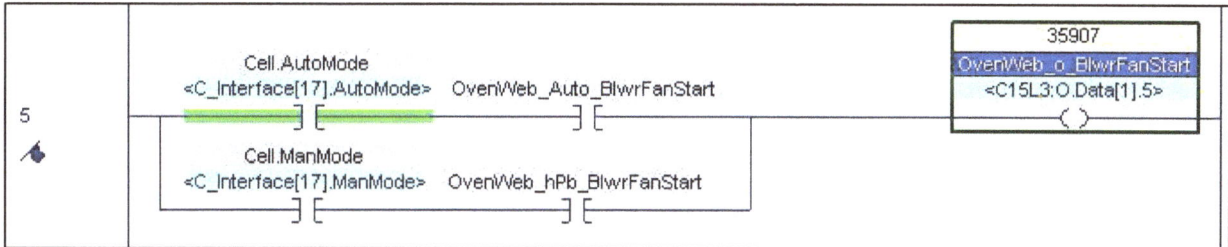

Figure 236 - Output Control Rung

Figure 236 shows that the Oven Web Blower output (the coil) is not energized. Since the AutoMode contact is energized, a cross-reference of **OvenWeb_Auto_BlwrFanStart**, is executed.

Step 1: Searching for the coil brings up this rung:

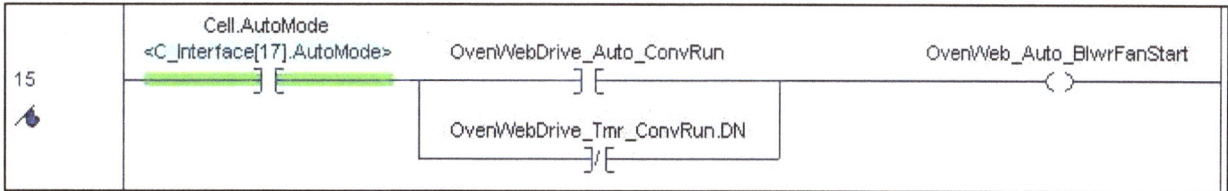

Figure 237 - Auto Command for Blower (Search Result Step 1)

Step 2: It appears from this logic that the blower starts when the conveyor runs.

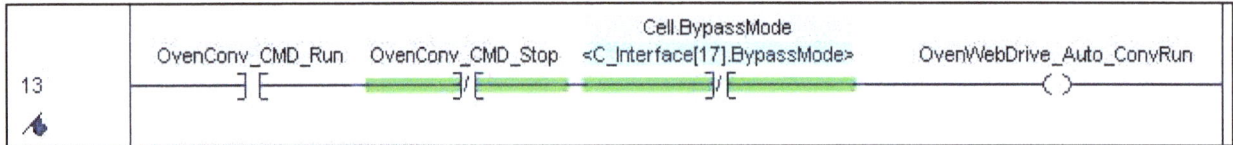

Figure 238 - Auto Command for Conveyor (Search Result Step 2)

Step 3-5: Searching for the CMD bit in Figure 238 further leads to the rungs below:

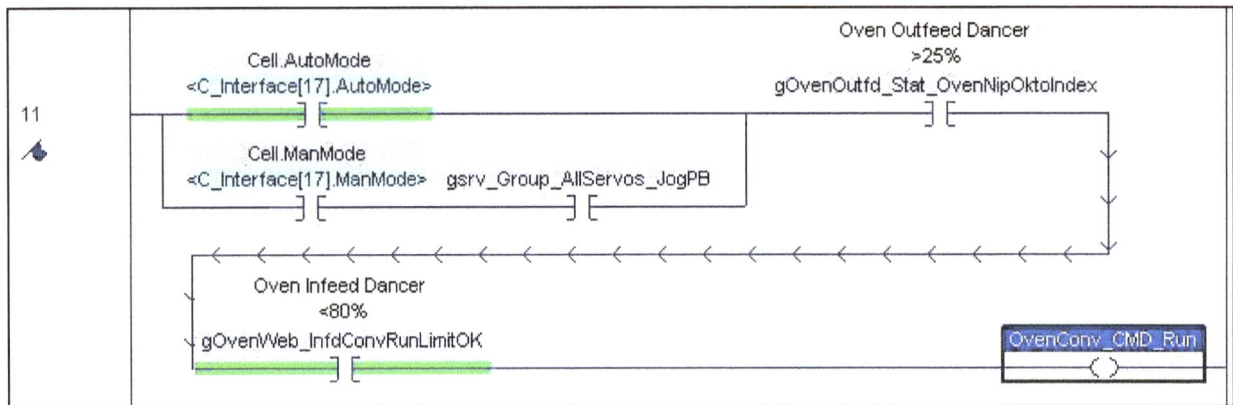

Figure 239 - Oven Conveyor Run Command (Search Result Step 3)

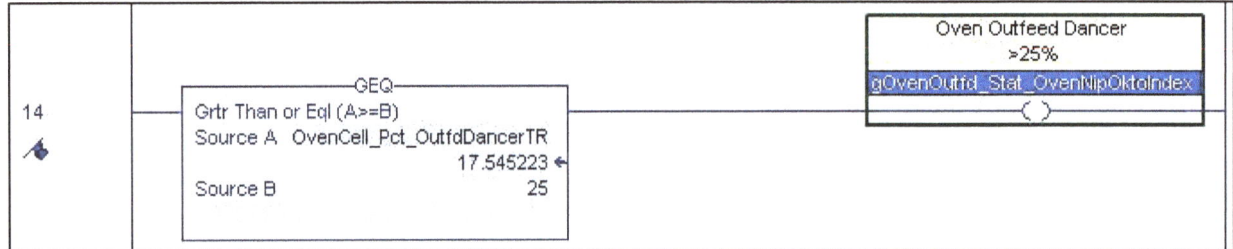

Figure 240 - Outfeed Dancer Position (Search Result Step 4)

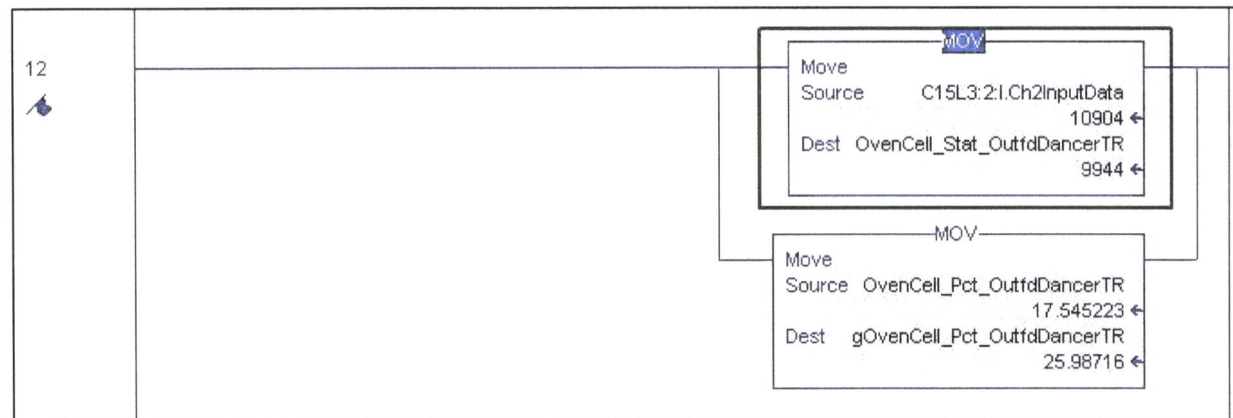

Figure 241 - Analog Input for Dancer Position (Search Result Step 5, Final Step)

Figure 240 shows a comparison instruction, which will not have a coil associated with it. Still, a comparison returns a true or false result, like a coil, and the place where that number is generated will be the cause of the result. This is sometimes called a **destructive** bit or value. If you look carefully at Figure 235, you will notice a column indicating whether the tag value is destructive or not. This is usually the value you are looking for.

Figure 241 shows that a physical address C15L3:2:I.Ch2InputData (an analog input value) is moved into the variable that we are looking for. Cross-referencing and searching will always end at either a **physical input point**, or at an address with no coil or address that has been changed by the program. This last would mean that the signal comes from outside the controller, such as an HMI, SCADA, or even a signal from another PLC.

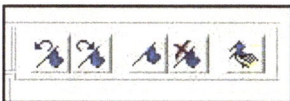

Figure 242 - Allen-Bradley Bookmarks

Did you notice the little flags next to the rungs? Allen-Bradley's ControlLogix platform has a toolbar called "Bookmarks" that allow a programmer to mark rungs and then index through them. Very handy!

Siemens also has a useful tool that allows one to look at which registers have been used. This can also be used to find addresses with no symbol or symbols with no address. Notice that several bits have been assigned that are also Word addresses (MB1026 & MB1027). This can be a useful to spot address interferences.

Figure 243 - Siemens Usage Table

Additional information on electrical troubleshooting of PLCs is covered in a later section.

Exercise 28

1. Forcing an input applies voltage to the physical input. True: _____ False: _____

2. Forcing an output applies voltage to the physical output. True: _____ False: _____

3. In the diagram at right, if Q98.4 "PP03" is forced, what will the state of Q98.5 "PP04" be? (Assuming that input I56.4 is off)

    _____

    _____

    _____

    _____

Network 43 : PP03

Comment:

```
  I56.4
  Return
   Flow                              Q98.4
  Switch                             PP03
  "RFS"                             "PP03"
   ─┤ ├──────────────────────────────( )──
```

Network 44 : PP04

Comment:

```
  Q98.4                             Q98.5
  PP03                              PP04
 "PP03"                            "PP04"
  ─┤ ├──────────────────────────────( )──
```

4. If the force on Q98.4 is removed and I56.4 is forced, what will the state of Q98.5 be?

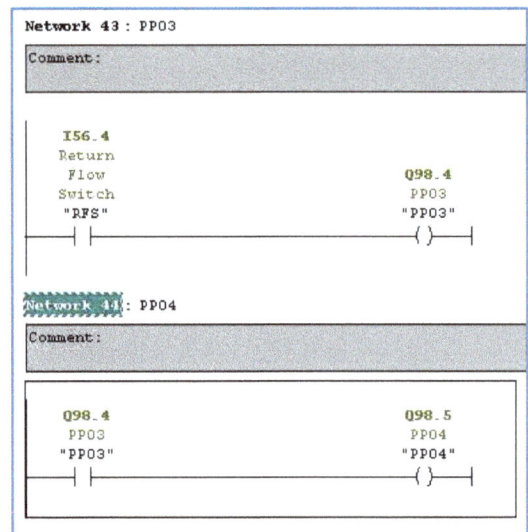

    _____

    _____

    _____

    _____

Figure 244 - Exercise 28, Questions 3, 4

5. To trace a signal through a program, what types of instructions should you look for?

    _____

    _____

    _____

6. What two types of addresses will represent the end of a search or cross reference?

    _____

    _____

    _____

    _____

Systems

All the components and controllers described in the previous sections need to act as a cohesive unit to produce products. Computers, PLCs, sensors, motors, and other devices are all part of the bigger picture of manufacturing.

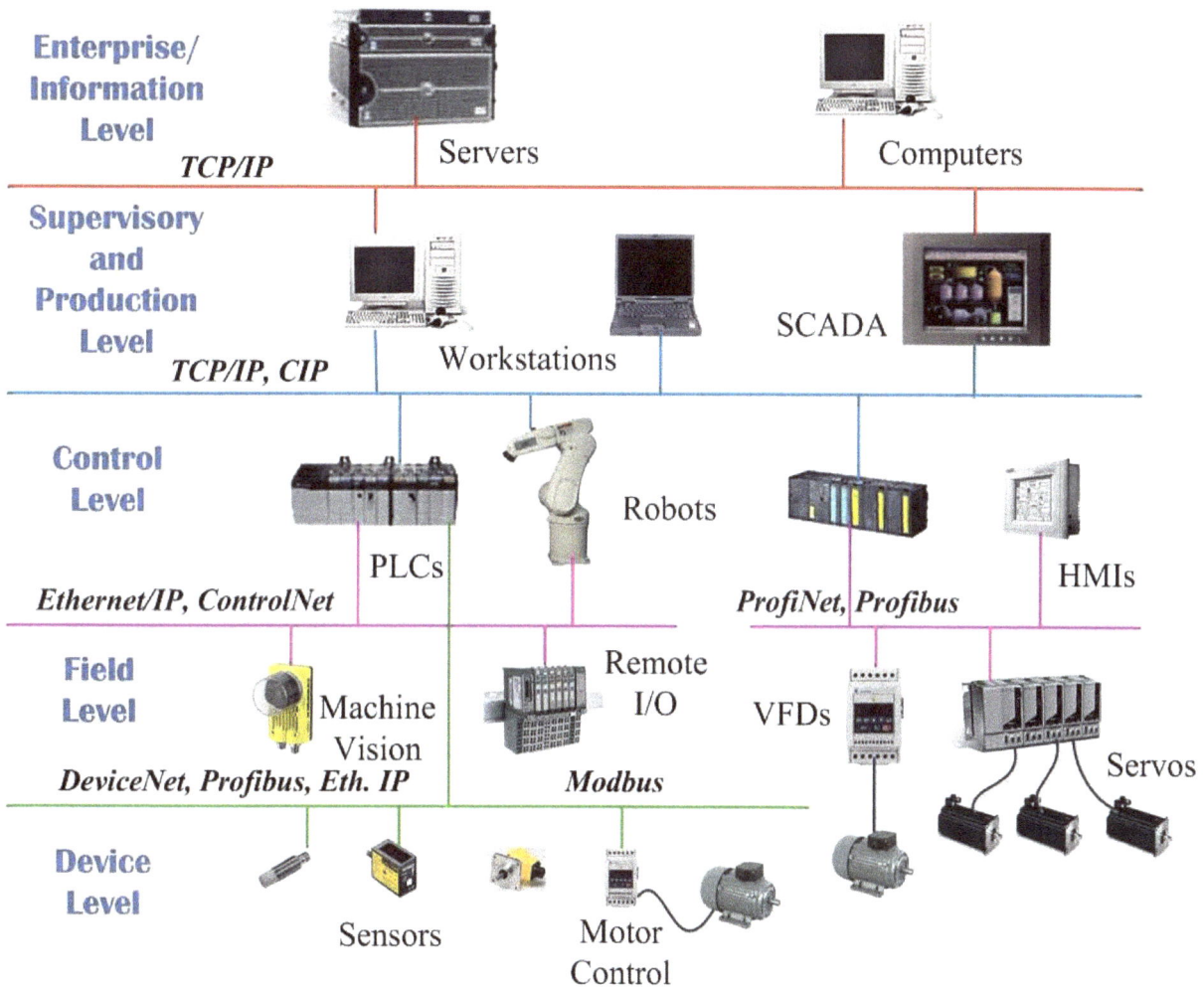

Figure 245 - Plant Control Hierarchy

Figure 245 illustrates how different controllers and elements of the system interface and bridge the various networks. All the networks below the control level need to respond quickly to changes, while the communications at the higher levels are concerned with transferring and saving large amounts of information.

Systems can be further broken down into individual machines, which can then be analyzed as a workcell, assembly, or mechanical powertrain. Analysis and troubleshooting then begins by isolating problems by looking at probable sources and eliminating other options.

Machines and Subsystems

Following are just some of the machines and subsystems that could be part of a manufacturing line. The common thread with all of these is that they have their own controllers that interface with the overall system.

CNC Machines

Computerized Numerical Control is the automated control of machining tools using a computer designed for that purpose. Multiple tools are often combined into a single cell that operates on a set of sequential instructions. G-Code is the most common CNC programming language; the instructions tell the motor axes where to move, how fast to move, and what path to follow. The most common method is for a lathing or milling operation to remove material until the workpiece is finished. As part of this operation either the workpiece or the tooling may be manipulated, and an automatic "tool changer" may be used to swap out cutting or burnishing tools.

Loading and unloading of parts is often done with an independent material handling system or robot. CNC machines can perform work on various metals or hard plastics, or even wood!

Figure 246 - CNC Machine (Courtesy of LaborMac)

Molding and Forming

Injection molding machines, extruders, and thermoformers are common stand-alone machines that produce plastic and rubber parts with heat. Control is sometimes done with a PLC, or with a dedicated controller. As with CNCs, these are often loaded and unloaded with robots.

Molding and forming processes involve the application of heat and pressure to the material, and molds or dies are used to form the parts. After removing the part from the mold, remaining material called "flash" must be removed. This may be done manually or through a process such as tumbling with an abrasive.

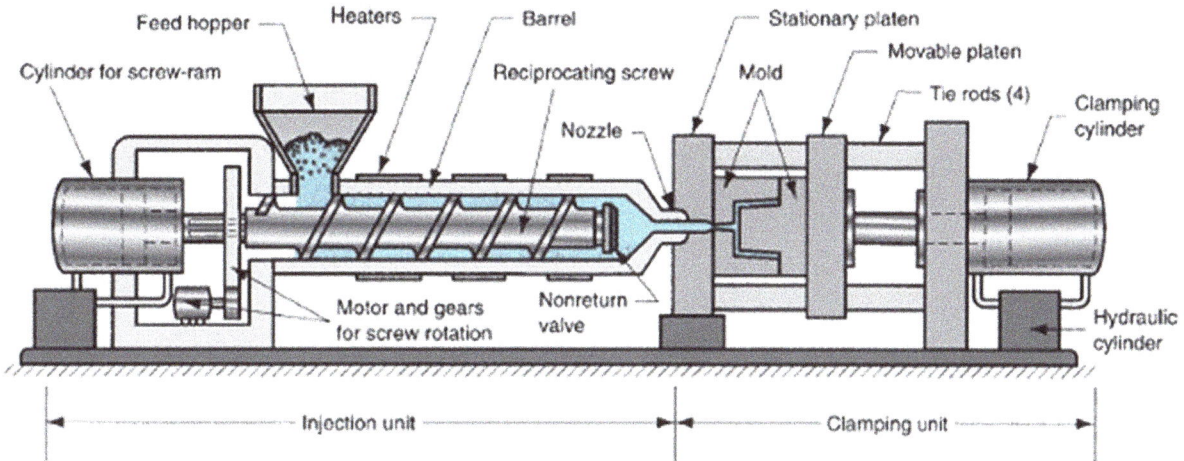

Figure 247 - Injection Molding Process

Presses

Presses are used to form parts at high pressure. Hydraulic presses are very common for large parts, but smaller presses may use servo motors for precise position and torque/pressure control. They may be controlled with a PLC or with a dedicated press controller.

Figure 248 - Hydraulic Press

Additional operations may be performed within the pressing operation such as hydroforming, which injects pressurized fluid into tubes so that it expands to fill the mold. This operation can also be performed with sheet metals using a bladder or in direct contact with the material.

Multiple dies may also be used within the same press, where parts are transferred from one station to another. This operation is known as transfer stamping, or a transfer press.

Notice the hydraulic power unit or "power pack" next to the press in Figure 248. This system is similar to that described in the hydraulic section of this book.

Material Handling

Material Handling systems may include conveyors, pick-and-places, escapements, pushers, and diverters. A PLC or other device is often used separately to control traffic independently of the manufacturing process.

Figure 249 shows some of the different common types of conveyors used in industrial material handling applications.

Belts can be made of plastic, rubber, or metal: rollers may be metal or various plastics also. Rubber cords are also sometimes used to move lighter objects inside of machinery.

Maintenance of conveyor systems includes cleaning, aligning, lubricating, and other motor-associated techniques. There are often steel chains linking motors to rollers or pulleys that need to be adjusted or tensioned.

There is also usually removable guarding over belts and chains. It is very important to replace guarding after maintenance activities.

Figure 249 - Types of Conveyors

Pick and Place mechanisms are used to move individual parts from one location to another. In their simplest form, they consist of a horizontal actuator, a vertical actuator, and some type of gripping or vacuum device to grasp the part. Actuators may be pneumatic, electrical, or a combination of both. Figure 250 shows a simple pneumatic pick and place used to move an object from one place to another.

More complicated servo operated pick and place mechanisms may have 3 or more axes and are actually considered gantry robots. They may use machine vision systems to locate and orient parts and are often used for palletizing.

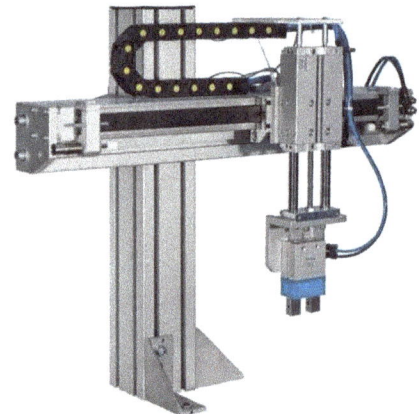

Figure 250 - Pneumatic 2 Axis Pick and Place

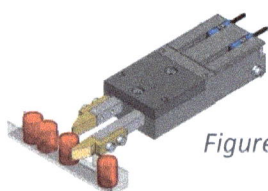

Escapements are used to singulate parts and feed them one at a time. They may be packaged as a single pneumatic device as shown in Figure 251 or take the form of a mechanical assembly of some kind.

Figure 251 - Pneumatic Escapement

Automated Storage and Retrieval Systems (ASRS)

Modern logistics requires computer-based systems to track part locations and automated load handling to store and retrieve items. Databases are used to track the location of items in real time and interface with controllers that move pallets, bins, and packages through the system.

Sometimes abbreviated AS/RS, Automated Storage and Retrieval is used in manufacturing, distribution, libraries, retail, and wholesale. It is part of the larger supply chain industry that includes trucking, railroad, air, and ship transportation of items. Equipment used in these systems includes conveyors and other material handling devices listed previously, but also uses large gantries with lifts, Vertical Lift Modules or **VLM**s, Automated Guided Vehicles or **AGV**s, and horizontal carousels. Robots on horizontal or vertical gantries are also common.

VLMs can be built to a height that matches available overhead space. They are sold in different configurations and sizes depending on the maximum load capacity. A VLM is comprised of trays on the front and back with an inserter/extractor system running down the center. They are a very efficient method of storage since they take up much less space than systems that require human access. Trays are automatically retrieved and delivered to pick windows for picking or filling. When complete the inserter/extractor takes the tray back to its location.

Horizontal carousels are a series of bins with shelves supported by an oval track. Shelves can be adjusted to resize changing inventory, and more inventory is directly accessible to operators than with VLMs. Multiple horizontal carousels are often arranged into workstations called "pods". These are then integrated with batch pick stations where multiple items can be accessed using **pick-to-light** systems.

Large gantries on rails can also be arranged to run on rails or wheels down aisles in warehouses to store and retrieve pallets and larger bins. Overhead cranes are also sometimes used for large items.

Pallets, bins, and packages are often labeled with barcodes or RFID tags. This allows individual items to be transferred without human intervention within the system and ensures that there is a backup to the database/location inventory system. Machine vision with optical character recognition is also commonly used.

Physical guarding and electronic safety systems with floor scanners and light curtains are used to prevent injury to operators. Controllers may be safety PLCs, SCADA-based systems with many separate controllers, or entirely computer-based systems with remote I/O.

Packaging Machinery

Packaging machines are usually stand-alone machines built by specialized OEMs. They use raw materials such as film, cardboard, or various formed products to arrange and wrap products for shipping or consumer use. They may use rolls of film, unfolded cardboard boxes, or stacks of Styrofoam or plastic cases.

There are many different standard types of machines used within such a system, and they often interface directly with the material handling controls.

Following are some common types of machines used in packaging:

Stand-alone Machines:	Subsystems:	
Sleeve Wrapper	Tape Machine	Barcode Readers
Case Packer	Adhesive Applicator	RFID Readers
Palletizer	Strapping/Banding machine	
Bagger	Labeler/Printer/Applicator	
Shrink Wrapper	Sealer	
Stretch Wrapper	Filler	
	Weigher/Scale	

Other types of devices used in the packaging industry include accumulating, collating and unscrambling machines and devices, part feeders such as bowl and step feeders and others.

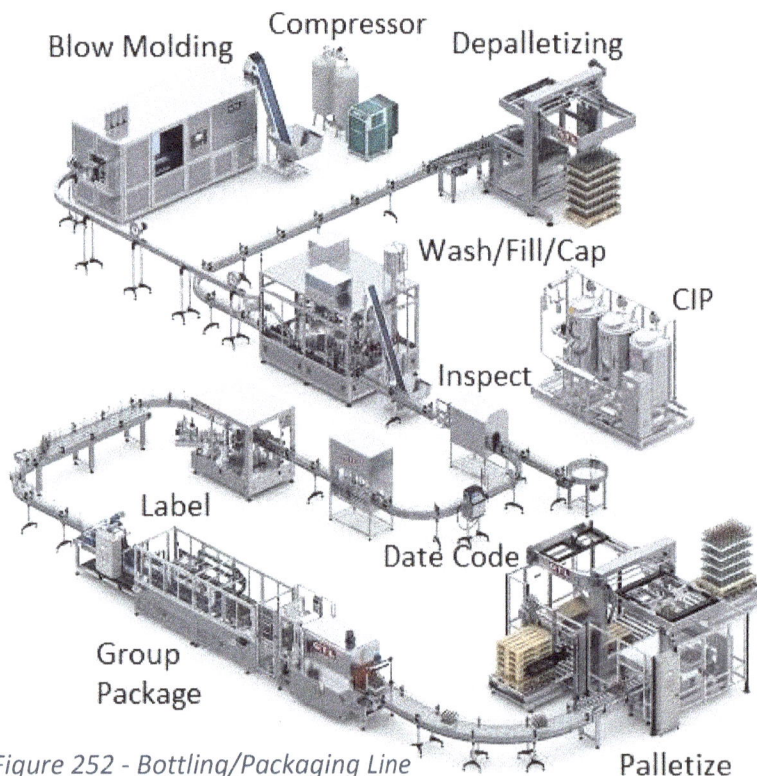

Figure 252 - Bottling/Packaging Line (Courtesy of GTL Packaging)

Figure 252 shows a bottling and filling line. The beverage industry is a major user of packaging equipment. The blow molding machine produces plastic bottles, while the depalletizer can optionally introduce glass bottles. After filling, bottles are inspected, labeled, grouped into cases, and then palletized for shipment. **CIP** is an abbreviation for "**Clean in Place**", which circulates cleaning fluids to sanitize the equipment internally. An additional packaging machine not shown here is a stretch wrapper to wrap the finished pallets.

Converting

Converting equipment involves taking raw material and slitting, forming, or rolling it into modified forms. Materials such as paper, plastic film, metal, foil, and cloth are often produced in long continuous sheets that are rolled up for convenient handling and transportation. These continuous sheets are known as **webs**, which are generally dispensed off an **unwinder** through various processing stations. An example of converting might be to take a web of plastic film, fuse the edges, and cut it into lengths, converting it into plastic bags.

Other converting processes include **slitting** (cutting a roll into smaller widths), **sheeting** (separating a roll into separate sheets), **coating** (applying a substance to the web surface), **laminating** (fusing multiple webs together) and **printing**. Additionally, materials are often die-cut out of multi-layered webs, ultrasonically welded or bar-sealed.

Web alignment is an important part of a converting operation. Moving webs have a tendency to track off center and wander out of alignment. Non-contact sensors are used to detect the edges of the web, and actuators are used to shift the web back to its desired path. The actuators might be pneumatic or hydraulic cylinders, or servomechanisms.

Tensioning of webs is also critical for many operations. This is a closed-loop operation also, using strain gauges to detect how tightly the web is stretched, and varying the speed or position of nips and rollers.

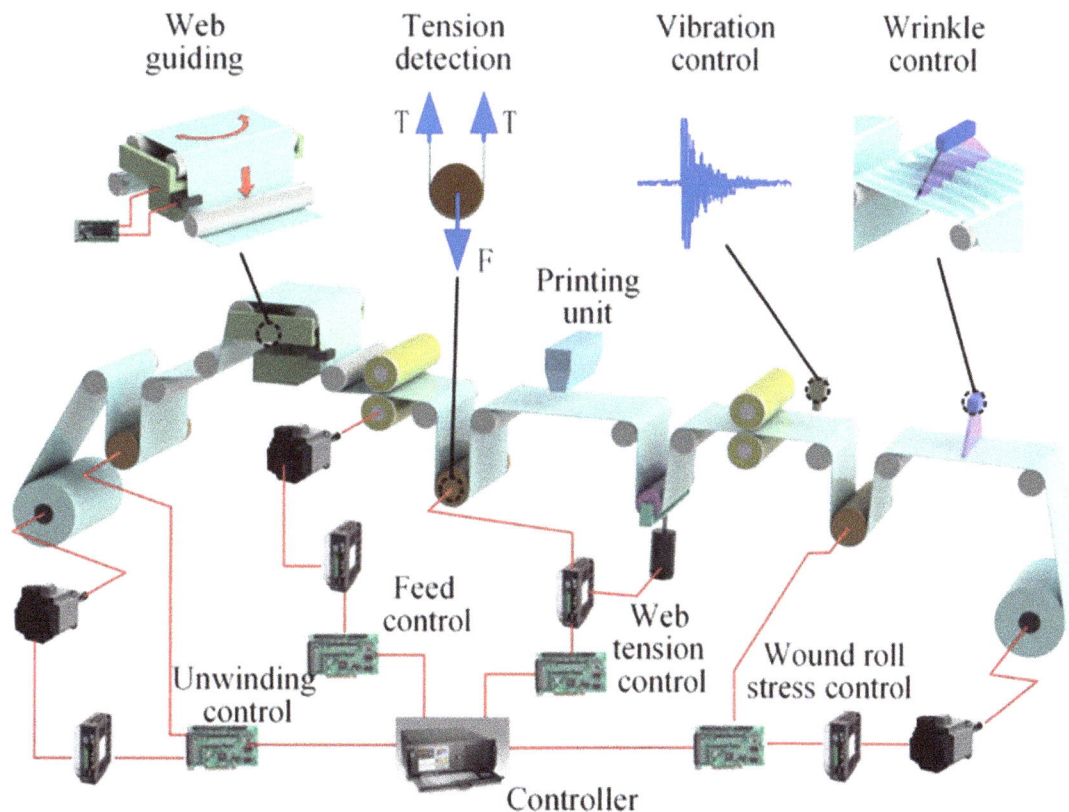

Figure 253 - Web Converting Line

Process Control

There are many different types of continuous processes used in industry. Web converting is a continuous process, as are many utilities such as electrical production and wastewater processing.

The beverage industry uses continuous and batch processing to produce the juices, sodas, and bottled water mentioned in the packaging of Figure 252. Batching is the process of making a tank or hopper of mixed materials then processing it into smaller packages or containers.

Figure 254 - Batching Tanks (American Beverage Depot)

The picture in Figure 254 is from American Beverage Depot in Miami, Florida, a plant where the author has done extensive work. These tanks are used to create batches of syrup used for boxed or bottled juices and sodas.

The processes in making these batches involve measuring tank levels, weighing solids such as powders, measuring temperatures in the pasteurizers, and measuring flow rates from one tank or process to another. There many different types of sensors used in these processes, such as pressure and flow transmitters, thermocouples and RTDs, pH sensors, and more.

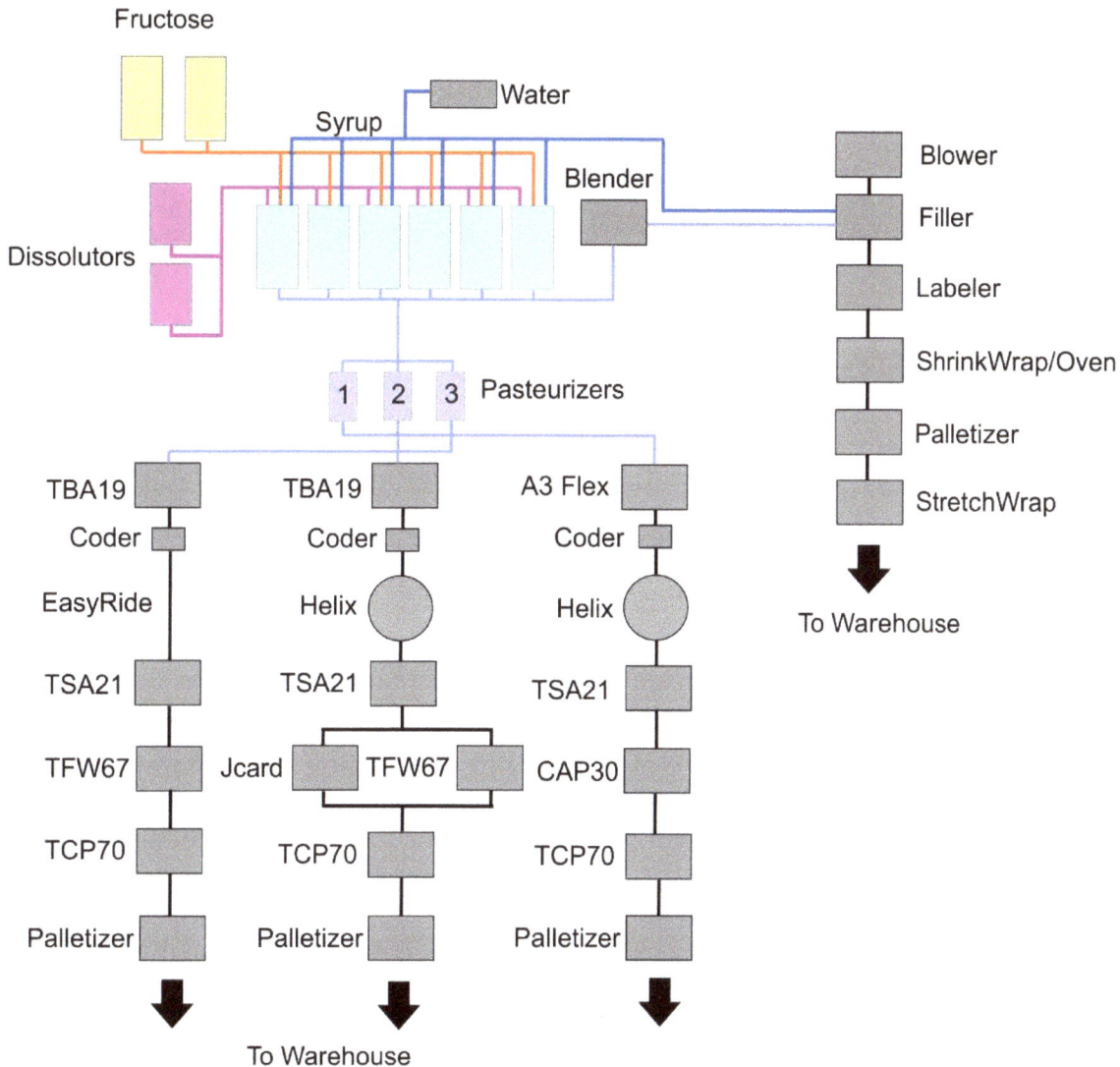

Figure 255 - ABD Process Flow

Figure 255 shows a simplified diagram of process flow through the plant. The gray objects represent discrete packaging equipment. One of the important concepts to understand here is that there are hundreds of individual subsystems that are connected by an overall SCADA system.

Not shown in this diagram is the CIP (Clean-In-Place) system that circulates various cleaning solutions through the processing equipment. This system interfaces with the other systems through a network of piping with sensors to ensure proper connections.

The batching system and tanks are controlled by a large Allen-Bradley PLC, each of the three fillers (TBA19s and A3Flex) are controlled by GE PLCs, and the other filler, blender and CIP are controlled by Siemens PLCs. The bottle blower is also controlled by a Siemens 400 with its own computer interface. There is a master SCADA system connected to all of the PLCs that runs Inductive Automation's Ignition on a computer server.

Figure 256 - Piping Connections

The picture above shows part of a skid with piping connections to route product or cleaning solutions to different equipment. Notice the posts connected to the pipes that extend to sensors mounted in the plates; all connections must be verified before starting a batch. There are hundreds of these sensors in different areas of the plant. Tanks can all be routed through the different pasteurizers to different filling lines; these pipe connections are connected by operators and verified in the software.

These connections and the pipe routing are all documented in Piping and Instrumentation Diagrams (P&IDs), as described earlier in this book. They are also shown on SCADA screens as shown in Figure 257.

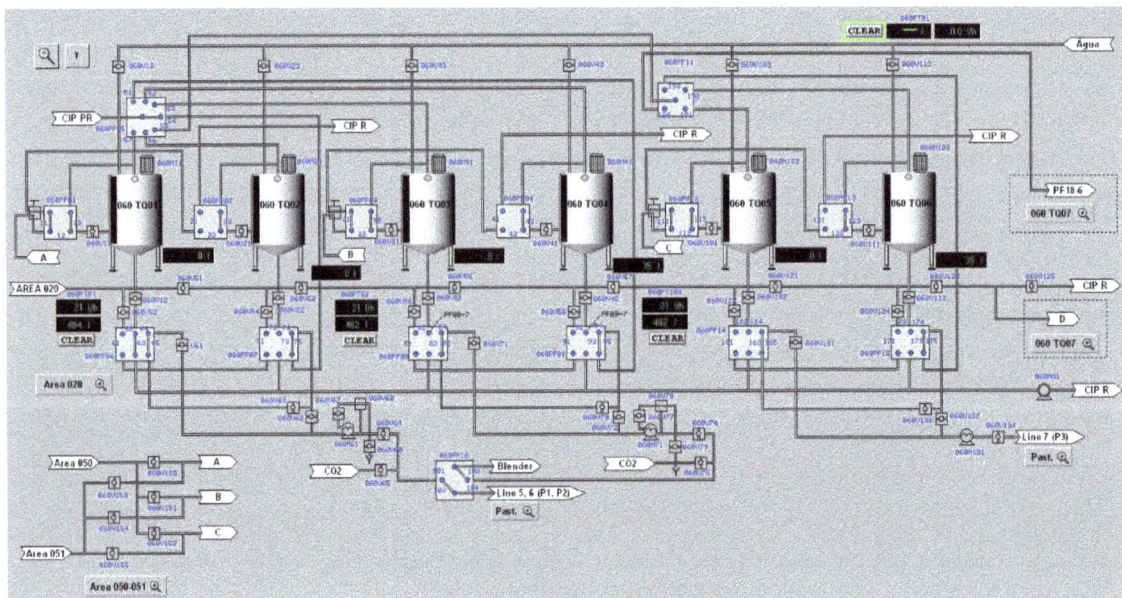

Figure 257 - Syrup Tank Piping

Visualization and control of processes is often done from workstations like the one shown in Figure 258, or from larger control rooms like the one shown later in the section on SCADA.

Recipe selection and batch setup is also done from the SCADA system, as well as data collection from all of the other machinery shown in Figure 255 previously.

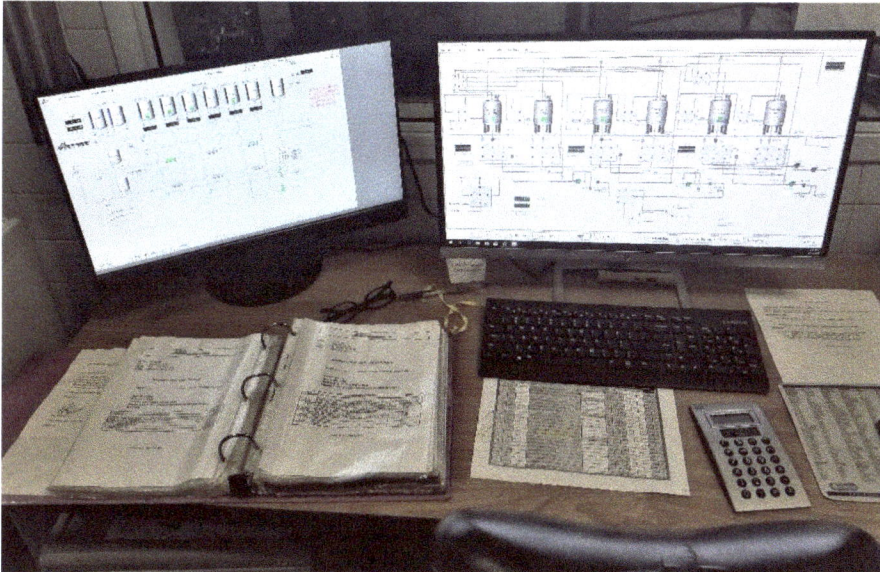

Figure 258 – Small SCADA Control Workstation

Maintenance of interconnected systems like this require extensive documentation including electrical schematics, P&IDs, and alarm message and diagnostics from the control equipment. There are also HMIs in many of the machine areas that display alarms and diagnostics.

In the United States, food and beverage processing falls under regulations administered by the Food and Drug Administration, or FDA. Much of the machinery is made of stainless steel and various polymers that are easy to clean and can withstand caustic chemicals. Personnel in areas where food and beverages are prepared must often wear hair and beard nets, gloves, and lab coats without buttons or pockets while working.

Food processing has similar aspects to the beverage industry in that products are often produced in batches. Handling of food products varies quite a bit, since products may take many forms from liquids to individual objects such as loaves of bread. Product is often frozen for storage and shipment in the food processing industry also.

Life Sciences, Medical Devices and Pharmaceutical

Also falling under FDA regulations are the manufacturing of drugs and medical devices. Many of these products are manufactured in **clean rooms**, which are classified according to ISO 14644-1 standards that specify a maximum number of particles above certain sizes that can be present in the manufacturing environment. Regulations for these rooms also specify what kinds of ceilings, walls, and flooring materials can be used, how many air changes must occur per hour or minute, and the types of HEPA filters that are used. Additionally, in the United States a federal standard US FED STD 209E applies.

Figure 259 - Clean Room Manufacturing

Products manufactured in these industries may be in liquid or capsule form in the case of pharmaceuticals or be complex electronic devices. Since the product directly affects the health or life of people, there are additional regulations for sensors and devices, lubricants and materials that can be used in machinery.

Software used in manufacturing processes must also be **validated**. FDA software validation is when an FDA-regulated company demonstrates and documents that their software can accurately and consistently produce results that meet predetermined guidelines for compliance and quality management. The FDA does not tell companies specifically how to validate their software but requires companies to document how they intend to do it and provide evidence for having done so.

One of the best practices that applies to FDA compliant software validation is to tie it to **change management**. This means validation is triggered every time there is a change in the control software or in a previously validated component. Because of this, control components need to use the exact same product and firmware revision, and any change in programming will require re-testing of the machinery to ensure that the quality results of the final product haven't changed. In typical manufacturing environments such as automotive or consumer products, technicians make changes to programs often, adding code or changing parameters. However, in medical device and pharmaceutical manufacturing controllers are usually locked down so that changes can't be made without re-validation.

Utilities

Electricity production and distribution, water and wastewater handling are process control type industries that don't involve packaging or delivery of discrete products.

Electrical power generation involves creation of electricity using hydroelectric, fossil fuels, natural gas, or nuclear power sources. In the United States, power generation is regulated by the U.S Department of Energy. Power plants contain one or more **generators** which rotate magnets around copper coils converting mechanical force to electricity. Like large electrical motors, electricity is created in three phases each 120 degrees out of phase. Most power plants use thermal energy (heat) to drive turbines that rotate the generator. Heat creates steam which turns the turbine and is condensed back into water.

Process control in a power generation plant involves monitoring of large rotating equipment, temperature regulation and of course monitoring of frequency, voltage, and amperage on a large scale. Much of this is done from a control room, again using SCADA software.

Power demand fluctuates throughout the day and across regions. To keep the power grid balanced at all times, generation operators must dispatch enough power to supply the demand. Base-loading power generation plants operate around the clock to satisfy the basic demand, while peak-loading plants gradually come online as demand peaks. Coal and nuclear plants, because they take up to 12 or more hours to start up, are often used for base power, while natural gas plants, which start faster but have higher fuel costs, are often used for peaking demands.

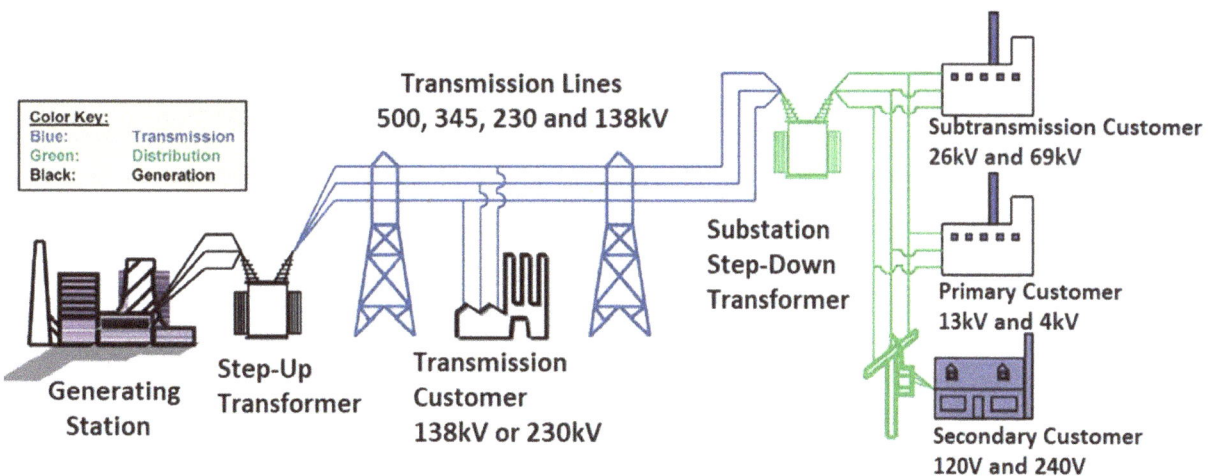

Figure 260 - Electricity Generation and Transmission Supply Chain

Power plants use "step-up" transformers to drastically increase voltage to the transmission system level. Facilities that house the equipment and conversion infrastructure are called substations.

Transmission lines interconnect to each other across different regions and are administered by Regional Transmission Organizations (RTOs) or Independent System Operators (ISOs).

There are various types of substations in the bulk power system, containing transformers, protective equipment such as relays and circuit breakers, switches for controlling high-voltage connections, and electronic instrumentation to monitor system performance and record data. Following are different types of electrical substation:

1. **Step-Up Substation:** Links a generation plant to the transmission system.
2. **High Voltage Substation:** Connects high voltage transmission systems to each other.
3. **Step-Down Substation:** Connects a high voltage transmission system to a sub-transmission system.
4. **Distribution Substation:** Connects transmission or subtransmission networks to medium voltage distribution networks.
5. **Converter Substation:** Connects non-synchronous AC transmission networks to systems with a different frequency.
6. **Switching Substation:** Acts as a circuit breaker in transmission and distribution networks for disconnecting parts of the network for maintenance.

Electrical power control centers are often located at substations but can be located anywhere with computer access to the equipment being controlled and monitored. As mentioned previously, these are SCADA systems connected to RTUs, DCSs, PLCs or other smaller controllers.

Water Treatment involves chemicals, filtration, heat, or biological treatments to ensure that water is suitable for human consumption and use. Water treatment in municipal facilities is usually combined with **Wastewater Treatment**, which is a process used to remove contaminants from wastewater and return it to the water cycle.

Domestic wastewater or sewage is generally treated in municipal or regional facilities. Additional industrial or agricultural treatment plants may be used to treat water from specific processes or activities.

Networks of pipes and pump stations are used to transport wastewater, which may contain both domestic sewage and urban runoff (stormwater) to the treatment plant. Treatment involves several stages of physical and biological processes.

Before passing sewage to the primary treatment, it is passed through a bar screen to remove large objects like cans, rags, sticks, and plastic packets or bags. A mechanical rake is used to collect the solids which are disposed of in landfills.

Sedimentation involves settlement or flotation of solids by gravity. Stones, grit, and sand are collected in a grit channel while less-dense solids are carried forward to the next stage. Gravity separation using an optimum flow rate allowing dense solids to settle in "primary settling" or "primary sedimentation" tanks. These are sometimes combined with skimmers called **clarifiers** that remove floating grease or soap scum and solids like feathers, wood chips, or condoms.

Secondary treatment is the removal of biodegradable organic matter. One method uses microorganisms that consume dissolved or suspended materials such as sugars, fats, molecules from human waste and food waste, soap, and detergent. The organisms reproduce to form cells of biological solids. Another method is to use **flocculation**, a chemical process where a clarifying agent is added to promote clumping of material into larger "flocs" or flakes for easier separation. A by-product of this process is sludge, which is often dewatered and dried to reduce disposal cost.

Additional treatments use chemicals to remove organic pollutants or kill bacteria or microbial pathogens by adding ozone, chlorine, or hypochlorite. This breaks down complex compounds to simpler compounds like water, carbon dioxide, and salts.

Polishing, also called "fourth stage" treatment, minimizes chemical reactivity and adjusts the pH of the water. Carbon filtering, filtration through sand or fabric filters are also used in this stage.

Figure 261 - Wastewater Treatment Plant

Additional reverse osmosis and de-ionization (RO/DI) treatments are often performed on both wastewater and water before either returning it to the water system or delivering water to consumers.

Pumping stations with instrumentation and backup pumps are often located at sites within the piping system. These are sometimes called "**lift stations**" on the wastewater side.

After collecting water from reservoirs, wells or other sources, water is treated at water purification plants. Some plants pre-treat incoming water with chlorine to minimize fouling organisms on pipes and tanks, but this has largely been discontinued.

Flocculation is used on incoming water with coagulants like aluminum sulfate to remove particles. This is done with a rapid mixing process while adding either metal salts or organic polymers. Sedimentation is then used again to settle floc to the bottom of sedimentation basins. Sludge is removed from the basin mechanically or done periodically in a manual cleaning process. Water is then filtered and disinfected with various chemicals including chlorine and ozone. Ultraviolet light is also used in some disinfecting processes.

Fluoride is also sometimes added with the goal of preventing tooth decay, and water may be conditioned to reduce the effects of hard water. This is done by adding chemicals or soda ash to precipitate excess salts.

After purification, water is stored in tanks or towers. Water pressure is usually required to be 45-80 psi at the point of use, this is usually done using pressure tanks located at high points throughout the distribution area. Pressurized water moves from the tanks to the water mains, where pumps or pressure reducing stations are used to maintain the proper pressure in the system.

As with electrical utilities, SCADA systems are used to monitor and control all of the different controllers and subsystems within a water utility. There can be thousands of devices within these networks, and alarms and historical data can be used to help diagnose problems and maintain the system properly.

Maintenance activities in water systems involve servicing of pumps, which in turn use motors, seals, and pressure regulation. Water systems are spread over large geographical areas, so the power feeds to stations and devices are all independent. Communications to devices in the control systems also often require intermediate networks such as wireless or cellular networks or wired commercial connections. Troubleshooting of these utility systems requires many of the techniques and concepts described in this book.

Manufacturing facilities also often have smaller self-contained water treatment units, especially in the beverage industry or pharmaceutical manufacturing. These usually use dedicated controllers for monitoring and control. Units like these usually have good documentation for operation and servicing.

Other utilities not described here include Natural Gas and communications services for phone, cable, TV, or internet.

Exercise 29

1. At what level of control is SCADA located? _____

2. List a few of the types of conveyors used in industry

   _____

   _____

   _____

   _____

3. What are some of the operations performed in the converting process?

   _____

   _____

   _____

   _____

4. List some of the types of machinery used in a packaging operation:

   _____

   _____

   _____

   _____

5. What does the abbreviation CIP represent in the process or beverage industry?

   _____

6. What standards are used to classify clean rooms? _____

   _____

7. What is software validation? _____

   _____

8. What do large utilities have in common for control systems?

   _____

Visualization and Notification

Visualization of machine processes and systems uses operator interfaces of various types, from touchscreens to computers. There are also other methods of alerting operators or other interested parties of abnormal machine conditions or production information. Audible alarms such as horn or buzzers, remote alerts on phones by text or e-mail, and reports that are generated periodically are all part of the bigger picture of the manufacturing system.

HMIs (Human-Machine Interface)

An HMI is a dedicated device that accesses control software to either perform functions or display information. They often communicate to a specific controller such as a PLC using ethernet, serial communications, or proprietary protocols as described previously in the communications part of this book.

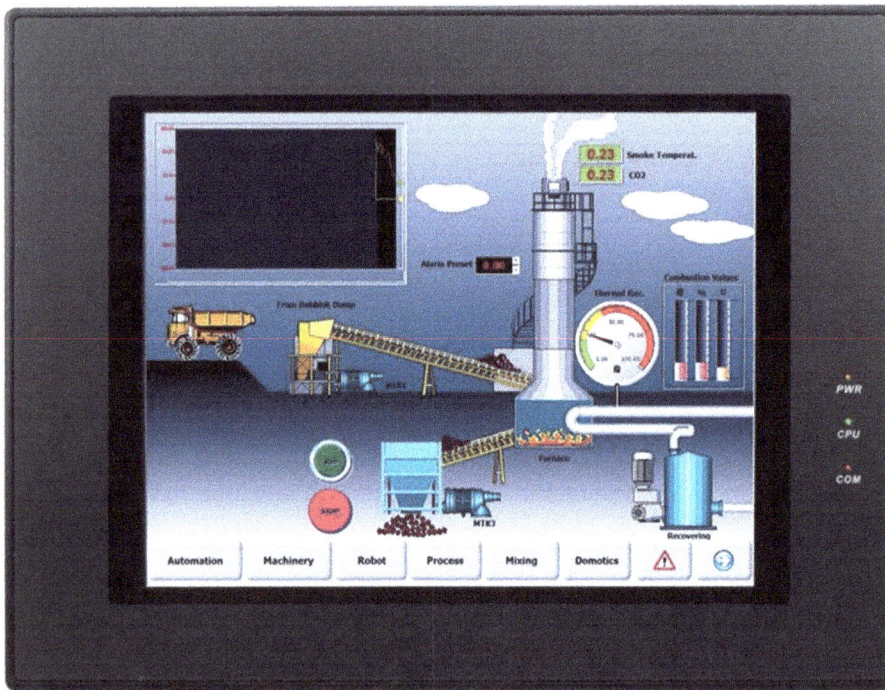

Figure 262 - Touchscreen HMI

There are simple HMIs that display only text, but as the cost of hardware has decreased and technology has improved, most are now colored touchscreens with a lot of functionality.

Figure 262 shows a small touchscreen HMI with visualization of a process and a couple of control devices. In addition, alarms may be displayed and there are navigation buttons at the bottom to change screens. Objects on the screen are linked to addresses or tags in a PLC to interface with the system.

Not all HMIs are touchscreens. Some HMIs use function keys with membrane keypads to select objects on the screen or enter numerical data. Touchscreens can be susceptible to environmental conditions in plants that make it advantageous to not touch the display surface.

Unlike a computer, HMIs do not usually have much memory and can't run other computer applications. Outside of an alarm history and some recipe storage, HMIs are generally simple interface devices.

The objects shown on the screen are generally selectable from a library within the software. Objects can be resized and vary from simple pushbuttons and pilot light indicators to numerical displays and text messages. Bitmaps or pictures can also be imported and added to the libraries.

Alarms are special events that allow messages to pop up in a window or display in a banner. They also can be archived in a history list with time and date information as mentioned previously. Other information such as when the alarm was acknowledged or cleared can also be shown.

Animation is often used to change an object's visibility, position, color or size on a screen based on an additional tag in the PLC.

HMI capabilities vary widely based on the manufacturer. More on development software for HMIs and SCADA is discussed later in this book.

SCADA (Supervisory Control and Data Acquisition)

SCADA is software that is deployed on a computer. This may be a laptop, desktop or industrial device and often involves multiple displays and processors.

SCADA is more of an architecture than it is a specific piece of software. It is differentiated from simple HMIs in that it involves computers, networked data communications, and graphical displays. Rather than interfacing with a specific machine or line, a SCADA system may interface with an entire plant.

Figure 245 at the beginning of this section showed a SCADA system at the Supervisory and Production level of the plant communication hierarchy and an HMI at the Control level, but this is not necessarily always true. Components of a SCADA system may be located at various levels.

Figure 263 - Control Room with SCADA

SCADA is often used at utility plants for electrical and wastewater, as well as in the petroleum and chemical industries. Figure 263 at left shows part of a control room with a variety of displays for different processes.

In addition to interfacing directly with equipment, SCADA computers often handle scheduling, historical data, alarms, and generation of reports. **DCS** (Distributed Control Systems), **RTUs** (Remote Terminal Units) and PLCs are commonly networked together with the SCADA system rather than the computers directly accessing sensors and device level components, but SCADA systems are very flexible and are deployed in different ways.

RTUs are used to connect field devices to the SCADA architecture. They don't generally perform logic functions themselves but can sometimes perform simple autonomous functions. They respond to instructions from SCADA and can save values for retrieval if disconnected. They often have backup batteries or power supplies and can implement energy saving measures like switching off I/O modules when not in use.

Figure 264 - Remote Terminal or Telemetry Unit (RTU)

DCS controllers are used in large facilities with a lot of analog devices, typically to control closed loops. When a large number of continuous control loops are present, an advantage of a DCS is that the plant can continue operation if any part of the DCS fails.

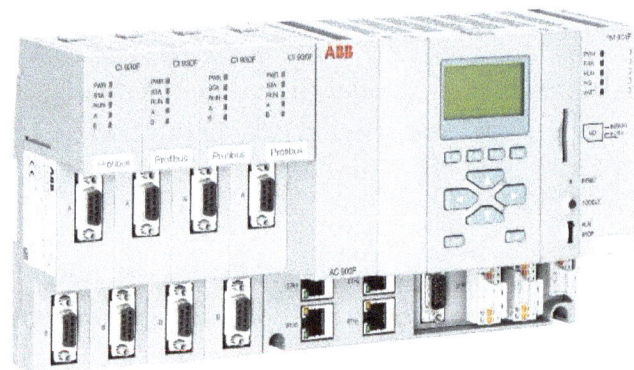

Figure 265 - Distributed Control System (DCS) Controller, ABB

Major PLC and DCS manufacturers usually also have software for SCADA, but there are also major independent software companies that produce it. Software licenses may be based on the number of tags used, or by the number of servers or devices.

Following is a list of some of the major SCADA software companies used in industrial plants:

FactoryTalk View SE (Allen-Bradley/Rockwell)   **WinCC** (Siemens)
Cimplicity, Proficy (GE Digital)   **WonderWare** (Aveva)
Ignition (Inductive Automation)   **Factory Studio** (Tatsoft)
Genesis64 (Iconics)   **WebStudio** (Indusoft)
Edge (Litmus)

Other controls companies that produce SCADA software include Schneider, ABB, Emerson, Mitsubishi, Honeywell, Yokogawa, and National Instruments.

Exercise 30

1. What are some of the differences between HMIs and SCADA?

   _____
   _____
   _____
   _____

2. What is "animation"? _____

   _____

3. What do the acronyms "RTU" and DCS" represent?

   _____

4. What is the basis for pricing SCADA software?

   _____

Operational Equipment Effectiveness (OEE)

In order to properly maintain equipment, it is necessary to understand its current condition and identify possible areas of improvement. An important tool for this is **OEE**.

OEE is a measurement of how well a machine or system is operating is used compared to its full potential. An OEE of 100% means that only good parts are produced (quality), at the design speed (performance), and for the full time with no interruptions (availability).

Measuring OEE is a standard procedure for most lines and machines and is part of the **TPM** (Total Productive Maintenance) concept developed in Japan by Seiichi Nakajima. This in turn is a part of **Lean/Six Sigma** principles used in manufacturing. By measuring OEE, manufacturing processes can be improved, and waste reduced or eliminated.

While OEE is measured against scheduled operations, a closely related measurement called Total Effective Equipment Performance (**TEEP**) is calculated against calendar hours, 24 hours a day for every day of the year. This term for calendar hours is called 100% loading.

Calculations for OEE and TEEP

The OEE of a manufacturing unit are calculated as the product of three separate factors:

- **Availability:** percentage of scheduled time that the operation is available to operate. Often referred to as uptime.

  The formula for availability is operating time/scheduled time.

  Example: A line is scheduled to run for an eight-hour shift, minus a 30-minute break. The scheduled time is therefore 480 minutes -30 = 450 minutes. The line has an unscheduled breakdown that takes an hour to fix. The actual operating time is then 390 minutes. 390 minutes/450 minutes = 0.866, or **86.7%**

- **Performance:** speed at which the Work Center runs as a percentage of its designed speed. Designed speed is considered the speed that a machine was built for and hopefully run at during initial Site Acceptance Testing (SAT).

  The formula for performance is (parts produced * designed time per part)/actual operating time.

  Example: The line is designed to produce 40 parts per hour, or one every 90 seconds. 242 parts are produced during the 390 minutes that the line ran. 242 parts * 90 seconds = 21780 seconds or 363 minutes, 363 minutes/390 minutes = 0.931, or **93.1%.**

  Another way to look at this: the designed production rate is 0.666 parts/minute, 390 minutes were available to produce parts, so the parts produced should have been 390*0.66=260. 242 parts produced/260 potential parts = **93.1%.**

Note that this calculation does not consider rejected or bad parts.

- **Quality:** Good Units produced as a percentage of the Total Units Started. It is commonly referred to as the first pass yield (FPY).

  The formula for quality is good parts/total parts produced.

  Example: Out of the 242 parts produced during the shift, 21 are rejects. 242 total-21 rejects =221 good parts. 221 good parts /242 total parts = **91.3%**

- The final formula for OEE is therefore (Availability)*(Performance)*(Quality).

  0.867*0.931*0.913 =0.737 or **73.7%**

To calculate the TEEP, the OEE is multiplied by a fourth component:

- **Loading:** percentage of total calendar time that is actually scheduled for operation.

  The formula for loading is scheduled time/calendar time.

  Example: The factory operates 5 days a week, 24 hours a day for 3 shifts. The loading factor is then 5 days/7 days = 0.714 = **71.4%**

  The TEEP is the bottom-line calculation for how effective the scheduling is, taking into account OEE. TEEP for the line is then 0.737*0.714 =0.526 or **52.6%**

These are not difficult calculations, and the data is easily obtained from the machine programming. It becomes more complex when evaluating an entire production line.

Each of these three components, Availability, Performance and Quality, can be individually targeted for improvement. Availability is improved by reducing downtime, i.e., proper maintenance. Performance is improved by ensuring that the maximum optimum speeds are attained for each assembly or component. Quality is optimized by ensuring the manufacturing process itself is repeatable and built properly.

It is highly unlikely that the overall OEE can approach 100%; an OEE of 85% is considered very good.

Six Big Losses

To be able to better determine the sources of the greatest loss and to target the areas that should be improved to increase performances, these categories (Availability, Performance and Quality) have been subdivided further into what is known as the 'Six Big Losses' to OEE. They originate from Total Productive Maintenance (TPM), a maintenance philosophy designed to integrate equipment maintenance into the manufacturing process.

The six big losses are categorized as follows:

Overall Equipment Effectiveness	Recommended Six Big Losses	Traditional Six Big Losses
Availability Loss	Unplanned Stops	Equipment Failure
	Planned Stops	Setup and Adjustments
Performance Loss	Small Stops	Idling and Minor Stops
	Slow Cycles	Reduced Speed
Quality Loss	Production Rejects	Process Defects
	Startup Rejects	Reduced Yield

Figure 266 - The Six Big Losses

The reason for identifying the losses in these categories is so that specific countermeasures can be applied to reduce the loss and improve the overall OEE.

Equipment Failure includes the breakdown of machinery due to tooling or equipment failure, or any unplanned maintenance. Unplanned stops could also be due to a lack of operators or materials.

Setup and Adjustments account for required changeover operations, cleaning, and planned maintenance. Required quality inspections also fall into this category.

Idling and Minor Stops can be due to misfeeds, material jams and periodic quick cleaning. These stops are usually a matter of seconds or minutes in length and may be due to upstream or downstream equipment speed issues. Other terms used in industry that are related are **Starved**, where a machine is waiting for product from upstream, or **Blocked**, where the machine can't release product because the downstream machine is not ready.

Reduced Speed can be due to manual reduction because machinery tends to break down at higher speeds, or because operators can't keep up with the production rate. Machinery may also adjust its own speed based on the sensed environment around it. This can be due to starved and blocked conditions as described previously.

Process Defects account for defective parts produced during steady-state operations. Common reasons include incorrect equipment settings, operator errors, bad materials, and lot expiration in pharmaceutical plants.

Reduced Yield can occur after equipment startups because of faulty setup or changeover, incorrect settings, or equipment that inherently creates waste as it begins operation, such as webs and packaging materials.

These quality losses include product sent back for rework.

Exercise 31

1. What are the three components required to calculate Operational Equipment Effectiveness (OEE)?

    _____

    _____

    _____

2. A machine is scheduled to run for 2 eight-hour shifts during weekdays. On Tuesday it runs for 850 minutes. What is its Availability? _____

    The design speed for the machine is one part/minute. 742 parts are produced that day. What is the machine's Performance? _____

    21 parts are rejected from the line and sent to rework. What is the Quality? _____

    What is the OEE for the machine? _____

    What is the calculated TEEP for the machine? _____

3. What category does a lack of operators to run equipment fall under?

    _____

4. If a line is stopped periodically so that product quality checks can be made, how should the lost time be categorized?

    _____

5. List some of the causes for products to be rejected:

    _____

    _____

    _____

    _____

    _____

The Industrial Internet of Things (IIoT) and Industry 4.0

The Internet of Things, abbreviated IOT, is the network of physical objects—a.k.a. "things"—that are embedded with sensors, software, and other technologies for the purpose of connecting and exchanging data with other devices and systems over the Internet.

The industrial version of this is the use of interconnected sensors, instruments, and other devices networked together with computers' industrial applications, including manufacturing and energy management. This connectivity allows for data collection, exchange, and analysis, potentially facilitating improvements in productivity and efficiency as well as other economic benefits. The IIoT is an evolution of a distributed control system (DCS) that allows for a higher degree of automation by using cloud computing to refine and optimize the process controls.

A natural extension of this is the use of microprocessor-based data acquisition devices to collect data and transmit it locally to a data collector, which may be a computer or PLC. New technology is constantly evolving that makes these devices less expensive and with more capabilities. Examples of these include **SoC (System on a Chip)** units such as Arduino and **SBC (Single-Board Computers)** like Raspberry Pi.

A change that has occurred with the advent of new devices interfacing with computers on the plant floor is the intersection of **Information Technology (IT)** and **Operational Technology (OT).** IT is an abbreviation long used to describe the field of computers and software used in the business world, but OT refers to the networking of operational processes and industrial control systems, including PLCs, HMIs, and SCADA. This is sometimes referred to as **IT/OT Convergence**, where technicians, maintenance and support staff need to learn cross-discipline technologies.

Industry 4.0 is a term that describes the ongoing automation of traditional manufacturing and industrial practices, using modern smart technology. Large-scale machine-to-machine communication (M2M) and the internet of things (IoT) are integrated for increased automation, improved communication and self-monitoring, and production of smart machines that can analyze and diagnose issues without the need for human intervention. Often described as the Fourth Industrial Revolution, the focus being the complete removal of paper documents in the manufacturing environment.

Along with these evolving technologies and capabilities comes a lot of new terminology. Fortunately, along with the improvements in technology comes the ability to research terms on the internet.

Search engines are your friend.

Tactics

In the Machine/System Theory part of this book, it was mentioned that observing a machine in operation or reading its documentation is a good way to understand what is supposed to do. In much of the following material the different types of machinery and components were discussed, along with other "Things You Need to Know".

Using your powers of observation, including senses, was also discussed. Many of the aspects of machinery that could be sensed by humans are also detectable by sensors, and many of these were listed and described.

This section discusses more on the tactical elements that are needed to solve problems with machinery. These include troubleshooting and machine improvement.

Troubleshooting

What is meant by troubleshooting? Troubleshooting is a systematic approach to problem solving that is used to find and correct issues with complex machines, electronics, and software. It is often applied to machinery that has stopped working or is not operating as expected.

Determining how a machine is supposed to operate involves the observation and documentation-reading skills mentioned earlier but may also require talking to the person who spends the most time with the machine, the operator.

There are some important concepts to understand when trying to determine the cause of a problem. One important relationship is between **correlation** and **causation**, just because one event occurred after another event does not mean that the first event caused the second. It could be a coincidence, or some other unseen event could have caused both. This is often stated as "correlation does not imply causation".

Ultimately there are three elements to solving problems.

1. Identify and document what the problem is.

2. Determine why it happened.

3. Determine a solution to prevent it from happening again.

Following are some techniques that can be used in troubleshooting systems and equipment:

Start with the Simple

It is often useful to examine the simplest or most obvious explanations first. People often complain when technical support people or manuals first ask the user if the receptacle has power or if the device is plugged in. It can save a lot of time to check for simple explanations like power, circuit breakers or fuses before moving on to more difficult or less obvious causes.

Look the machinery over and see if there are lights showing or obvious jammed parts. Keep in mind that the jam could have been caused by something else, and that there may be other sources of power that power the devices. Ask yourself: What has changed since the last time the machine worked properly?

Begin from a Known Good State

Starting a machine from a known good state, such as a home position with no parts, can help identify problems. A well-known example of this is rebooting a computer; this clears the current RAM memory of the processor and re-starts the operating system.

This is especially important with assembly machinery. Ensuring that a good part runs all the way through all of the steps of the process makes sure that the tooling, sensors and actuators are all positioned correctly and their speeds are correct.

Substitute Components

Sometimes called "shotgunning", troubleshooters could check each component one by one, substituting known good components for each suspected bad part. This is certainly not the most efficient way of solving the problem, and there is a risk that the thing that originally caused the failure can also make the new good component fail. Substitution can be effective if used in a well thought out way, but this should be a last resort. You could end up destroying multiple expensive components!

Use Checklists and Flowcharts

Creating a checklist, flowchart, or table of procedures in advance can help. This creates an organized sequence that technicians or operators can follow when troubleshooting equipment. These lists can be kept on a computer or even on the machine HMI so that they can be easily accessed.

Keeping records of previous problems and their solutions can help in creating these lists. The lists should be updated as new events occur.

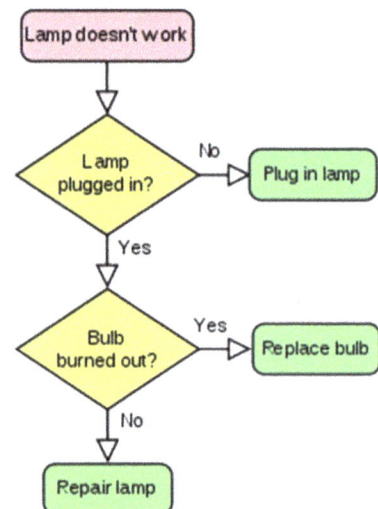

Figure 267 - Troubleshooting

Dependencies can be used to create these lists or charts also, often a component depends on another system in order to operate correctly. Identifying these can help when creating flowcharts.

Reproduce Symptoms

If you can re-create an error it becomes easier to isolate and resolve its cause. If the problem can be reproduced consistently, this works well. Many problems can be intermittent in nature though, making them difficult to reproduce. Identifying environmental causes such as heat or humidity at the time of the occurrence can be useful here.

Split the System

A technique called "half-splitting" can be helpful in limiting the choices for bad components. Dividing a series of connections or nodes in half can identify where a voltage or communication signal is lost. It also works well on systems that are made of a series of sequential functions.

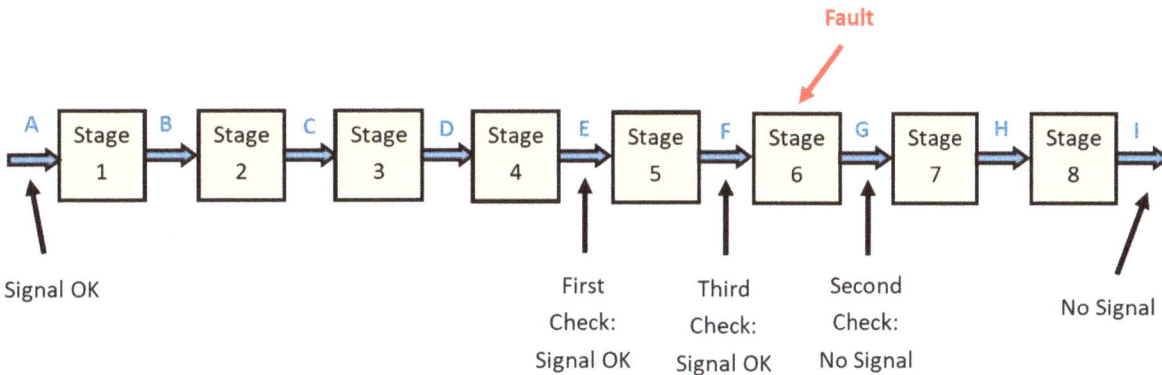

Figure 268 - Half-Split Methodology

Root Cause Analysis

Root Cause Analysis, or **RCA**, is the process of discovering the origin of problems in a system. The purpose is to identify solutions for the base problem, rather than just treating symptoms and putting out fires. To this end there are templates and techniques that can be used for this.

One method of finding root causes is the **Five Whys** technique. This is a series of questions that can be asked to drill down or iterate to underlying causes.

The following example of this technique illustrates this:

Problem: The machine won't run.

1. Why? – It has no power. (First why)

2. Why? – The circuit breaker tripped. (Second why)

3. Why? – The circuit drew too much power when the motor got stuck. (Third why)

4. Why? – A part fell into the drive chain. (Fourth why)

5. Why? – The chain cover was left off after servicing. (Fifth why, a root cause)

Of course, you could use more or less "whys" if necessary, but this is the formal method defined in the Toyota Production System (TPS), where many Kaizen and Lean Manufacturing techniques come from.

Factors contributing to defect XXX

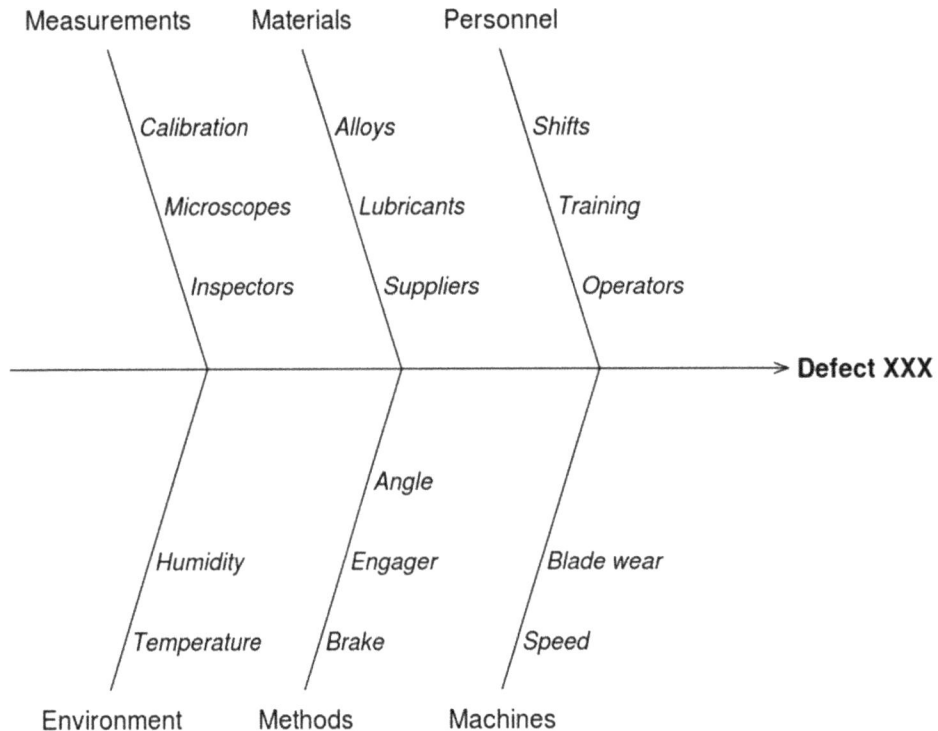

Figure 269 - Ishikawa or Fishbone diagram

A common tool used to identify possible causes of problems or defects is the Ishikawa diagram, named after the originator. These are also known as fishbone, herringbone, cause and effect, or even "Fishikawa" diagrams! This method is graphical, which can be useful when brainstorming. Additional branches can be drawn from other branches if required. A disadvantage is that the diagram can become easily cluttered if there are many possible causes.

Intermittent and Multiple Symptoms

Some of the most difficult problems to solve are those that are **intermittent** in nature. Hours or days can go by in between occurrences, and there doesn't seem to be a pattern.

Some of the best techniques to use in solving these problems involve trying to reproduce the symptoms as mentioned previously. When defining the problem itself, it is important to list all the suspected causes and components involved. If causes like heat or humidity are on the list, artificial stresses like heaters or humidifiers can be used. If things like vibration or mechanical stresses are suspected, vibration monitors can be installed, and physical movement of assemblies attempted. All of these should be well documented to prevent doing the same thing over and over.

Race conditions in PLC programs or other errors are also possible, though this would fall under the category of "bad design". Properly written and tested control programs should not have these kinds of problems.

Sometimes a problem such as a bad component is found and technicians assume the problem has been solved, only to discover that the bad component caused another component to be damaged. Technicians and operators also sometimes make adjustment to machines such as motor speed or pneumatic flow controls because of component wear or damage, finding that when the component is repaired or replaced, the machine doesn't run properly anymore. This falls into the category of **multiple symptoms or causes**. Accurate machine documentation for setpoints and adjustments, and good recordkeeping of repairs and modifications can go a long way toward minimizing these issues.

Exercise 32

1.   What is the relationship between Correlation and Causation?

_____

2. What are the three basic elements of solving problems?

_____

_____

_____

3. What is the purpose of Root Cause Analysis?

_____

_____

4. List some of the techniques to keep in mind while troubleshooting a system

_____

_____

_____

_____

_____

Tools and Techniques

Tools that can be used in maintaining and troubleshooting equipment include physical items and devices, software, and written materials. The most important tool of all of course is your **brain**; this is the tool that guides a troubleshooter or technician to all the other tools!

Along with the tools of the trade, it is important to learn their proper use. This section also covers some of the common tasks in industrial use.

Mechanical Tools and Techniques

Measurement

Physical measurement of components and tooling can be important in ensuring that systems operate as designed. While there are various large systems used for precise measurement such as Coordinate Measuring Machines (CMMs), this book will discuss handheld measurement tools.

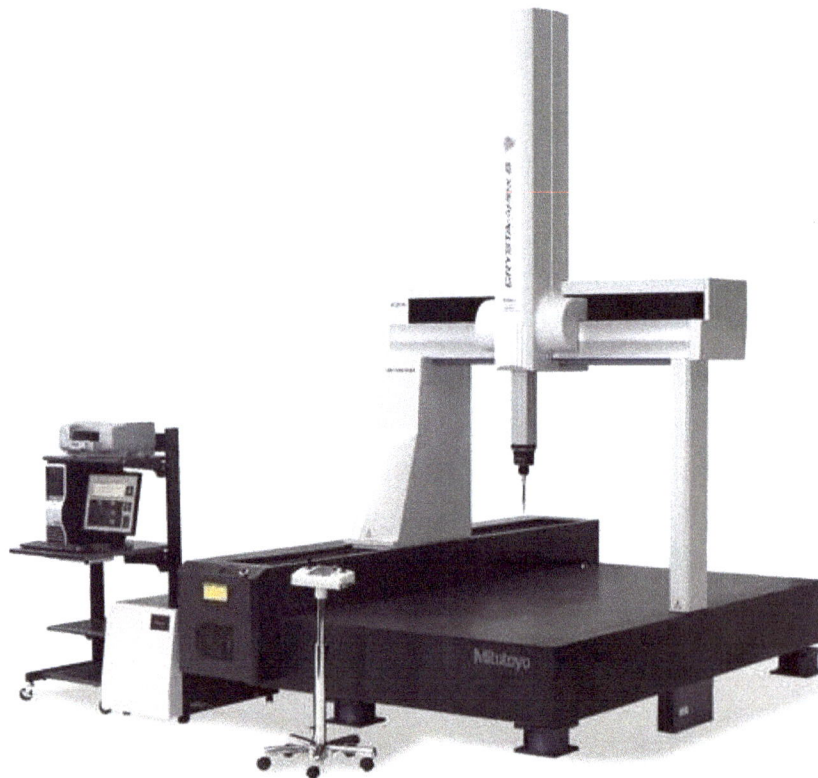

Figure 270 – Mitutoyo Coordinate Measuring Machine (CMM)

Calipers

A caliper is a device used to measure the dimensions of an object by moving its measuring face to a surface, then reading a gauge or display when it stops. Physical configurations include inside and outside calipers, so named by the dimension it is meant to measure, and divider calipers which are sometimes used to mark locations in metalworking.

The most common types of calipers in common use for measurement are vernier, dial, and digital calipers. **Vernier** calipers use a sliding scale that is read by aligning marks manually and visually interpreting the position; these are the least expensive of the three options. **Dial** calipers have a round dial that performs much the same function. The needle must be read in relation to the marks on the dial. The dial usually rotates once per inch, 1/10 inch, or millimeter. The dial reading is then added to the coarse reading of the scale.

Digital calipers have a readout as shown in Figure 271. They are more expensive than either vernier or dial calipers but have a higher resolution and therefore are more accurate. They can also be scaled to either metric or inch-based measurements.

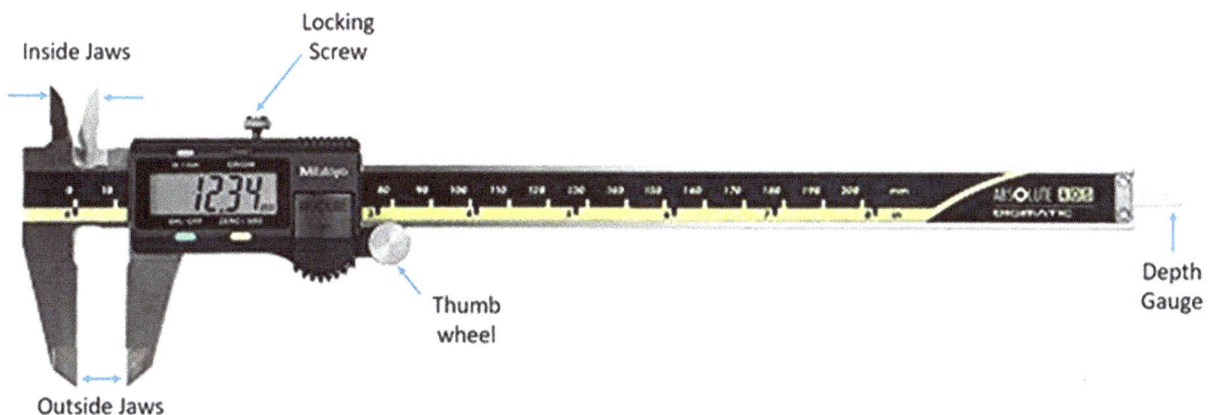

Figure 271 - Parts of a caliper

Digital calipers also often have a serial output, allowing measurements to be interfaced with a computer or recorder. Electronics also allow a "reading hold" feature where dimensions can be read later after the measurement is made in a non-visible area.

Inside and outside jaws allow for measurement from outside diameters and lengths to inner diameters of holes or pipes. The thumbwheel makes the caliper easy to use with one hand, and the locking screw can aid in creating a simple go/no-go gauge. The depth gauge probe at the end is useful for measuring tapped or blind holes.

Micrometers

A micrometer uses a calibrated screw, placing the object to be measured between the spindle and anvil. The spindle is moved by turning the thimble until the object lightly touches both faces. A useful feature is the ratchet which ensures the micrometer is not over tightened, which can make the measurement inaccurate.

As with calipers, there are digital versions in addition to the vernier type shown in Figure 272.

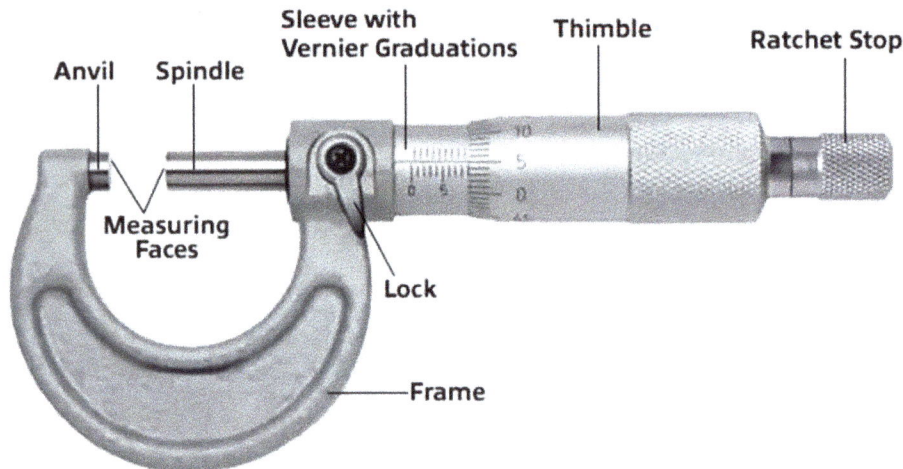

Figure 272 - Vernier Micrometer

Micrometers tend to be more accurate, with greater resolution than calipers, but can be more difficult to use.

Tapes and Rulers

A simple tape measure or ruler is a standard tool any technician should keep in their tool kit. When using these basic measuring devices, it is important to view the scale straight on and if marking or scribing to do so accurately.

Proper use of a tape measure or ruler requires precision in interpreting the position of an object relative to the markings on the scale. Larger measures or increments have longer lines to make the scale easier to read.

Tapes and rulers come with different scales depending on whether inch or metric measurements are needed. Some come with both scales.

If the goal is simply to duplicate the length of parts, marks can be made on paper or on the part itself to measure against.

Torque Drivers and Wrenches

Torque drivers and torque wrenches are calibrated to provide a specific torque to a screw, nut, or bolt. Beam type wrenches have a deflecting pointer as shown in Figure 273 that shows the force exerted by the wrench. There are also dial types that perform a similar function.

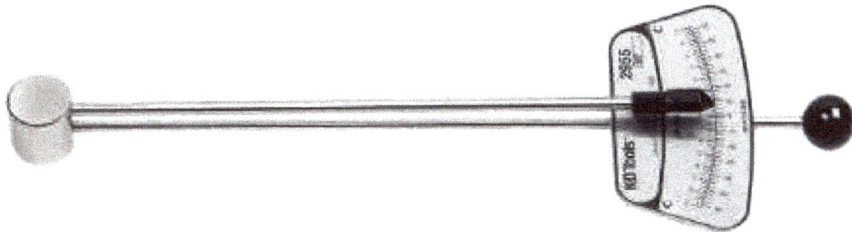

Figure 273 - Beam torque wrench

Other ratcheting "click" torque wrenches allow a preset force to be set. When the allowable torque has been reached the driver will ratchet, limiting the amount of torque to the fastener.

Electronic indicating torque wrenches use a strain gauge that converts the torque to a scaled value. When used in a pneumatic or other powered torque driver, the force can be limited to a setpoint.

Exercise 33

1. What are the three most common types of calipers used for measurement?

    _____

    _____

    _____

2. Which are considered more accurate: micrometers or calipers?

    _____

3. In addition to measuring, what is the purpose of a torque wrench or driver?

    _____

Hand Tools

Wrenches and Drivers

Most good mechanics will have a good set of **crescent wrenches** and **socket wrenches** in both standard and metric sizes. While an adjustable wrench can be flexible, maintenance technicians sometimes disdainfully call them "nutwreckers".

Allen wrenches are also necessary in a mechanical tool kit, either as individual items or in a folding set. There are also a variety of other types of bits available for **ratcheting drivers** such as the hex, Torx, and screwdriver bits shown in Figure 274.

Standard **screwdrivers** are also necessary for small fasteners and removing covers. Mechanical technicians often have a set of electrical hand tools including screwdrivers for wire terminals.

Pipe and strap wrenches can also be useful for turning larger diameter cylindrical objects.

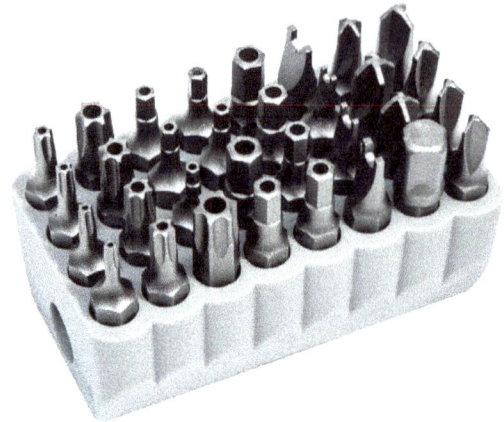

Figure 274 - Hex and Torx Bit Set

Pliers and Cutters

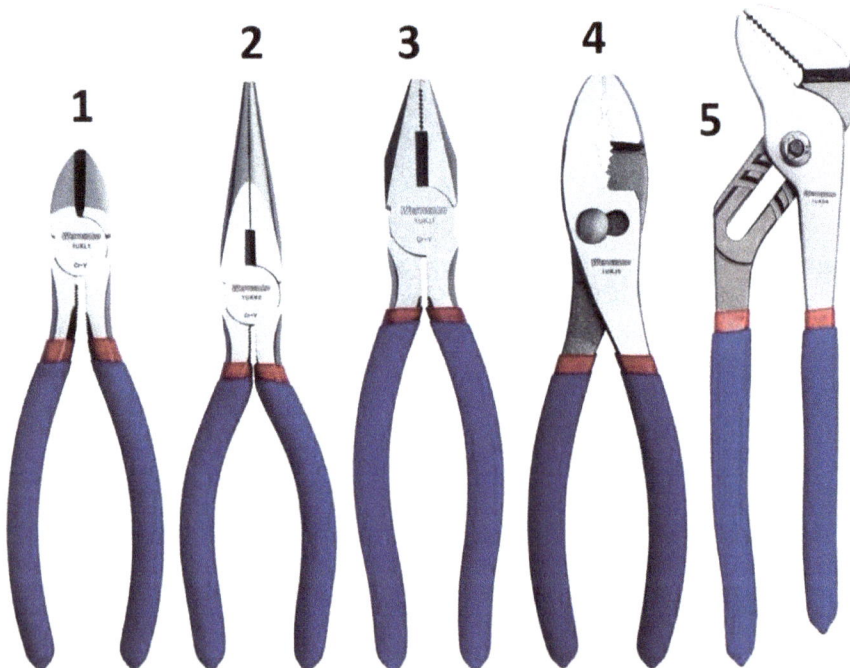

Channel locks, cable or diagonal cutters, and even a hacksaw can be useful hand tools for a technician. Figure 275 shows a variety of cutters and pliers. From left to right:

1. Diagonal Cutters
2. Needle Nose
3. Linesman
4. Slip Joint
5. Channel Locks

Figure 275 - Cutters and Pliers

Taps and Dies, Screw Extractors

Taps are used for threading holes, while dies are used to thread rods. They come on standard and metric sizes, and with different thread pitches. Tapping sets often come with drill bits in the correct size for the tap.

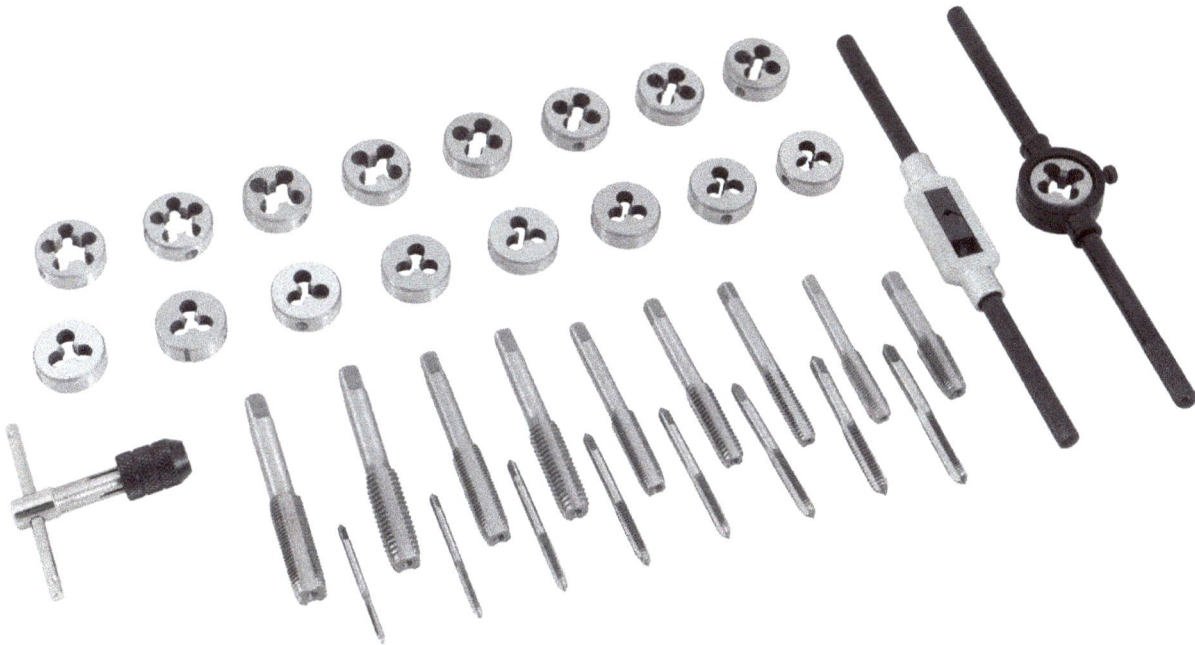

Figure 276 - Tap and Die Set

Tapping is much more common than threading in the typical maintenance arena for the purpose of using screws or bolts to attach things to each other. Using oil or tapping fluid as a lubricant is advised when tapping. The smaller taps in Figure 276 are used with the smaller tap wrench on the left, while larger taps work with the T handle on the right.

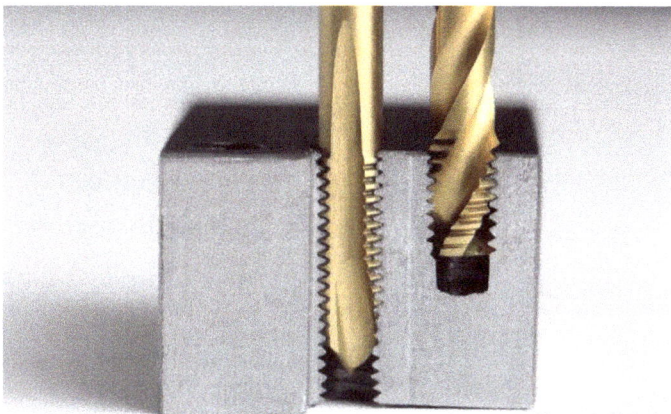

There are different types of taps as shown in Figure 277, but all have flutes (the indented areas on the sides) to help clear metal shavings. Tapping a hole can be done by hand or using a drill, it often is done in multiple stages, backing the tap out of the hole several times to clear the flutes. Tapping is a slow and careful process.

Figure 277 - Through hole (left) and blind hole (right)

No matter how carefully screws are inserted, the drive (the part where you insert the screwdriver) can get messed up. This is where **screw extractors** can be handy. They can also be used to remove broken screws or bolts. Like taps, they come in different sizes. Sometimes called "easy outs", they are spiral bits that turn in the opposite direction of the screw or bolt (typically counterclockwise). Some have different tooling on each end of the bit; the end opposite the bit is used to prepare the screw head by grinding the hole to the proper shape.

Figure 278 -Screw Extractors

Power Tools

Common hand-operated power tools used in industrial maintenance include drills, saws, and grinders. With all of these come wearable tooling, bits for drills, blades for saws and abrasive disks for grinders.

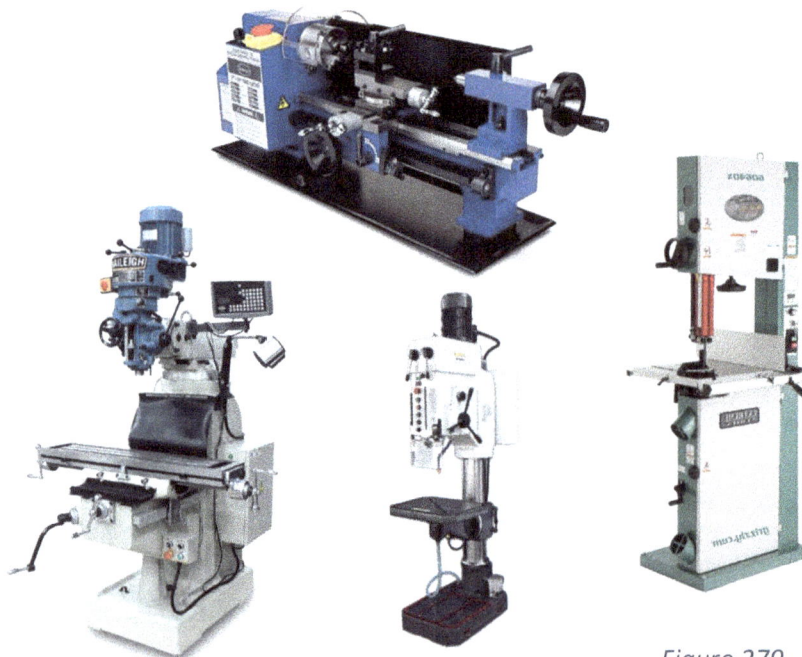

Additional tools used in the industrial environment include mills, lathes, presses, shears, band and table saws and many others. This book doesn't go into detail on these other than brief mentions in the following sections. Each requires further training on operation and safety.

Figure 279 - Larger Shop Tools. Clockwise from Top: Lathe, Bandsaw, Drill Press, Vertical Mill

Powered tools may use a power cord that plugs into an electrical outlet, be battery powered, or use an air connection in the case of pneumatic tools.

Drills

Drills may only provide rotational force to a bit to drill a hole, but impact drivers use a hammer and anvil mechanism to apply rotational impacts. This creates more torque to a fastener than a standard rotational drill. Most impact drivers have two hammers, but some have three. These tools are specified in impacts per minute, or **IPM.**

Hammer drills provide forward impacts instead of rotational and are generally specified in beats per minute or **BPM.** They are usually used for masonry or stone. Some drills combine the rotational impacts and hammering for tough jobs, these are known as rotary hammer drills, and use a different mechanism for impacts than the rotary and hammer types. The hammering function can be turned on and off with a switch.

Figure 280 - Bosch Corded Hammer Drill

Drills usually have a finger operated trigger that can control the speed of rotation, as well as a switch or button to change the direction.

Drill chucks can either be tightened by hand or by using a chuck key; some are available that do both. The type that uses a key is called a **Jacobs Chuck**, while the keyless type is sometimes known as an **Albrecht Chuck** after the company that invented it.

There are different sizes of chuck available depending on the diameter of bit being used. These can be switched out or replaced on a drill press.

Figure 281 - Drill Bits

Drill bits are designed to drill holes in common materials including metal, wood, plastic, ceramic, or concrete. Bits can be further subdivided to be used on steel, aluminum, cast iron, or sheet metal. Bits are generally categorized by the diameter of the hole being drilled. Figure 281 shows various types of bits used in maintenance. The leftmost bit and the others that are the same diameter for their full length are generally used for metal. They are sometimes called "twist" bits. The materials used are different depending on the type of metal. Cobalt is often used for harder metals as it dissipates heat quickly, while black oxide and titanium bits are often used for softer metals. High-Speed Steel (HSS) is another common material used for drilling wood, light metals, wood, and PVC.

The second bit from the left is called an auger bit. It has a screw tip that helps draw the bit into the material and locate the tip. The two wider bladed bits are called spade bits and are usually used for wood or softer materials.

The second bit from the right end is called a step bit and is used to drill holes of multiple sizes in sheet metal. It is sometimes used in panel fabrication. The last bit is used for deburring holes or countersinking screws. Some bits have a countersink bit in combination with a standard spiral bit.

Materials such as plastics, glass, and ceramics use special bits not shown in Figure 271. It is important to choose the correct bit for the material it is being used on.

Technique: It is important when drilling holes to hold the drill as vertically as possible, perpendicular to the workpiece. Fixturing or clamping the workpiece to ensure it doesn't move is also critical. Using a drill press will ensure that the bit is vertical.

Speed is also important. Drilling too fast creates heat, which reduces the life of the bit. Don't apply too much pressure either as the bit can break.

Lubrication helps carry heat away from the workpiece and bit, and can prolong the life of the bit. It can also be useful to start with a smaller bit and work your way up to a larger diameter.

Using a **pin punch** to create a dimple in metal before drilling can improve accuracy in hole location and prevent the bit from "walking". It is also a good idea to deburr the hole when done, removing sharp edges.

Most importantly, wear safety glasses when drilling! Bits or metal shavings can damage your eyes! Gloves may also be needed to prevent injury to the hands.

Saws

Hand operated saws include portable band saws, sometimes called "portabands", jig saws, and circular saws. There are also larger floor or table mounted versions such as upright band saws, "roll-in" saws that operate on a slanted surface so that the weight of the saw pushes the blade through the material, table saws which are basically circular saws mounted to a slotted table, and "chop saws", which are circular saws with a pivoting blade. Additionally, tabletop scroll saws have a thinner blade than a jig saw and can be used to cut sharper curves.

Saw blades can also be angled to create beveled or mitered cuts. Jig saws and circular saws often have this feature.

Figure 282 - Left to Right: Jig Saw, Portable Band Saw and Circular Saw

Blades for these saws are generally made of tempered high-grade tool steel that is slightly flexible. Circular and chop saws also have blades made of abrasive material such as aluminum oxide or silicon oxide grains laid over a fiberglass mesh. These blades can be used for metal, concrete, or masonry.

Generally speaking, finer toothed blades are used for metals and harder materials. The more teeth on a blade, the better quality and finish of cut will occur. TPI is the number of teeth per inch, usually softer materials such as plastics or wood will use blades with a TPI of 6-20, and metals will have a TPI between 14 and 36. Lower TPI numbers remove material faster but create a rougher cut.

There are also different shapes for teeth, depending on the speed and material. Teeth may be triangular or hook shaped and are usually arranged in alternating directions. Some blades even use several different sized teeth, removing more material but also providing a better finish to the material. Blades also have different thicknesses; it important to take this into account when measuring. Remember to subtract the thickness of the blade from the material cut. The thickness of the slot left when cutting material is called the "**kerf**".

Technique: As with drill bits, heat is created by the metal blade cutting through material. It is important not to apply too much pressure pushing the saw blade through the material, as this creates heat, which reduces its life and can even break the blade.

Residue can build up between blade teeth. It is important to clean blades occasionally to reduce "blade drag", which further heats the material. Lubricants can also be used to reduce heat but can also increase drag or even contaminate the material being cut.

It is very important to choose the right blade for the material being cut. Selecting the wrong blade can damage both the workpiece and the blade, and even cause metal pieces to fly across the workstation. As with the use of any tool, wear safety glasses always and gloves when necessary.

Saw blades can create sparks. It is critical to ensure that flammable or explosive substances are not located near the cutting area. Abrasive saw blades and grinding wheels are designated either ferrous or non-ferrous. Ferrous blades are ideal for cutting metals that contain iron, such as stainless steel or cast iron. Choose a non-ferrous blade when cutting softer metals like aluminum or copper.

Grinding, Polishing and Sanding

Handheld grinders are often used to finish off rough edges, deburr parts, and smooth welds. They can also be used to create unique finishes on metal parts. Manual grinding can often be done with a simple file, but using a powered tool is often more efficient, especially when a large amount of material needs to be removed. Angle grinders are the most common hand-held grinding tool, these usually use 4-1/2" grinding wheels.

Tabletop rotating grinders for metal are called bench grinders. These may also have buffing wheels or brushes to put a shine on metal or plastic surfaces as shown in Figure 283. Bench grinders are often used for sharpening tools. They have two wheels on them, often with a coarse and fine grit.

Figure 283 - Angle Grinder (Left) and Bench Grinder (Right)

Belt sanders are another common tabletop tool for polishing or finishing material. These often also have a circular sanding pad on them as shown in Figure 284. Belts come in different sizes as shown in the illustration.

Figure 284 - Belt Sanders

Notice the tabletop surfaces that allow the workpiece to be held in a proper position. These tables are often adjustable to allow the workpiece to be angled.

There are also handheld sanders for finishing flat surfaces.

Belts and pads come with a variety of abrasive roughness, known as "grit". This refers to how many particles can fit through a 1-inch filter, so a lower grit number has larger, coarser particles. Grit sizes vary from about #24 (very coarse) to more than #7000 (extremely fine). Another measurement of roughness uses numbers preceded by a **p** or **k**, indicating the number of grains per square centimeter.

Different materials are also used depending on the surface it is applied to. Zirconia and ceramic sanding belts last longer and remove metal faster than aluminum oxide or silicon carbide, but aluminum oxide is often used for hard metals like iron or steel, and silicon carbide is used for softer metals like aluminum or brass.

Techniques: As with cutting and drilling operations, grinding and sanding creates heat. This can become very apparent when holding a piece of metal against a running belt with bare hands.

It is also difficult to use fixturing or clamps when sanding or polishing, this creates the opportunity for the workpiece to fly off and hit something, or someone... use great care around others, and as always, wear safety glasses.

Grinding metal creates even more sparks than saws do, so ensure there are no flammable objects or vapors present.

Consistent smooth movement is the key to creating a good finish with belts or rotating abrasives. Using a less abrasive (higher grit) belt or disk takes longer but can help ensure a better finish to the workpiece.

Exercise 34

1. What are the two methods of holding a hand-operated tap?

    _____

2. Make a list of some of the hand or powered tools you need for your work:

    _____

    _____

    _____

    _____

    _____

    _____

3. What is another name for a screw extractor? _____

4. What are three sources of power for powered tools?

    _____   _____   _____

5. What is the cause of excessive tool wear for drill bits, taps and saw blades?

    _____

6. What are the two designations for abrasive saws and grinding wheels?

    _____   _____

7. What is the name for the thickness of a saw blade, or width of its slot?

    _____

Electrical Tools and Techniques

Multimeter

The most important measurement tool an electrician uses is a multimeter. This device measures resistance, voltage or current in an electrical circuit. It is portable, battery-operated, and there are many different price levels and additional capabilities available.

Resistance measurement requires a power source, usually a battery, to create current flow through the resistance being measured. This makes it very important that _power is removed from the circuit under test_. Also, the resistance should be isolated from the rest of the circuit to ensure a proper measurement. Since the voltage of the power source is known, current is measured in the meter and used to calculate the resistance. On older analog meters, there were range settings for resistance ranges, most modern digital meters do this automatically.

Voltage can be measured for both AC and DC. Most meters require the user to select between AC or DC measurement, but the range may be auto-selected.

Older multimeters were analog, with a needle pointing to a scale, but modern meters are typically digital with protective features to avoid damaging the meter.

Current measurements required the meter to be placed in series with the circuit under test, but some models come with a clamp which is looped around the conductor. It is very important to set the meter to the correct units of measure, meters can be destroyed when setting the meter to current or resistance and applying it to live voltage.

Older analog meters such as the Simpson 260 shown in Figure 285 use a needle to indicate the measured value on a scale. Because there is only one needle but different ranges can be selected, there are multiple scales on the meter. The needle position is determined by the amount of current flowing through a coil and reading the position of the needle can be imprecise.

Additionally, the correct range must be selected on the meter itself, and if the correct type and range of measurement is not selected it can be dangerous.

Figure 285 - Simpson Analog Multimeter

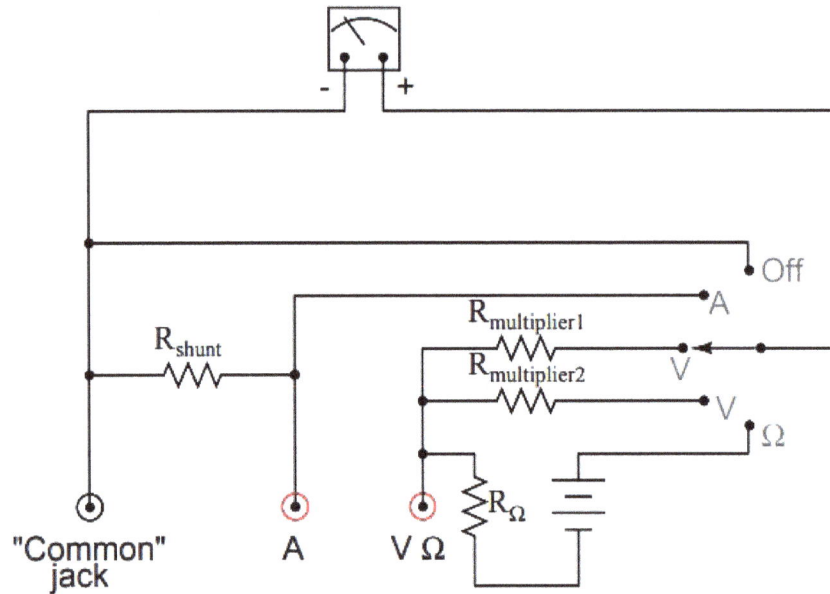

Figure 286 - Simplified Multimeter Circuit

Figure 286 above shows a simplified circuit for a multimeter. Though not shown, the Ampere (A) and Ohms (Ω) settings also have resistors placed in series with the needle coil. Ammeters in particular use a shunt in parallel with the meter movement to divert most of the current around the coil. A resistance is also placed in series based on the range setting; higher resistances are used with higher ranges. The circuit illustrates the hazards associated with selecting the incorrect range or parameter, placing the leads across a voltage with the setting on amperes can destroy the meter or injure the technician!

Resistance readings require a voltage source to create current through the resistor. This is in the form of a battery, which must be replaced occasionally. Since the available current depends on the state of the charge of the battery, a multimeter usually has an adjustment for the ohm scale to zero it. Place the leads together with the dial on the desired ohms range and then adjust the needle to zero. It is also important not to attempt to read resistance on energized circuits. An open circuit will read "**OL**", signifying Overload, or out of range.

Digital multimeters can include many different functions in addition to the standard resistance, voltage and current described previously. Tests can be performed on thermocouples, diodes, transistors, and capacitors. They can do frequency measurements and capture peaks. They are also much more accurate and safer than analog meters since a signal can be sampled before performing the test.

Figure 287 shows a digital multimeter from Fluke that records events and allows graphing. It has a real-time clock that time-stamps readings and allows data to be saved for analysis. It also has software available that help solve complex electrical problems and wireless capability. This allows readings to be monitored remotely.

A note in the manual for this meter reads as follows:

"To avoid circuit damage and possibly blowing the meter's current fuse, do not place the probes across (in parallel with) a powered circuit when a lead is plugged into a current terminal. This causes a short circuit because the resistance through the meter's current terminals is very low".

In addition, if a lead is plugged into either of the current terminals and the rotary switch is not set to the correct position, the meter makes a chirping sound and displays "Leads connected incorrectly". This is much safer than the analog meter described previously!

Figure 287 - Fluke 289 True RMS Data Logging Multimeter

The *i info* button even brings up more information about the selected function. These meters are quite expensive but can save time when used by a trained technician.

Additional Meter Techniques and Safety

Like machine safety, there are categories of electrical measurement safety. **Category I** includes basic protected electronic equipment, where measures are taken to limit transient overvoltages to a low level. This generally involves low voltage DC. **Category II** is single phase receptacle connected loads, this includes appliances and portable tools. **Category III** is three phase equipment and single-phase lighting equipment (277v), and **Category IV** includes three-phase at the utility connection. These categories are all for equipment under 1000V, detailed in IEC 61010.

The real issue for meter circuit protection is not just the maximum steady-state voltage range, but a combination of both steady-state and transient overvoltage withstand capability. Transient spikes can be caused by internal equipment failures or external effects like lightning. Therefore, power lines exposed to outdoor effects are considered the most dangerous, even if the voltage is only 120VAC.

A technician working on equipment in a CAT I location can be exposed to DC voltages much higher than the electrician working on a motor in a CAT III location, but the transient voltages are a lesser threat because the energy available to an arc is limited by both wire size and circuit protection.

Non-contact voltage detectors are a quick, inexpensive way to check for the presence of live voltage. Whenever possible work and testing should be done on de-energized circuits, but sometimes this is not possible.

The biggest hazard from electric shock exposure is from current passing through the chest, causing heart fibrillation. Therefore, when measuring hazardous voltages or current it can be useful to perform measurements with one hand. One probe can be connected to a neutral or grounded point via alligator clips or screw clamp, while the other probe is applied to the measured point. Both probes may also be held in the same hand. Always connect to the neutral or lower voltage terminal first and remove the hot lead first.

Hang or rest the meter if possible; try to avoid holding it in your hand to minimize exposure to the effects of transients.

In many plants and other working environments there are Personal Protective Equipment (PPE) and Lock-Out/Tag-Out (LOTO) requirements that protect the technician. PPE includes safety glasses or a face shield and insulated gloves. Watches, rings, and other jewelry should be removed, and flame-resistant clothing worn. Standing on an insulated mat is also best.

Meter probes and cables should be well insulated to protect the technician from hazardous voltage. Meters are often fused, but test leads with fuses are also available. Meters with recessed input jacks and test leads with shrouded input connectors should be used if possible.

Use the three-point test method, especially when checking to see if a circuit is dead. First, test a known live circuit. Second, test the target circuit. Third, test the live circuit again. This verifies that the meter worked properly before and after the measurement.

To determine what kind of meter you should own, analyze the worst-case scenario of your job and determine what category your use or application falls into. First, choose a meter rated for the highest category you could be working in. Then look for a multimeter with a voltage rating for the category matching your needs. Don't forget the test leads; they should be certified to a category as high or higher than the meter.

A "**Wiggy**" is a solenoid operated meter that checks for the presence of voltage. The original unit used a spring-loaded solenoid that moved a needle, but modern voltage testers often simply use a light.

A Megohmmeter or "**Megger**" is a special ohmmeter used to check the resistance of insulators. Older models use a hand operated crank to generate a high DC voltage, newer models use a battery. They are also often used to check motor windings.

Exercise 35

1. What three values can any multimeter measure?

_____   _____   _____

2. What two things need to be set or configured on a multimeter <u>before</u> performing a measurement?

_____   _____

3. What precautions should be taken before measuring resistance?

_____

_____

4. List some of the additional features a multimeter can have:

_____

_____

5. What are some of the safety precautions that should be followed when working with live voltage?

_____

_____

6. What is a "Wiggy"? _____

7. What is a "Megger"? _____

Electrical Diagnostics

There are many other "Bench tools" available for diagnosing electrical and electronic circuits. An **oscilloscope** displays analog waveforms for frequency analysis. These devices used to require glass cathode ray tubes (CRTs) to display the waveforms, but since the advent of LCD (liquid crystal) displays handheld and portable units have become available for field service applications.

Oscilloscopes, also called "scopes" or "o-scopes", graphically display varying signal voltages as a function of time. This allows signals to be compared and diagnosed for properties such as amplitude, frequency, rise time, distortion, or time interval.

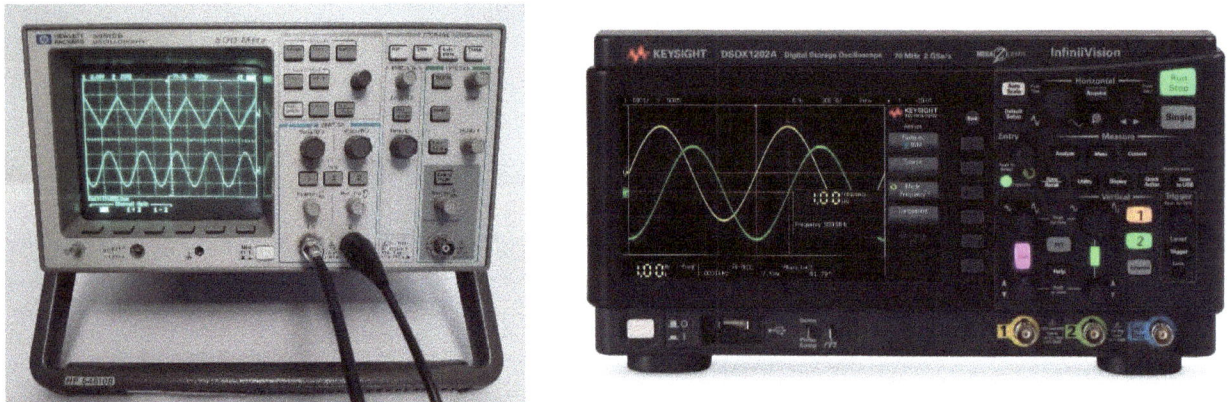

Figure 288 - Left: HP CRT Oscilloscope   Right: LCD Oscilloscope

The two oscilloscopes in Figure 288 are dual channel scopes. Additional channels allow signals to be plotted against each other such as voltage (V)and current (I), or two different points on a signal path. Newer scopes also capture data and import it into computer software for analysis.

Common uses for an oscilloscope in industrial automation are viewing waveforms for encoders and troubleshooting VFDs. The waveform for the output of a VFD should be sinusoidal with little or no distortion, flat-topped sine waves are often a sign that another non-linear load is attached to the same feeder circuit. In this case the problem is not with the VFD or the motor.

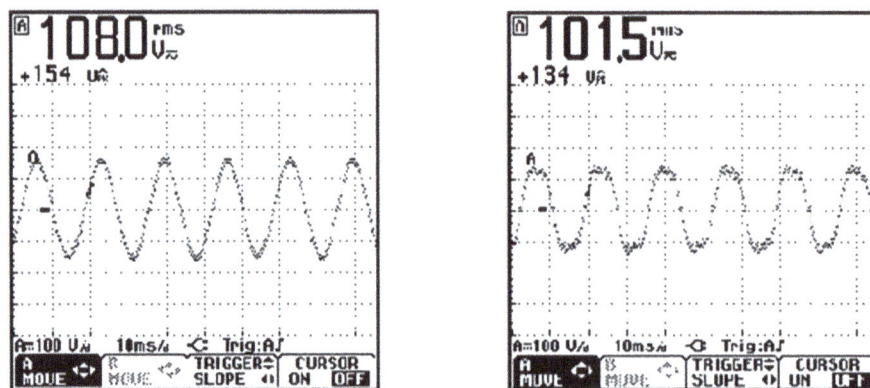

Figure 289 - VFD Outputs: Ideal Sinusoid (Left) vs. Flat-topped (Right), courtesy of Fluke

An oscilloscope shows not only amplitude but any distortion, disturbance or noise that might be affecting the waveform. This makes it an excellent tool for diagnosing and solving electronic problems.

Most of the safety concerns associated with multimeters also apply to oscilloscopes. Meter leads need to be well insulated and appropriate for the type of circuit under test. Leads are often coaxial cables.

Function Generators are used to inject signals of different forms and frequencies into circuits so they can be analyzed. Frequency, amplitude and wave shape are all adjustable parameters that can be set. Function generators that only produce sinusoidal waveforms are known as signal generators.

Waveforms that can be generated include sinusoidal, triangular, square, and sawtooth. More expensive units, sometimes called Arbitrary Waveform Generators or AWGs, can even generate more randomized signals for higher end design and test purposes.

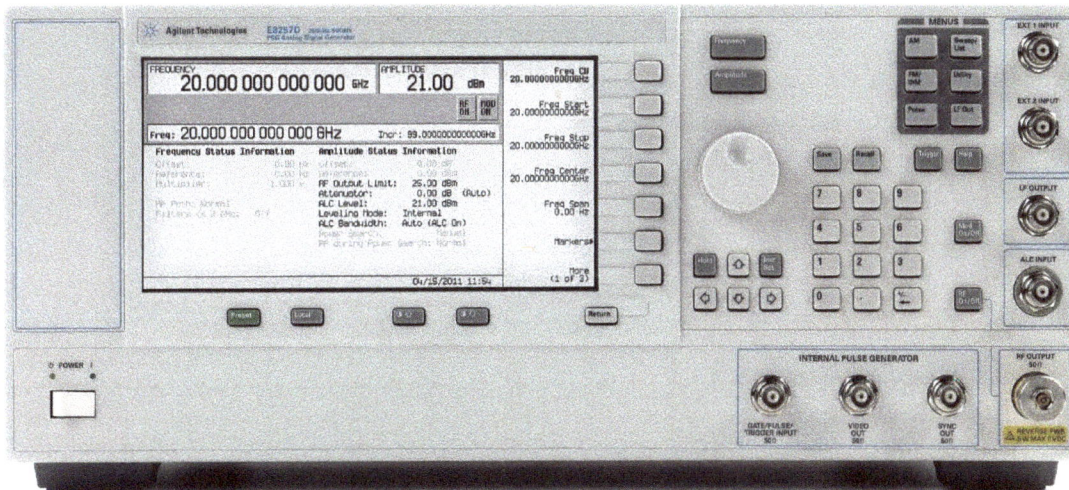

Figure 290 - Agilent Analog Signal Generator

Function generators are often used with oscilloscopes to diagnose circuits. They may also be used to generate Pulse Width Modulated (PWM) signals for motor control.

Spectrum Analyzers are used to measure the magnitude of an input signal versus frequency within the full frequency range of the instrument. The primary use is to measure the power of the spectrum of known and unknown signals. By analyzing the spectra of electrical signals, dominant frequency, power, distortion, harmonics, bandwidth, and other spectral components of a signal can be observed that are not easily detectable in time domain waveforms. These parameters are useful in the analysis of electronic devices, such as wireless transmitters. Oscilloscopes sometimes have spectrum analysis functions built in.

Thermal Imaging cameras use infrared detection to identify hot spots in control panels or build heat profiles for motors. A good practice is to capture good quality infrared images when a motor is operating under normal conditions, giving you a baseline measurement. Motors will list the normal operating temperature on the nameplate along with full-load amps (FLA).

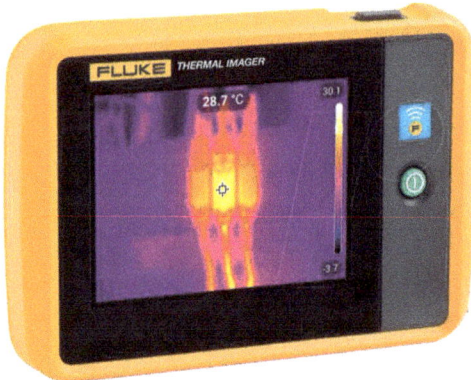

Figure 291 - Fluke PTi120 Thermal Imaging Camera

Temperatures can be captured for all critical powertrain components: motors, shaft couplings, bearings, and the gearbox. Small differences can indicate problems. As the load increases, the temperature will rise. If there is a problem, there will often be greater temperature differences at higher loads.

Abnormal thermal patterns or temperatures can be caused by high-resistance contact surfaces at switch contacts or connections, load imbalances, or improper sizing of conductors.

Failed components also typically look cooler than normally functioning ones. An example of this is a blown fuse. In a motor circuit this can result in a single-phase condition that can damage the motor.

Exercise 36

1. What is an oscilloscope used for? _____

_____

2. What does a function generator do? _____

_____

3. What is the purpose of a spectrum analyzer? _____

_____

4. What can cause abnormal thermal patterns or temperatures?

_____

_____

Electrical Hand Tools

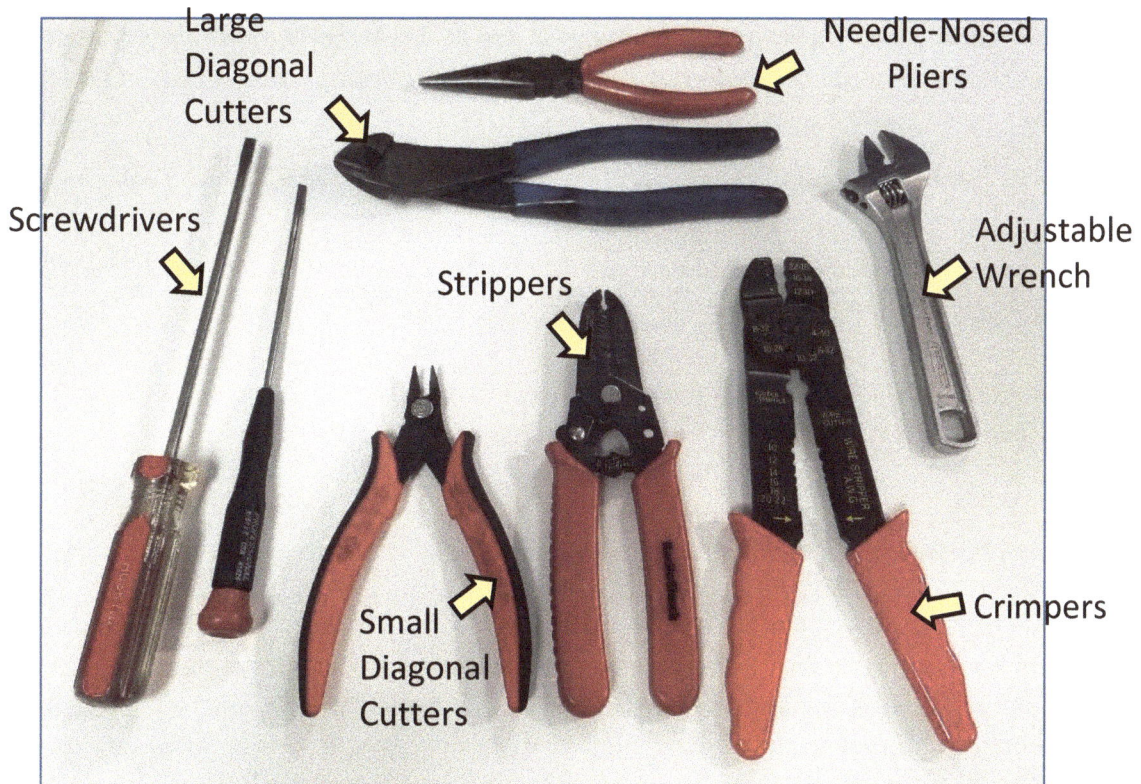

Figure 292 - Typical Electrical Hand Tools

Shown above are some of the basic hand tools needed for control wiring in an industrial environment. Larger wire gauges require cable cutters or a hacksaw. Other optional tools include vise grips, alligator clips, a soldering iron, and solder.

Shown below are some of the better, higher-priced tools available. Left to right: strippers, ratcheting crimpers, and ferrule crimpers.

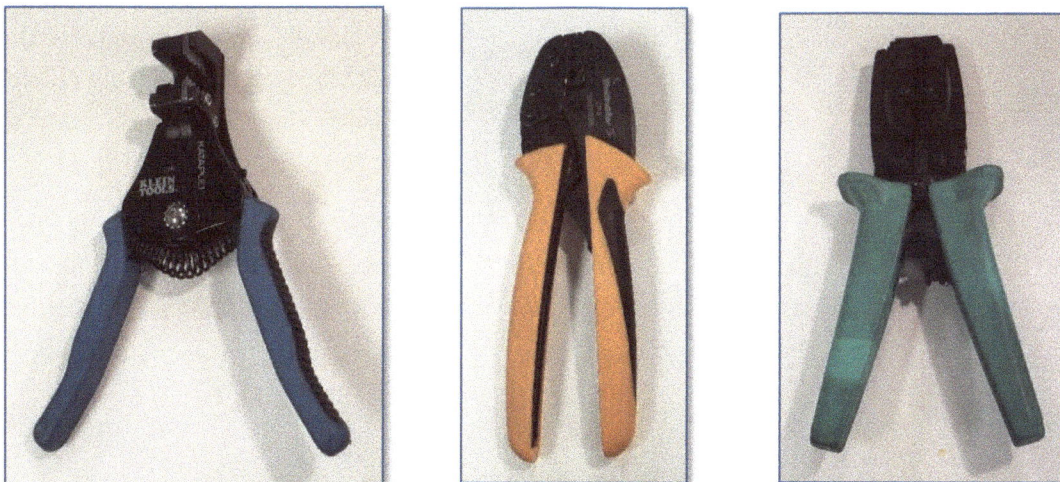

Figure 283 – Strippers and Crimpers

Electrical Fabrication Techniques

Many of the tools listed in the mechanical section are also used for wiring-related tasks. Drilling and tapping holes in metal backplanes is common when mounting components in an electrical enclosure.

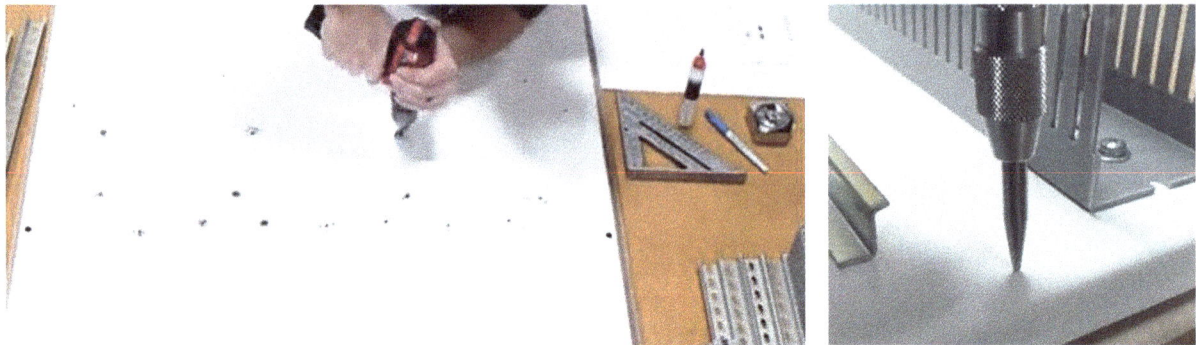

Figure 294 - Control Panel Preparation

Figure 294 shows some of the techniques used in preparing a control panel for components. On the left the backplane is being drilled and tapped after marking hole locations. Note some of the tools shown for marking and aligning holes. The right side shows a "pinger", or spring loaded punch, commonly used by technicians to mark hole centers. This makes sure that the drill bit doesn't drift while drilling.

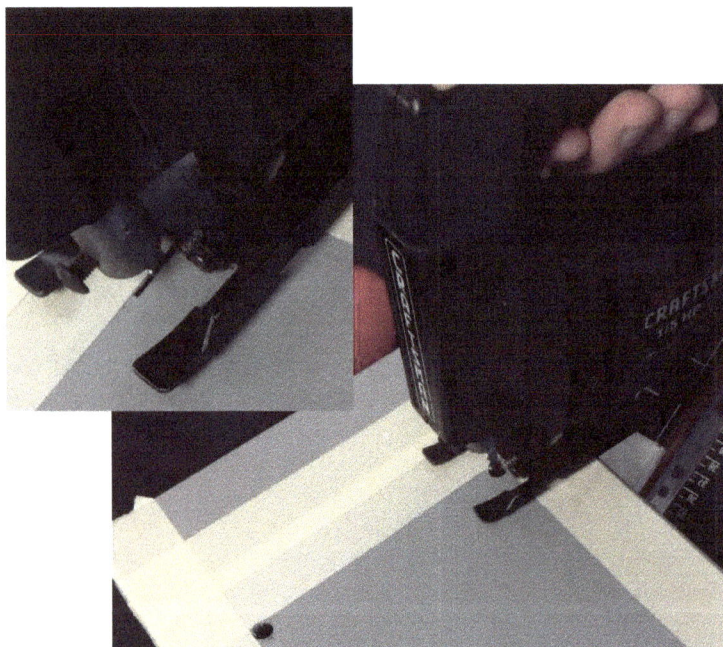

Figure 295 - Cutting a Hole for a Touchscreen

Larger holes for HMIs or touchscreens are usually done with a jigsaw as shown in Figure 295. After marking the opening the edges are taped to ensure the saw doesn't scar the edges of the opening.

Holes are drilled at the corners of the rectangle, large enough for the jigsaw blade to fit in. Jigsaw cuts can be turned slightly to ensure that the cut is at the edge of the tape.

After cutting the opening a file is used to smooth the rough edges of the cut. A handheld "deburring" tool can be used to treat the corners.

Knockouts are used to make holes for pushbuttons and conduit entries in control enclosures. While drills are useful for smaller holes and a stepper bit can be used to enlarge them, the hole will require a lot of deburring and metal shavings can get into delicate electrical equipment.

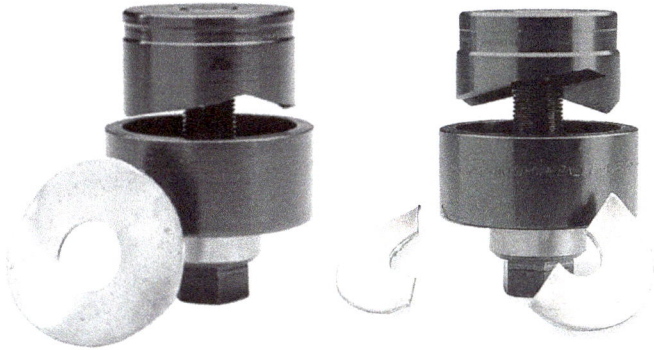

Figure 296 - Standard and Split Knockouts

The basic hand operated knockouts in Figure 296 are operated with a wrench. A hole is drilled in the enclosure slightly larger than the size of the threaded shaft, then the hollow part of the die set with the shaft and bearing is inserted into the hole. The cutting die is then threaded onto the shaft (known as a "draw stud"), and the wrench is turned until the hole is complete, pulling or drawing the die through the metal. It can be useful to use a lubricant or cutting oil to increase the life of the die.

Knockouts are available in standard conduit fitting sizes, but special sizes for 22mm and 30.5 mm pushbuttons are also important for technicians in a panel shop. It is important to note that the dimension listed for the punch may not be the actual hole size; a ¾" conduit punch actually makes a hole 1.1" in diameter for ¾ nominal size conduit. There are also knockout dies for square devices, such as 1/16 DIN controllers.

Figure 297 shows a hydraulic punch driver kit from Greenlee. Hydraulic drivers have a handle that is pumped to turn the shaft, these can save a lot of time. Battery operated units are also available.

It is important when operating a punch not to tighten the draw stud too quickly, especially for smaller diameter studs, as they can break.

Figure 297 - Greenlee Hydraulic Punch Kit

Ferrules are used to reinforce the ends of stranded wires. They are small, deformable metal tubes, usually made of tin-plated copper. The tube is compressed with a crimping tool before inserting it into terminals. Ferrules are sized for different wire gauges and ensure that a good connection is made. They are available in different lengths and sleeves come in a variety of colors. Some ferrules have wider openings to allow 2 wires to be crimped together.

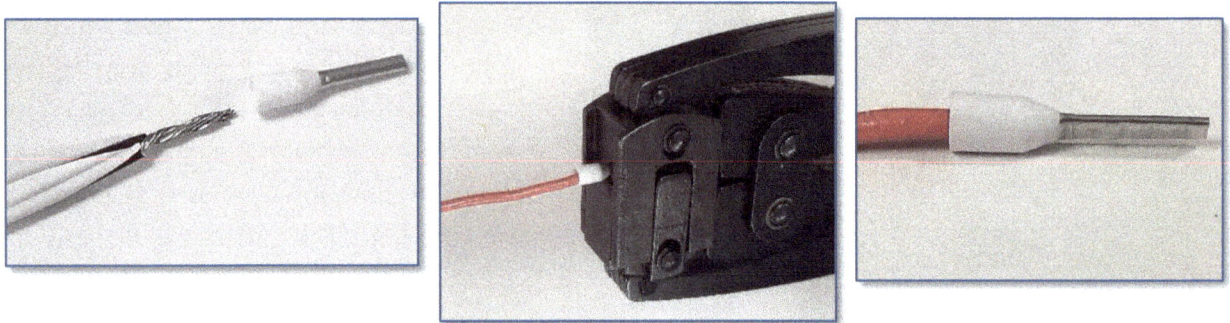

Figure 298 - Ferrules and Crimpers

The tool used to crimp the ferrule in the pictures above is an expensive ratcheting type, but decent results can be obtained by using pliers or vise grips.

Before terminating wires into terminals or terminal blocks they must be **labeled**. There are a wide range of options as shown in Figure 299.

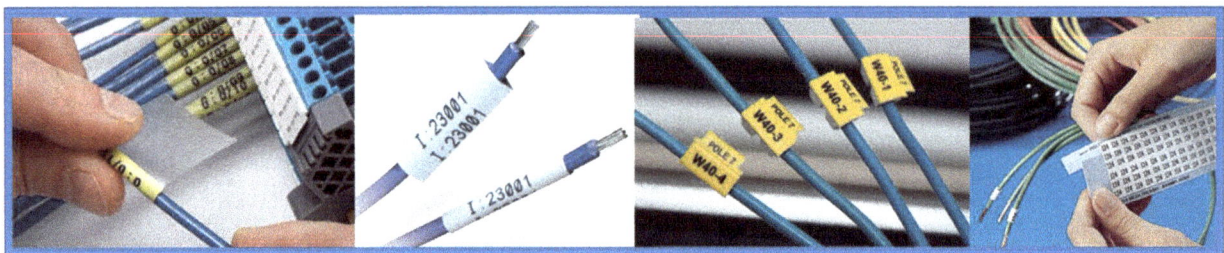

Self-Laminating   Heat Shrink   Clip-On   Adhesive

Figure 299 - Wire Label Types

Self-laminating labels have a clear cover that wraps over the print. This ensures that the ink doesn't smear. Heat shrinkable labels are slipped over the end of the wire and then hot air is blown over the label to shrink it for a snug fit. Clip-on labels are more easily removed from the wire if needed, while the simple adhesive labels are available in sheets or small booklets.

The adhesive labels are not typically used for control wiring but are sometimes used in the field by electricians. The first three types can be printed using special printers designed for the purpose. The first two types can also be printed on sheets in a regular inkjet or LaserJet printer.

Figure 300 - Brady Label Printers: Left - S3000,   Center - BMP21 Handheld,   Right - A5500

Figure 300 shows several printers made by Brady, a major manufacturer of labeling solutions.

Electricians and electrical technicians in an industrial environment can be called on to perform many different types of tasks. They work with large gauge wire and cables when servicing motors and power distribution, and small or fine wire when working with controls. While many drawings include full details on wiring, sometimes a technician needs to make decisions on the size and type of wire to use in a specific application. The **National Electrical Code** (**NEC**) as specified in NFPA 70 provides more details on wiring techniques and regulations. Appendix A in the back of this book shows the ampacity for different gauges of wire, and Appendix B lists the full load current and fuse or breaker sizing for motors. These are meant to be a reference for technicians, but the National Electrical Code supersedes all tables in this book.

***NFPA** is the acronym for the **National Fire Protection Association**. It is an international nonprofit organization that publishes codes and standards to minimize injury, death and property loss.

Below are listed some of the recommended colors for wiring in an industrial plant in the United States:

24 VDC:	Positive – Blue	Negative/0 – Blue/White Stripe
24 VDC Sensors:	Positive – Brown	Negative/0 - Blue
120 VAC:	Line - Red Neutral – White Ground – Green	
208/240 VAC:	Phase 1 – Black Phase 2 – Red Phase 3 – Blue Neutral – White	
277/480 VAC:	Phase 1 – Brown Phase 2 – Orange Phase 3 – Yellow N - Gray	
Externally powered circuits: Yellow		

Exercise 37

1. List some of the tools an electrician or electrical technician should have in their tool kit:

   _____

   _____

   _____

   _____

   _____

   _____

   _____

2. What tool is usually used to make a hole in an enclosure door for a touchscreen?

   _____

3. What is the purpose of a ferrule? _____

4. List some of the types of labels available for wire:

   _____   _____

   _____   _____

5. What is the main source that should be consulted concerning electrical standards and regulations? _____

6. A 10 Horsepower, 240 VAC motor needs to be wired to a motor starter in a control panel. What gauge and color wire should be used? (Use Appendix B in the back of the book)

   _____

Understanding Schematics and Documentation

To diagnose problems and work on automated equipment in an industrial facility a technician needs to be able to read and understand electrical and pneumatic schematics or drawings. One problem with this is that there are several different types of drawings with different standards for each!

NEMA and IEC Symbology

For electrical schematics, the United States generally follows NEMA (National Electrical Manufacturers Association) symbology while much of the rest of the world has adopted the IEC (International Electrotechnical Commission) recommendations.

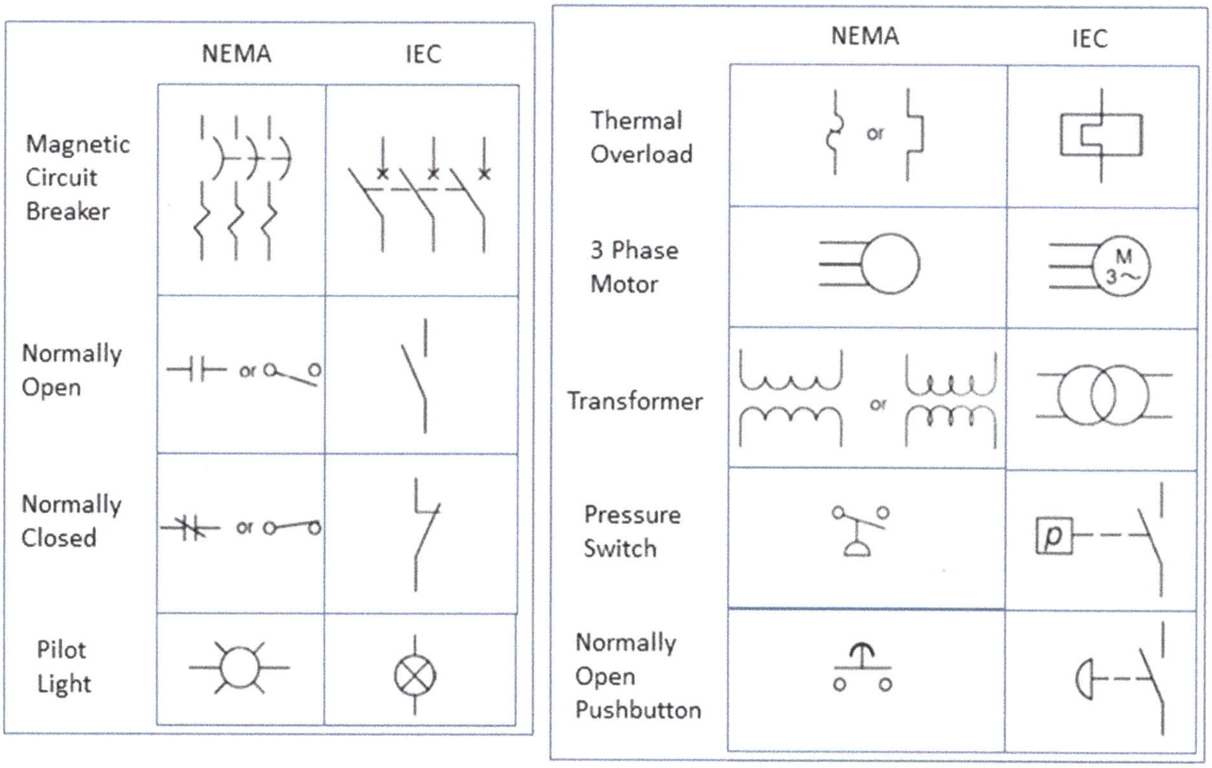

Figure 301 - NEMA vs. IEC Symbols

Figure 301 shows some of the differences between NEMA and IEC symbols. Because of the global nature of machinery production, technicians often need to learn both.

In addition to the symbols, device designations can also differ across different drawing sets. Fortunately, most electrical schematic drawing packages include a key or list of designations and symbols. Device designations include an abbreviation (PRX for proximity switch, MTR for motor, etc.) and a numerical identifier. The number can refer to several different things; sometimes it refers to the assembly number as described in the P&ID section of this book. In

OEM equipment, sometimes designers simply number sensors starting at 1, such as PRX1, PRX2.

The "Sweet Machine"

The following examples are taken from an Automation NTH trade show demo called the "Sweet Machine". It uses a robot to dispense candy. The project number is used as the filename for the electrical CAD drawings, and each sheet is saved as a separate file.

DRAWING INDEX

FILENAME	SHT	REV	DWGDESC	DWGDESC2	DWGDESC3
P2104014-000	000	0	TITLE SHEET		
P2104014-001	001		DRAWING INDEX		
P2104014-005	005	0	DEVICE DESIGNATIONS		
P2104014-006	006	0	WIRING NOTES		
P2104014-007	007	0	GROUNDING METHODS		
P2104014-008	008	0	DRAWING SYMBOLS		
P2104014-010	010	0	ONE LINE DRAWING		
P2104014-200	200	0	240VAC POWER DISTRIBUTION		
P2104014-230	230	0	EPSON VT6	ROBOT	CONTROLLER
P2104014-231	231		EPSON VT6	ROBOT	INTERCONNECT
P2104014-300	300	0	120VAC POWER DISTRIBUTION		
P2104014-400	400	0	24VDC POWER DISTRIBUTION		
P2104014-401	401	0	24VDC POWER DISTRIBUTION		
P2104014-405	405	0	24VDC POWER DISTRIBUTION	SAFETY DEVICES	
P2104014-409	409	0	24VDC POWER DISTRIBUTION	SAFETY DEVICES	
P2104014-410	410	0	UPPER	CONVEYOR	
P2104014-412	412	0	LOWER	CONVEYOR	
P2104014-420	420	0	LEFT ELEVATOR	CONVEYOR	
P2104014-421	421	0	RIGHT ELEVATOR	CONVEYOR	
P2104014-500	500	0	FRONT UPPER	GUARD DOOR	SAFETY CIRCUIT
P2104014-501	501	0	LOCKING	GUARD DOOR	SAFETY CIRCUIT

Figure 302 - Sweet Machine Drawing Index

The electrical drawings will usually have an index that lists the pages in the drawing set. Figure 302 shows the first section of pages in the index. There are 58 pages in this particular electrical drawing set. Sheet numbers are grouped in series, the main power and robot drawings are in the 200 series.

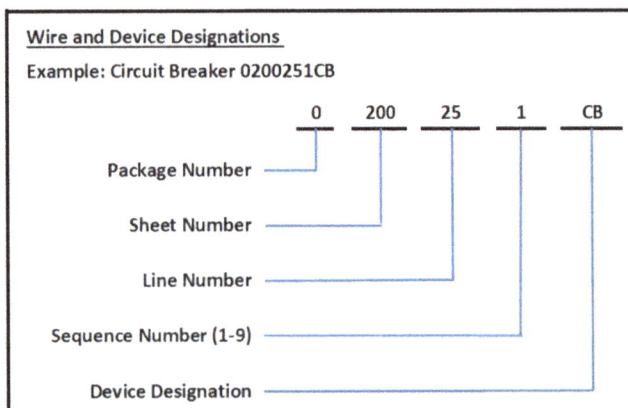

Wire and Device Designations

Example: Circuit Breaker 0200251CB

0 200 25 1 CB

Package Number

Sheet Number

Line Number

Sequence Number (1-9)

Device Designation

Figure 303 - Device Designation

Figure 303 shows a common method of using the drawing sheet number and line number where the device is drawn in the schematic set. This method is used in both IEC and NEMA drawing packages and makes it easy to locate the device.

The circuit breaker for the robot is designated 200251CB, the zero is omitted.

Example: Robot Circuit Breaker

The circuit breaker for the robot is found on sheet 200 which is the 240 power distribution sheet. The lower right corner of the drawing set is where the Title Block is located.

CUSTOMER: AUTOMATION NTH 282 Mason Road Lavergne TN 37086	PROJECT: AUTOMATION NTH SWEET MACHINE		
DRAWN BY: M. GOSLIN	TITLE: 240VAC POWER DISTRIBUTION		
ENGR APPD: J. BUCK			
DATE: 6-22-2021			
FILE NAME: P2104014-200		SHEET: 200	REV: 0

Figure 304 - Sheet 200 Title Block

Title blocks contain information about the customer, the people who drew and approved the drawings, revisions and more.

The device designations are usually found in the drawings, on tags or terminal blocks in the panel, and near each device in the field or on the machine.

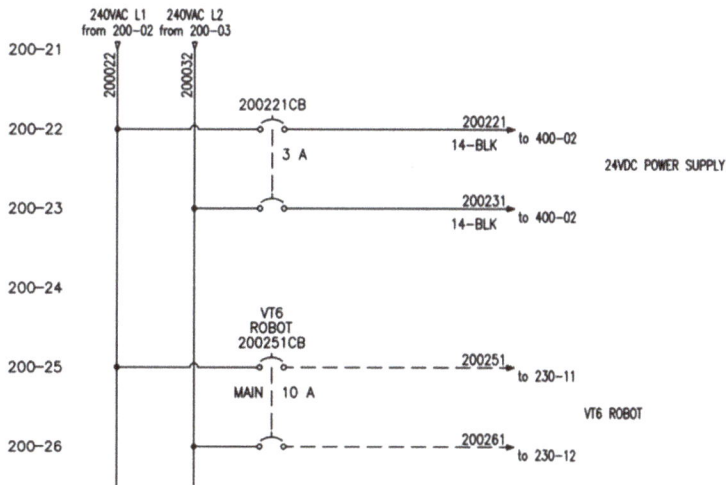

Figure 305 - Circuit Breakers in CAD Drawing (Sheet 200)

Figure 305 shows the area on Sheet 200 where the circuit breakers are located. In addition to the device number, other pertinent information can be found here. In this case, the amperage of the circuit breaker and the wire numbers are of particular interest to a technician.

Wire numbers are designated using the same method, using the sheet and line where the wire originates. Off-page connectors are also shown to follow the power flow.

Figure 306 shows a picture of the circuit breaker with a device label at the top on the backplane. Wire labels are also shown wrapped around the incoming and outgoing conductors. Both ends of wires should always use the same number.

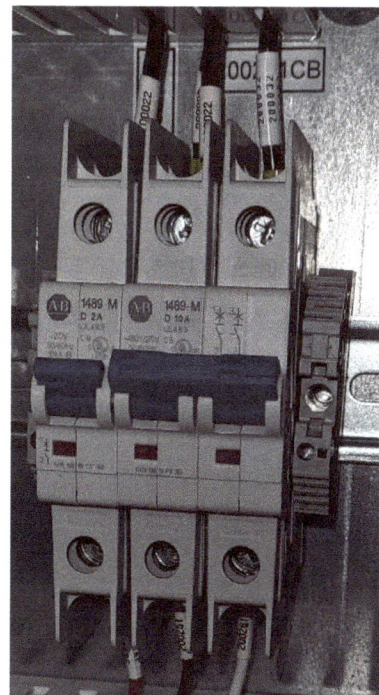

Figure 306 - Circuit Breaker in Panel

Sheet 820 in the Drawing Index is listed as Main Enclosure Interior Layout. Figure 307 shows the circuit breaker's location on the backplane, along with the number 134. This number references the Bill of Material (BOM), which is listed as Sheets 890 and 891 in the Index.

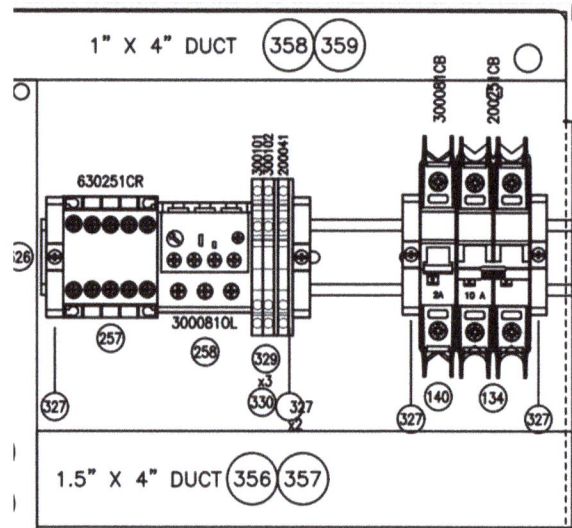

Figure 307 - Enclosure Interior Layout (Sheet 820)

Item 134 shows the manufacturer, part number and description of the circuit breaker. This can be used to order spare parts for the machine.

129	1	ALLEN BRADLEY	194R-N30-1753-PYN1	DISCONNECT SWITCH, NON-FUSED, 30A, 600VAC, 250VDC, 3 POLE
130	3	ALLEN BRADLEY	194R-30-MTL3	TERMINAL LUGS, MULTI-TAP, 14-4 AWG, FOR 194R DISCONNECT, L(
133	1	ALLEN BRADLEY	1489-M2D030	MINIATURE CIRCUIT BREAKER, 2 POLE, 3 AMP
134	1	ALLEN BRADLEY	1489-M2D100	MINIATURE CIRCUIT BREAKER, 2 POLE, 10 AMP
140	1	ALLEN BRADLEY	1489-M1D020	MINIATURE CIRCUIT BREAKER, 1 POLE, 2 AMP
141	1	ALLEN BRADLEY	1489-M1C080	MINIATURE CIRCUIT BREAKER, 1 POLE, 8 AMP
146	1	PULS	CP20.241	POWER SUPPLY, 24VDC, 20A, 100-240VAC, 50-60HZ, 4.26/2.23A
148	2	ETA	EM12D-TIO-000-DC24V-40A	ELECTRONIC 40A SUPPLY MODULE WITH IO-LINK
149	4	ETA	REX12D-TE2-100-DC24V-1A	ELECTRONIC CB 2-CH 24VDC 1A-10A VARIABLE

Figure 308 - Bill of Material (BOM)

The information obtained from this example can be summarized as follows:

- Component location in electrical drawings by sheet number, also shows connected devices and ratings.
- Component location in enclosure by interior layout. Also used to find item in BOM.
- Component location in enclosure located by printed label
- Component part number in Bill of Materials

Additionally, manufacturer's "cut sheets" are sometimes included in operations or maintenance manuals.

Drawing Package and Template

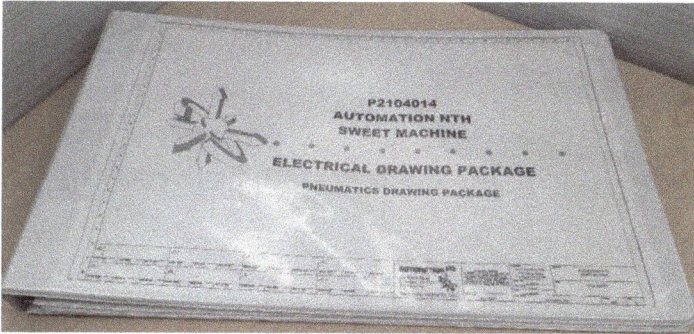

Figure 309 - Sweet Machine Drawing Binder

Electrical schematics are usually printed on 11" x 17" format paper (size B) in landscape format. This allows pages to be folded and put in a standard 3 ring binder or put in a larger binder as shown in Figure 309.

Sheet 000 of this drawing set is the title page, shown inserted into the binder cover in the figure above. Sheet 001 is the Drawing Index, then several sheet numbers are skipped to leave room for larger drawing packages.

Sheet 005 has a list of device abbreviations, including not only the physical items but also for Advanced/Returned, Normally Open/Closed and more. Sheet 006 has Wiring Designations as shown in Figure 301, along with color codes for wire, wiring practices and various notes for general wiring and special conditions.

Sheet 007 shows grounding methods for the system ground and equipment bonding, including detailed instructions on attaching ground wires to the stud on a backplane or welded to the enclosure.

Sheet 008 lists symbols used in the drawing package. As mentioned previously, NEMA and IEC symbology is different, Automation NTH standards use NEMA symbols.

The drawing set for electrical schematics includes these sheets in every package, so this serves as a starting template before customizing drawings for a specific panel.

Single Lines (SLDs) or One Lines

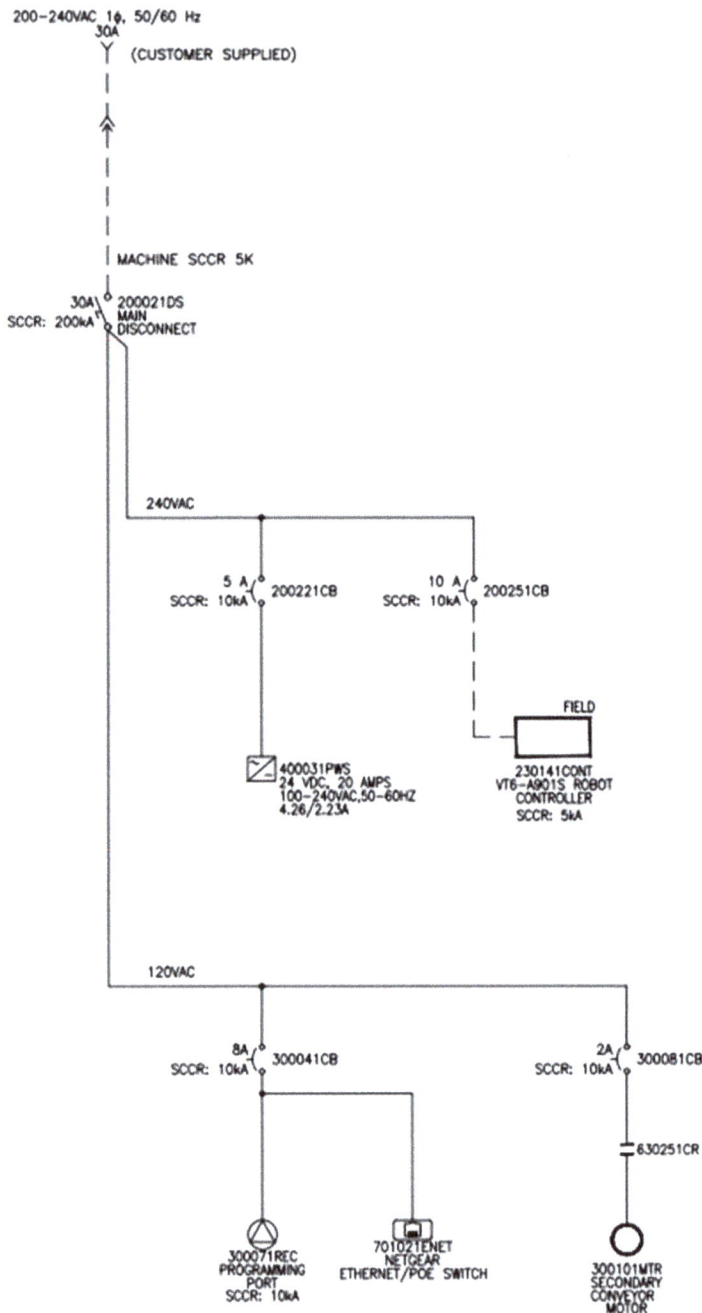

200-240VAC 1φ, 50/60 Hz
30A
(CUSTOMER SUPPLIED)

MACHINE SCCR 5K

30A 200021DS
SCCR: 200kA MAIN DISCONNECT

240VAC

5 A 200221CB
SCCR: 10kA

10 A 200251CB
SCCR: 10kA

FIELD

400031PWS
24 VDC, 20 AMPS
100-240VAC,50-60HZ
4.26/2.23A

230141CONT
VT6-A901S ROBOT
CONTROLLER
SCCR: 5kA

120VAC

8A 300041CB
SCCR: 10kA

2A 300081CB
SCCR: 10kA

630251CR

300071REC
PROGRAMMING
PORT
SCCR: 10kA

701021ENET
NETGEAR
ETHERNET/POE SWITCH

300101MTR
SECONDARY
CONVEYOR
MOTOR

Figure 310 - One-Line Drawing, Sheet 010

Single Line or **One Line** drawings are used to visualize the power distribution of the whole machine. Rather than showing every wire or phase, these drawings simplify circuits while showing all the major elements. Sheet 010 of the Drawing Index is titled One Line Drawing.

Abbreviated SLD, a single-line diagram or drawing is a primary resource to calculate short circuit currents and serves as a basic design and safety document. The National Electric Code NFPA 70 recommends that SLDs show all electrical equipment in the power system and gives all pertinent ratings for voltage, frequency, transformer impedance, available short-circuit currents and more.

SCCR is an abbreviation for **S**hort **C**ircuit **C**urrent **R**ating, which is the amount of fault current at a nominal voltage that an apparatus or system can safely withstand without causing a shock or fire hazard.

The SCCR for the entire control panel is determined by the lowest SCCR value for any component or branch circuit.

Power Distribution

The next sheet in this drawing package is the 240VAC Power Distribution page mentioned previously. The main disconnect and incoming power connection is designated in the upper left corner of the sheet, starting at the first line.

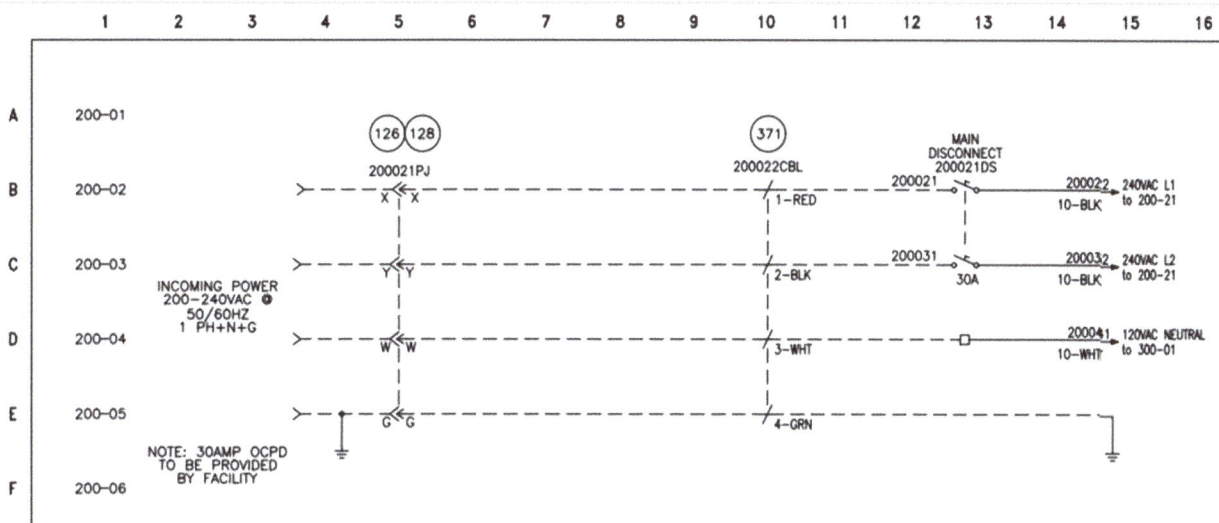

Figure 311 - Incoming Power and Disconnect

OCPD is an abbreviation for **O**ver**c**urrent **P**rotection **D**evice. The drawing states that this will be provided by the facility, presumably in the form of a circuit breaker.

The main disconnect for this panel is shown just to the left of the red handle in Figure 312. This disconnect is not fused, but a similar version for fuses is also available.

The color and gauge (#10AWG) of the incoming power is specified in the drawing. Notice that the Neutral wire for 120VAC is not connected through the disconnect for safety reasons. It is continued on Sheet 300.

Figure 312 - Main Disconnect

The devices next to the PULS DC power supply with the green lights are identified in the BOM of Figure 308 as "Electronic 40A Supply Module with IO-Link" (Item 148) and "Electronic CB 2-CH 24VDC 1A-10A Variable". More on this will be discussed later in this section.

120 VAC Power

Sheet 300 shows the distribution of 120VAC power. A neutral wire is provided as shown in Figure 311, it does not run through the disconnect.

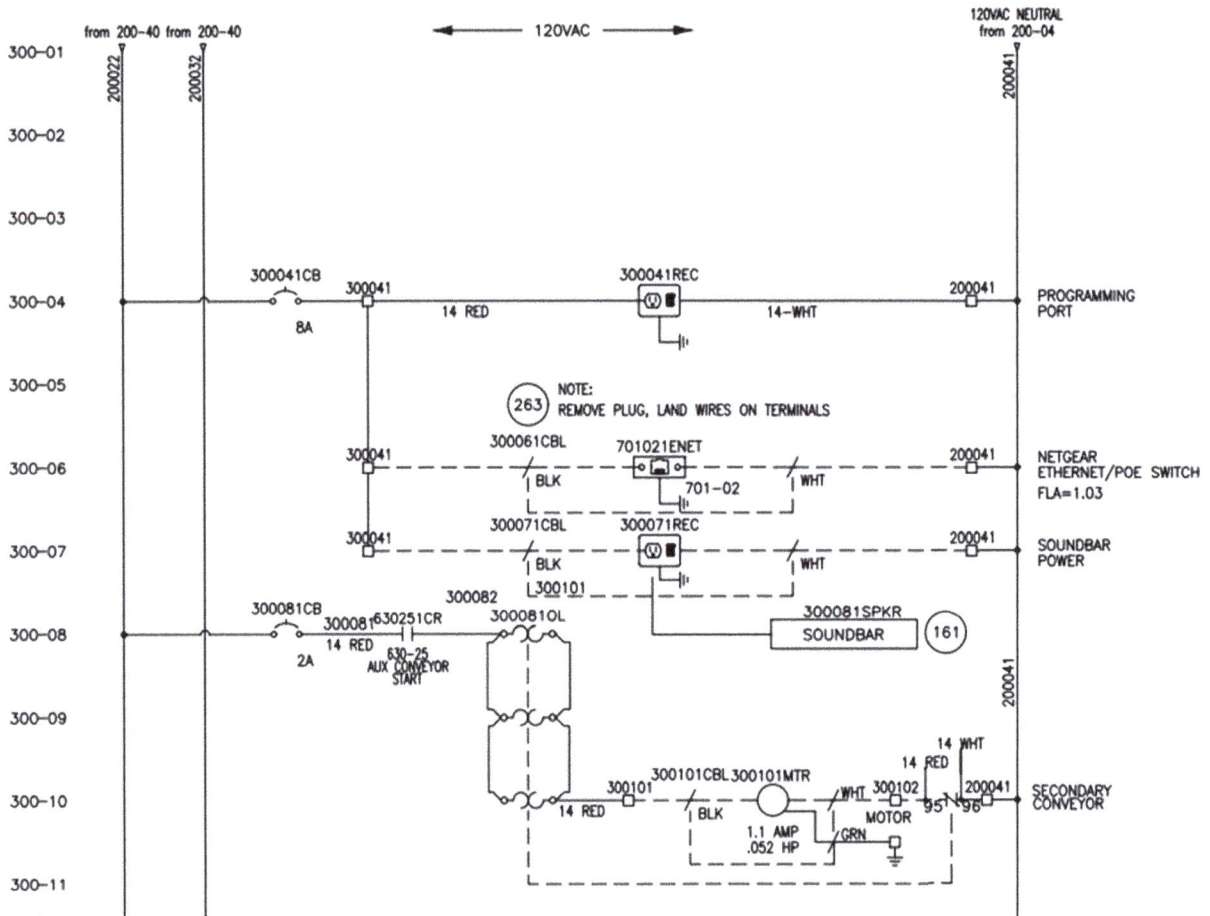

Figure 313 - 120VAC Power Distribution

Figure 314 - Programming Port and Receptacle

Line power is provided from one leg of the 240VAC supply through the two circuit breakers. One breaker supplies the programming port receptacle on the enclosure door, which is to be used ONLY to power a laptop or computer. This is not to be used for power tools! A NetGear ethernet switch and a powered Bluetooth speaker are also on the 8 Amp breaker 300041CB.

The small AC powered conveyor that runs through the "Black Box" is connected to the second 2 Amp breaker 300081CB. The 300081OL overload device is in the panel next to the contactor 630251CR.

IO-Link

IO-Link is an industrial communications networking standard defined in IEC 61131-9. It is used to connect sensors and actuators to either an industrial fieldbus such as Profibus or DeviceNet, or to an industrial ethernet network. A system consists of an IO-Link master and one or more IO-Link devices.

Cabling between the master and the devices consists of three or five conductor cables carrying 24VDC power and using either 4 or 5 pin connectors to connect to the ports on the master unit. The cable also carries data and can act as a simple digital input or output compatible with standard sensors, or can be configured to transfer process data, value status, device data or event data.

Figure 315 - IO-Link Network

Figure 315 shows an IO-Link Master connected to various input and output devices. The PLC communications link in this case is Ethernet/IP and is connected to an ethernet switch.

Remote I/O blocks, intelligent sensors, valve manifolds and "smart" stack lights can be connected to the IO-Link Master. The circuit breakers and controller act as one module to the IO-Link Master.

Node addresses are assigned using software from the device manufacturer. In the case of the Sweet Machine, the devices are from IFM Efector, and the online and offline parameter setting software is called **LR Device**.

Figure 316 shows the IO-Link Master on the Sweet Machine with 3 additional IO-Link remote I/O blocks next to it. The green cable in the Master is connected to the Ethernet/IP network in the main enclosure, which in turn is connected to the PLC.

Figure 316 - IO-Link Master and I/O Blocks

DC Power

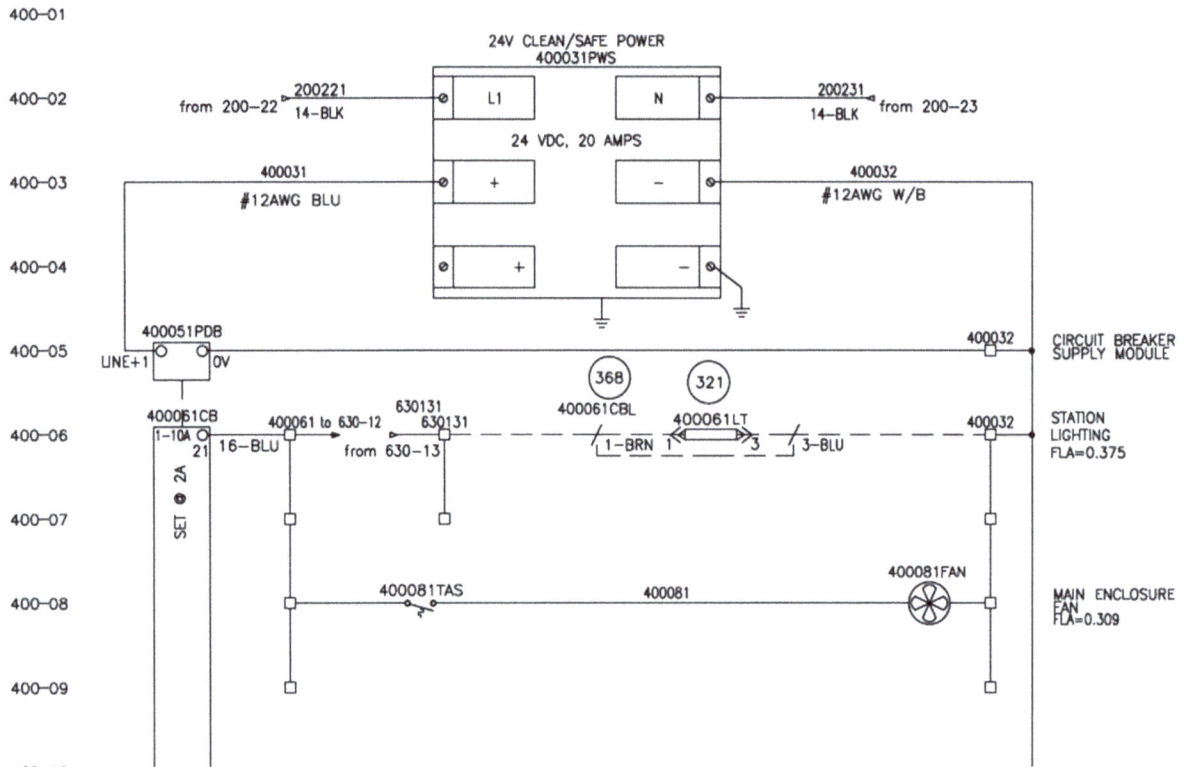

Figure 317 - DC Power Distribution

DC Power is distributed from the power supply through the electronic circuit breakers shown in Figure 318. There are 6 breaker circuits shown on sheets 400 and 401, each of the breaker circuit trip points can be independently set with messaging from the PLC. The breakers are configured with two circuits to each module as shown.

400031 is wired into the circuit breaker supply module and is distributed in the following circuits:

400061 – Station Lighting and Fan – 2A

400151 – Ethernet switch, HMI, Light Strip – 5A

400221 – PLC MOD and SA power, PLC Remote Rack, Safety circuit – 5A

400311 – IO-Link Remote Master Sensors, IO-Link Remote Actuator Supply – 5A

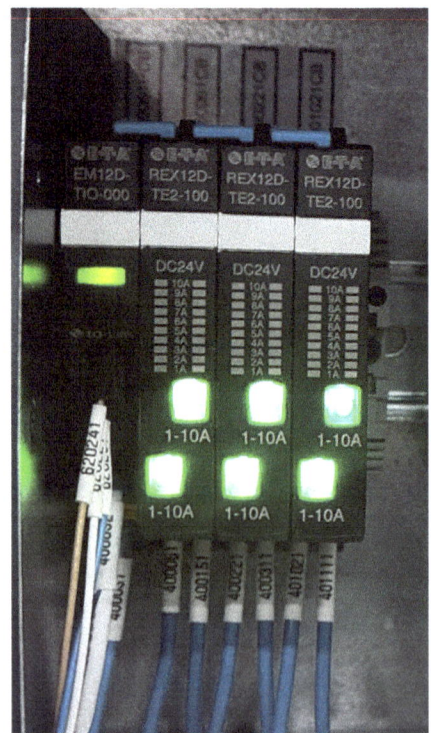

Figure 318 - Electronic Circuit Breakers

401021 – Upper and Lower Conveyor Power, Left and Right Elevator Conveyor Power – 6A

401111 – Robot Safety Circuit, Hazardous Power – 5A

The IO-Link Master itself receives power from circuit 400311 as shown in Figure 317.

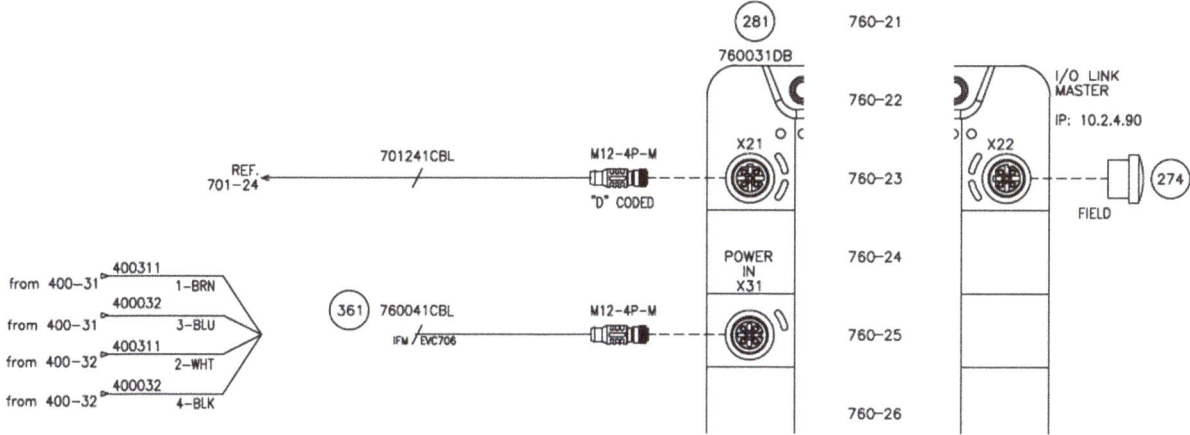

Figure 319 - IO-Link Master Power (Sheet 760)

Communications to the circuit breaker module is done via an IO-Link capable input module, 1734-4IOL. This module is part of the Point I/O rack shown in Figure 320. Since Point I/O is connected to the processor by Ethernet/IP, the full functionality of IO-Link is available.

Figure 320 shows the IO-Link Input module, notice the more typical standard 8 point input module (1734-IB8) next to it. Wiring connections to these modules use spring clips. A small screwdriver is inserted into the rectangular hole, the stripped wire is inserted into the round hole, and the screwdriver tip is removed, clamping the wire.

Figure 320 - 1734-4IOL IO-Link Input Module

Figure 321 shows the schematics for the circuit breaker connection. Pin 2 acts as both an input point and a communications link to the breaker, while pins 1 and 3 provide power to the small processor in the Supply Module.

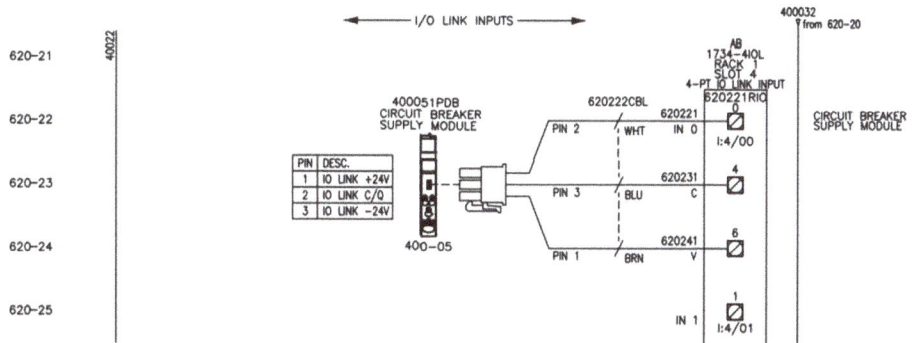

Figure 321 - Sheet 620 - Circuit Breaker Connection for IO-Link

Safety Wiring

In the **Safety System** part of this book, it was mentioned that **Safety PLCs** are sometimes used instead of individual safety controllers. The Sweet Machine uses a Compact GuardLogix 5380 Safety Controller. This is part of the CompactLogix family of PLCs and is programmed with RSLogix Studio 5000 software.

Figure 322 - Sweet Machine Safety PLC

Allen-Bradley safety components are easily identified by being red in color. The PLC itself is shown on the left in Figure 322, and the red modules to the right (1-3) are safety rated I/O modules.

Sheet 600 of the schematics shows the power connections for the PLC. The MOD power connection is used for system-side module power, including the processor itself.

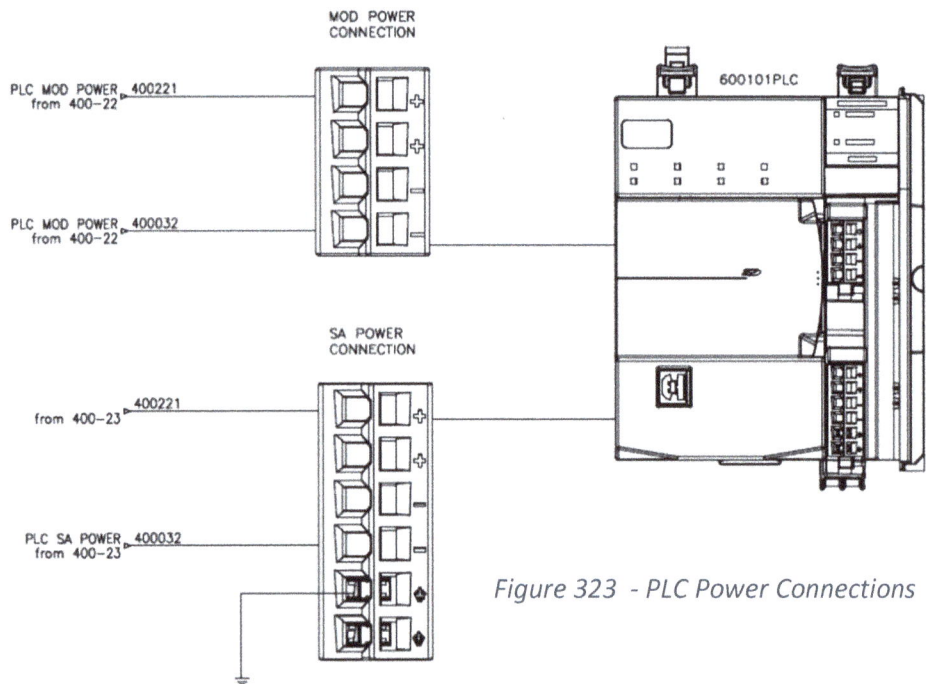

Figure 323 - PLC Power Connections

The SA (Sensor/Actuator) power connection provides field side power. Both provide power across a bussed connection to other modules connected in the rack.

The Sweet Machine does not have any I/O modules connected directly to the processor, but if they did, power for operating the modules would be passed through the MOD power bus connection. This is required to transfer data and execute logic. The SA bus is isolated from the MOD bus and can be separated further by using Field Potential Distribution (FPD) devices. These create multiple SA power buses, allowing power to be removed from modules using safety relays.

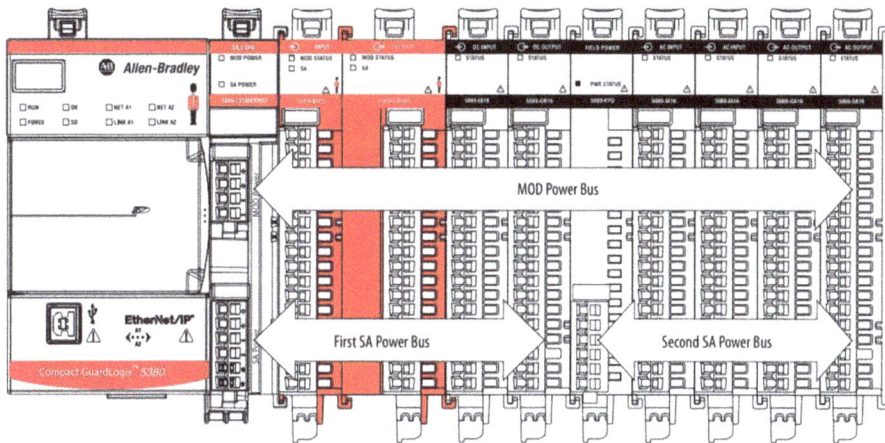

Figure 324 - PLC Power Buses

Safety devices are connected to the three red remote I/O modules shown in Figure 322. Modules 1 and 2 are safety rated input modules, and module 3 is a safety output module. As with the Category 3 and 4 safety circuits described in the Safety section of this book, all devices are dual channel and control-reliable.

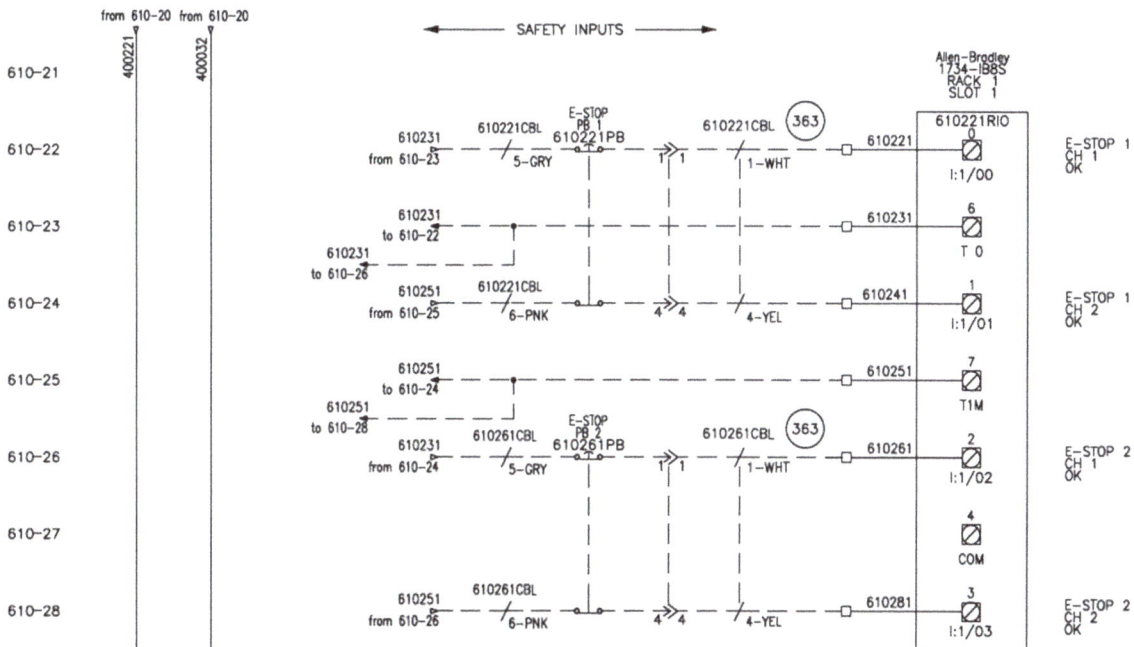

Figure 325 - E-Stop Circuits (Safety Inputs) Sheet 610

Like the safety controllers described in the safety circuit section, the channel circuit is not standard 24VDC voltage but a pulsed signal when configured as "Pulse Test". The pulse drops to zero volts every 144 milliseconds for a period of 525 microseconds. There are two **test outputs**, labeled T0 and T2, and two test outputs with muting capability, labeled T1M and T3M. The muting outputs allow testing of muting lamp status in addition to acting as a standard output, a continuous pulse test or as a power supply for safety devices.

Test outputs in combination with safety inputs allow short circuit, cross-channel, and open circuit fault detection. This works in combination with the safety outputs of module 3 to turn off the safety relays that supply power to output cards connected to actuators.

While safety input cards allow for the wiring and logic associated with dual channel input devices, the safety output card of module 3 allows for wiring of redundant output devices such as safety relays.

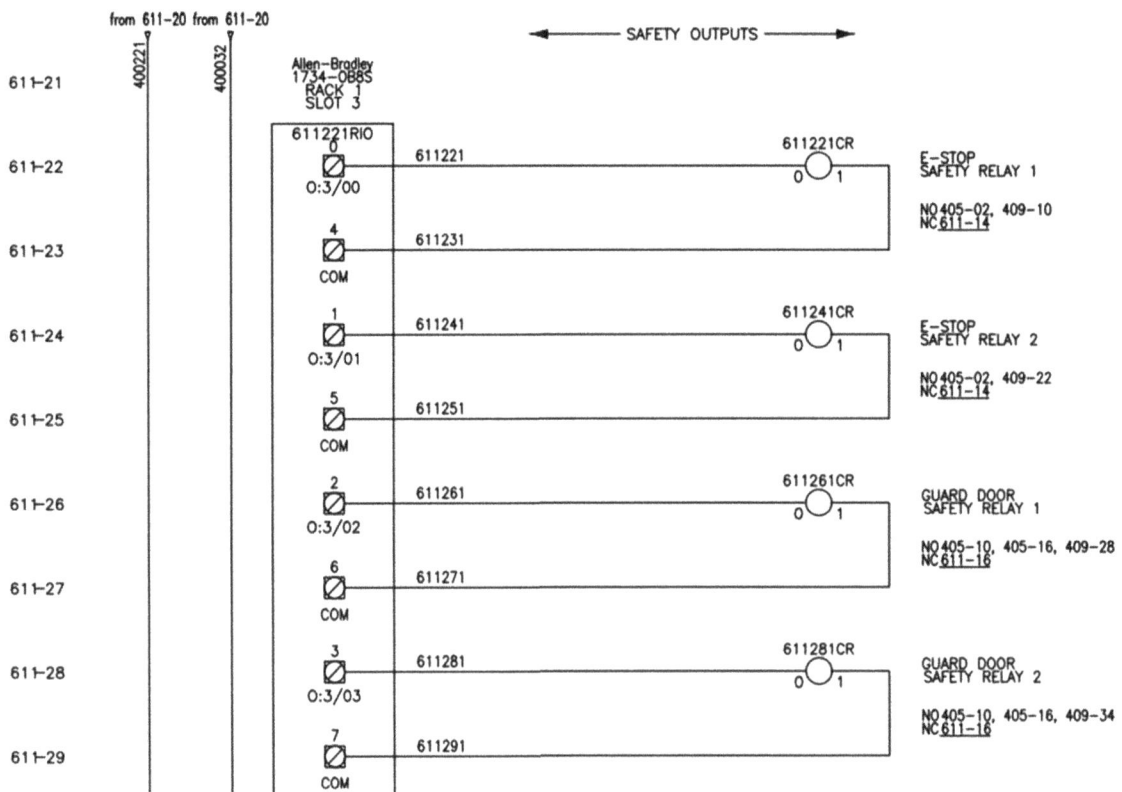

Figure 326 - Safety Relays (Safety Outputs) Sheet 611 Top Right

As with the Category 3 and 4 safety circuits described in the safety section of this book, since everything is redundant two coils are required for disconnecting power to devices. To make it easier for troubleshooting, notice that the page numbers of the contacts for the safety relays are listed under the names of the coils.

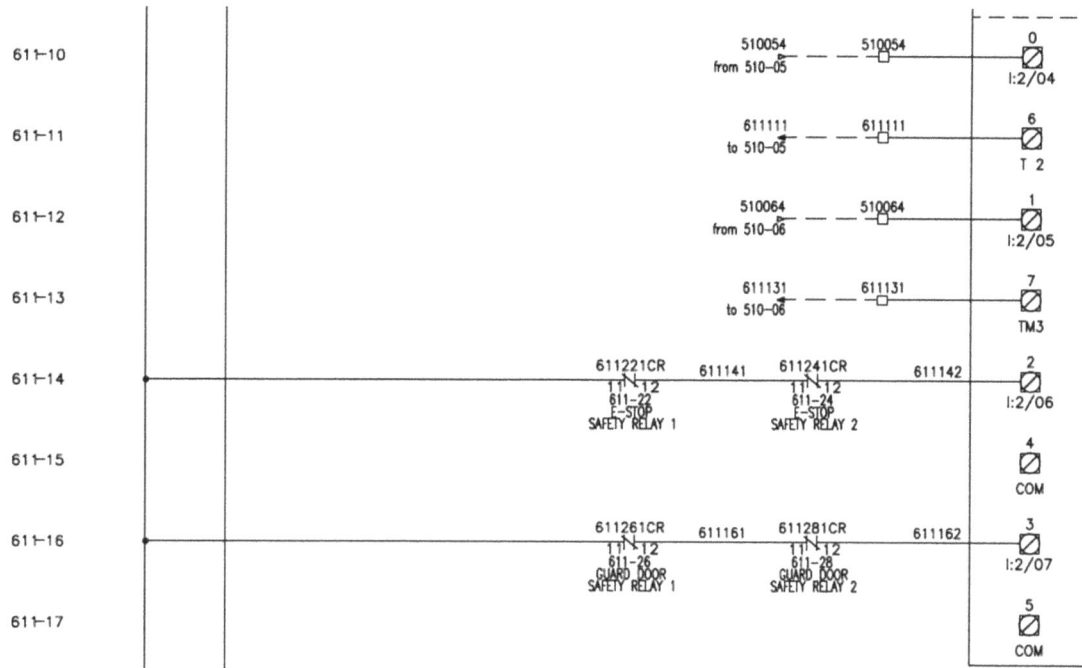

Figure 327 - Contact Monitoring for Safety Relays (Sheet 611 Bottom Left)

Unlike the E-Stop pushbuttons of Figure 325, the test outputs are not used for the monitoring of the coils. The power through the contacts is from 400221, +DC directly from the breaker.

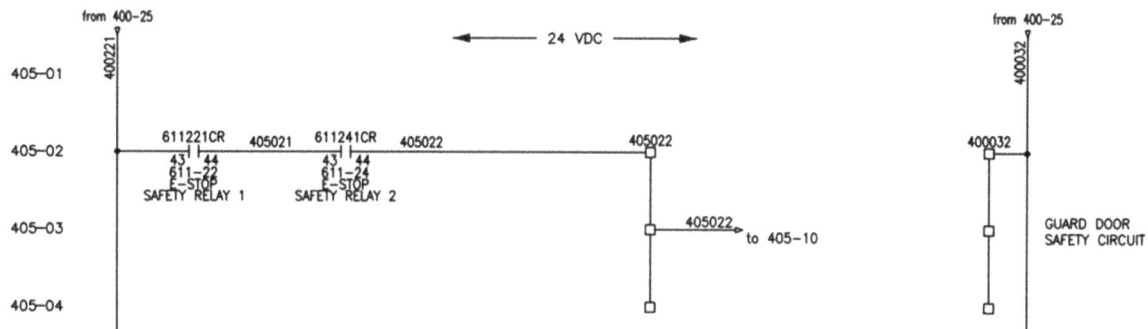

Figure 328 - E-Stop Safety Contacts Sheet 405

Figure 328 shows power flow through the safety relay contacts. If it was desired to stop hazardous motion with only the E-Stops, 405022 could be connected to the power feed of the output card. In the case of the Sweet Machine however, all hazardous power is also routed through the Guard Doors.

The Guard Door safety switches have redundant NC contacts like the E-Stop pushbuttons. There are two magnetic locking guard switches and eight regular safety switches on the sweet machine.

It can be difficult to understand the wiring of safety circuits in general because there are many links from one page to another. In addition, the guard switches used in this application are RFID coded. The coding can be made specific to each switch and actuator, but in the case of the Sweet Machine, all switches accept any actuator.

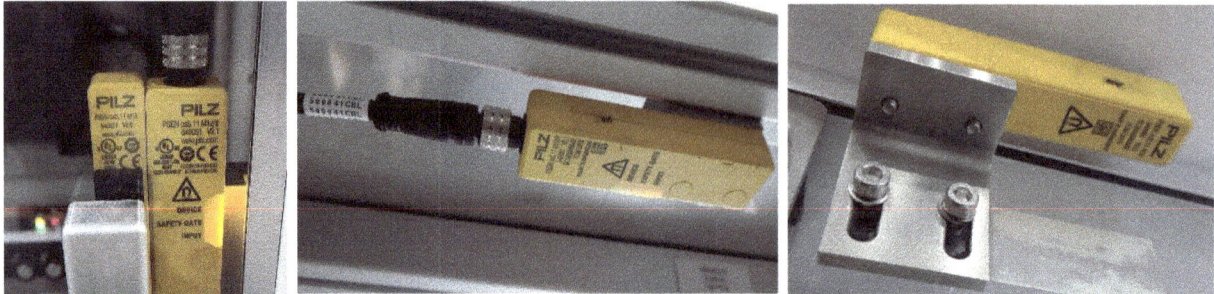

Figure 329 - Left: Door Switch and Actuator, Center: Switch, Right: Actuator

Power to each of the switches is applied independently of the safety circuitry on the A1 and A2 terminals. These **Pilz** brand safety devices all have small processors inside that perform functions like decoding the RFID address and diagnostics.

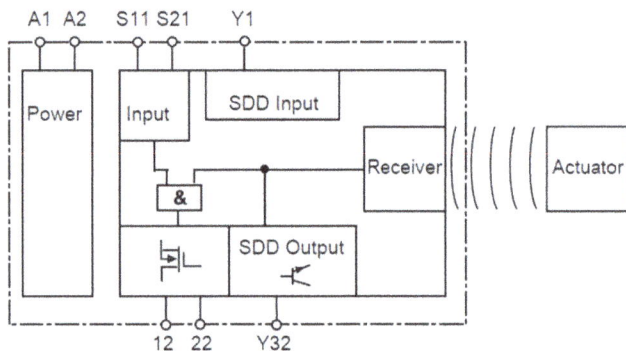

When analyzing a safety circuit, it can be useful to download manuals and diagrams for the devices to understand how they operate. Figure 330 shows a block diagram from the operations manual for the RFID safety switches. The terminal designations for safety devices is common across many manufacturers.

Figure 330 - Pilz Coded Safety Switch

The A1 and A2 terminals are designations for where operating power is applied to devices, A1 being +DC and A2 being -DC or 0 volts. S terminals (S11, S12) are safety inputs and are usually assigned to the incoming side of a series of safety devices. The 12 and 22 terminals are the corresponding output side. Y terminals are typically assigned to "electrically operated output devices", but the SDD input and output terminals for Pilz safety systems are used for **SDD** (**S**afety **D**evice **D**iagnostics) circuits.

These switches also have internal diagnostics to partially lock the switch if one safety input switches from high to low while the other remains high. There are LEDs on the switch that will flash to indicate a problem with the circuit.

The eight safety switches on the doors are listed below:

1. Front Upper Left
2. Front Upper Right
3. Front Lower Left
4. Front Lower Right
5. Rear Left
6. Rear Right
7. Left Elevator
8. Right Elevator

In addition, there are two locking switches to prevent access to the interior of the machine while the robot is operating. These are also Pilz devices that have 500N of magnetic locking force and must be energized to release. There is one at the top of each of the front and rear doors.

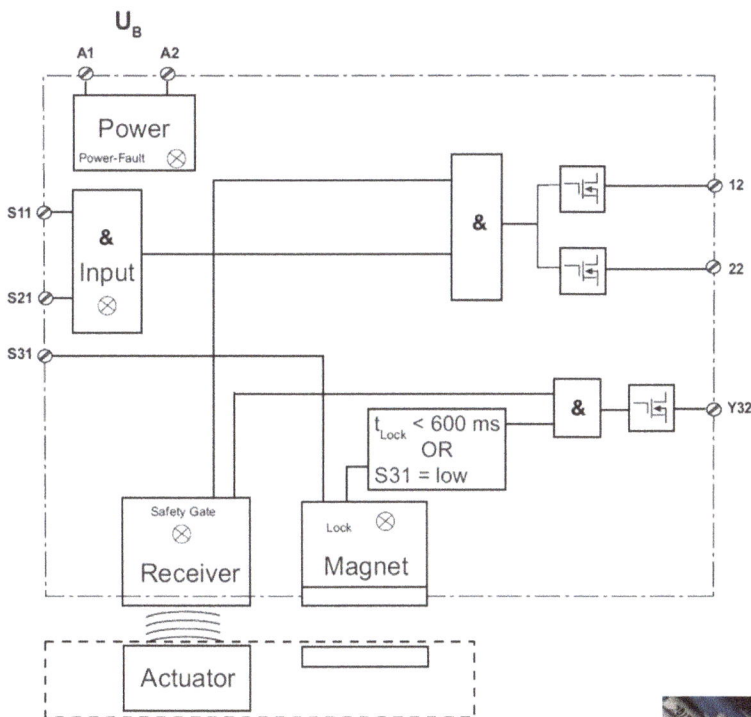

Figure 331 is a block diagram for the magnetic locking switches. The terminal designations are like those of the safety switch described previously with the addition of terminals to release the lock (S31) and monitor whether it is closed and locked (Y32).

All of the dual channel safety circuits are run in series with each other and terminated inside the control enclosure as shown in Figure 332.

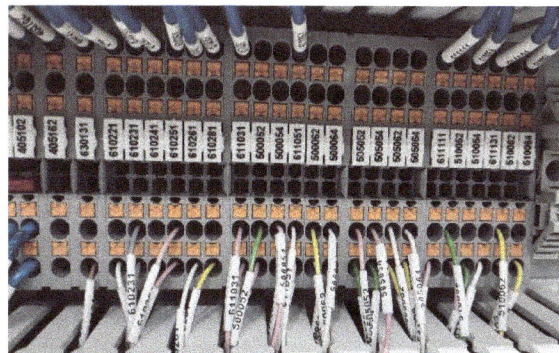

Figure 331 - Locking Safety Switch

Figure 332 - Safety Switch Terminals

Sheet 500 of the electrical schematics shows the front door safety switches in series.

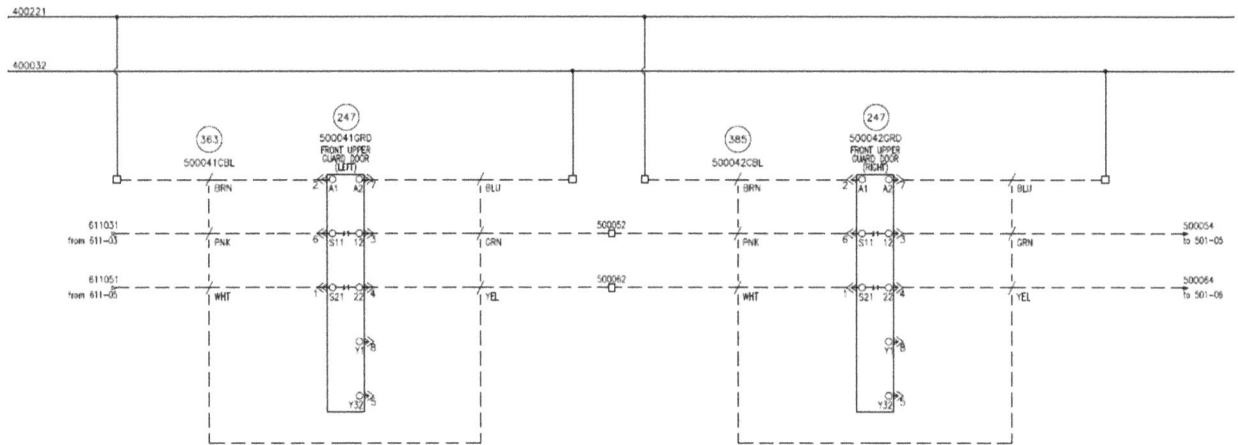

Figure 333 - Sheet 500, Upper Front Door Safety Switches (Top of Door)

Sheet 500 is continued on Sheet 501, where the locking switches are placed in series with the door switches. Both the front and back locking switches are shown.

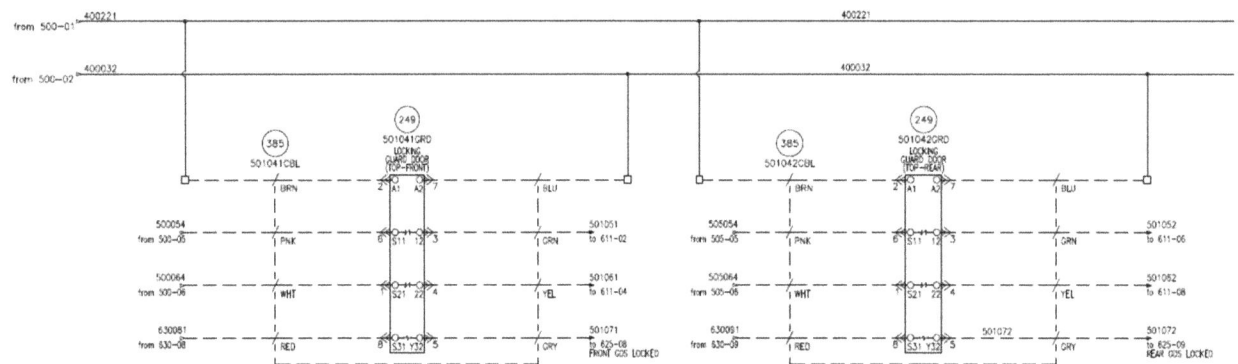

Figure 334 - Sheet 501, Locking Door Switches (Bottom of Left Door)

Like the Emergency Stop pushbuttons shown in Figure 325, the Guard Door circuits begin and end on a safety input card, in this case Sheet 611 which shows both slot 2 (safety inputs) and slot 3 (safety outputs).

Sheet 501 also shows the pinout for the locking door switch. This can be useful when using a multimeter to test electrical signals.

PIN	FUNCTION	TERMINAL	COLOR
1	INPUT, CH 2	S21	WHT
2	+24VDC	A1	BRN
3	OUTPUT, CH 1	12	GRN
4	OUTPUT, CH 2	22	YEL
5	SIGNAL OUTPUT/DIAGNOSTIC OUTPUT	Y32	GRY
6	INPUT, CH 1	S11	PNK
7	0VDC	A2	BLU
8	LOCK/UNLOCK	S31	RED

Figure 335 - Locking Door Switch Cable Pinout

Sheet 502 shows the front lower guard doors. The left and right doors are in series like the upper doors shown in Figure 333 but are not in series with a locking switch. The rear doors are wired in series with a locking switch and are shown on Sheet 505.

The left and right elevator doors are also wired in series with each other and are shown on Sheet 510.

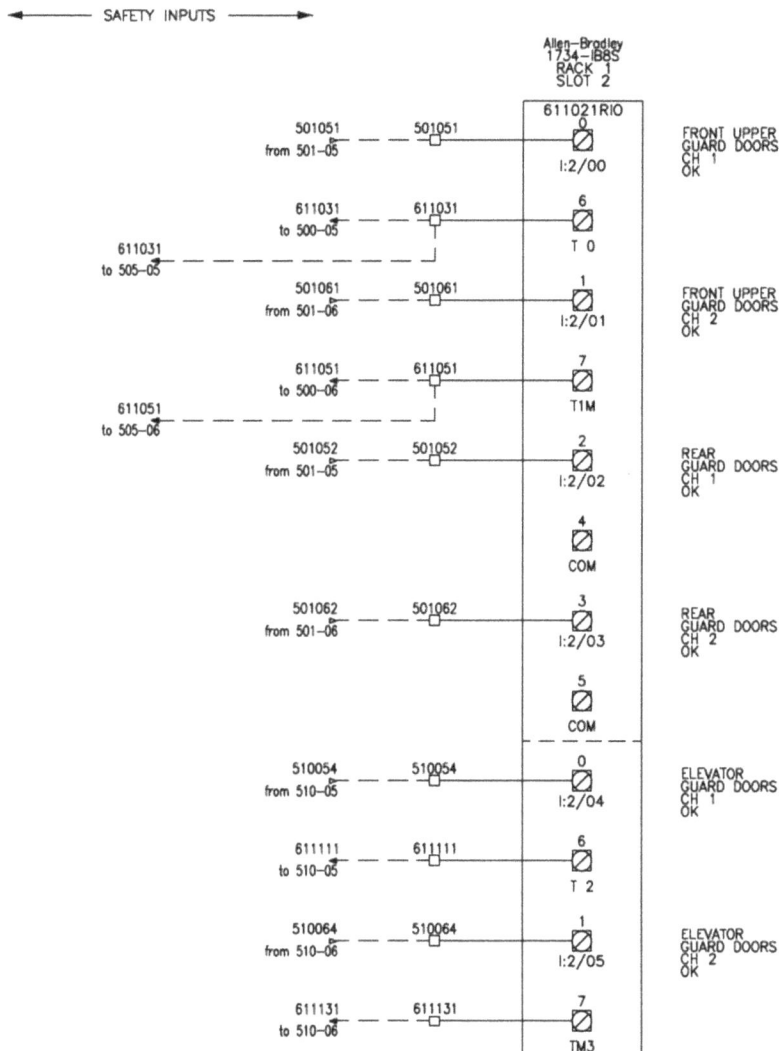

Figure 336 shows the ends of the "safety loops" for the guard doors. As with the Emergency Stops, test outputs are used to make the circuit control reliable, continuously testing the circuit with electrical pulses. Note that T0, T1M, T2, and TM3 all feed the loops for guard doors.

The front lower guard doors are terminated on Sheet 610 at I:1/04 and I:1/05 for the CH1 OK and CH2 OK signals, they use T2 and TM3 as test outputs.

As mentioned previously, the safety inputs and safety outputs of a safety PLC operate together to create a control reliable circuit that can meet Category 4 requirements according to ISO 13849-1. This is accomplished in two important ways:

Figure 336 - Sheet 611, Input Module 2 Guard Door Monitoring

1. Inputs are dual channel and constantly monitored with a "watchdog" circuit (i.e the test outputs)

2. Outputs are dual channel/redundant, safety circuitry is monitored to detect system failure and facilitate a safe shut down.

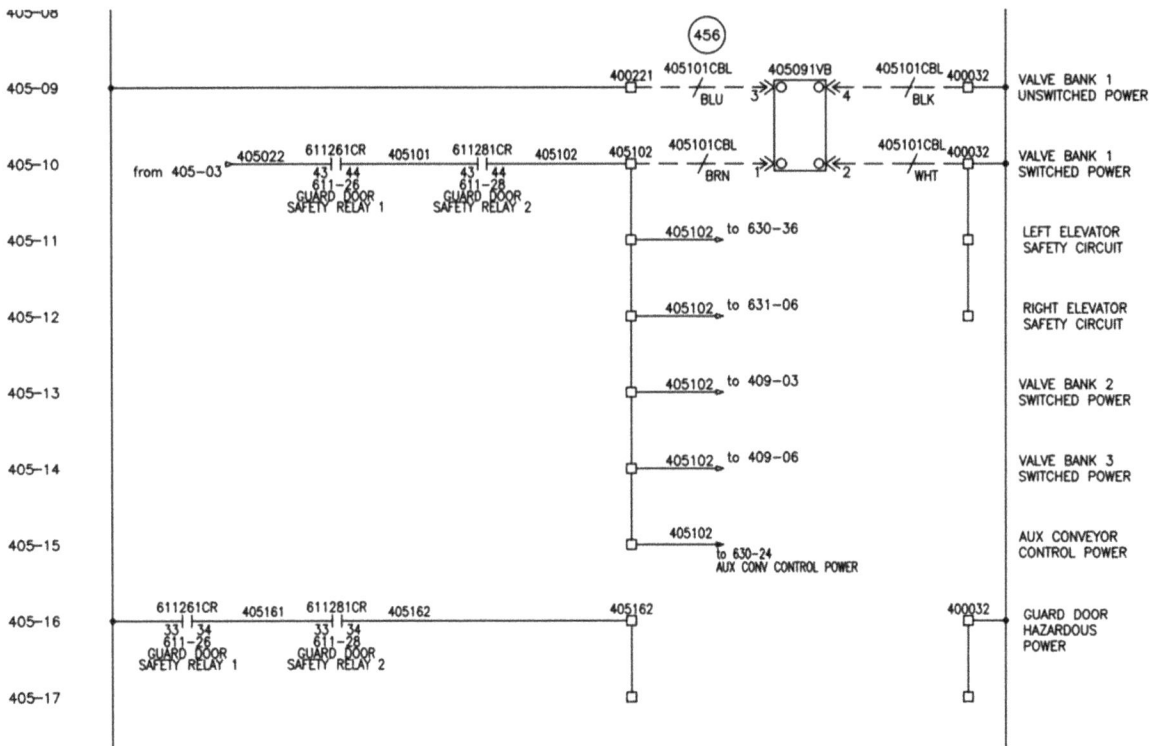

Figure 337 - Switched Power Feed, Sheet 405

Power from the E-Stop Safety Relays is fed through the Guard Door Safety Relay contacts, feeding all hazardous motion components. This includes valve banks, conveyors, elevators, and any other moving devices, except for the robot.

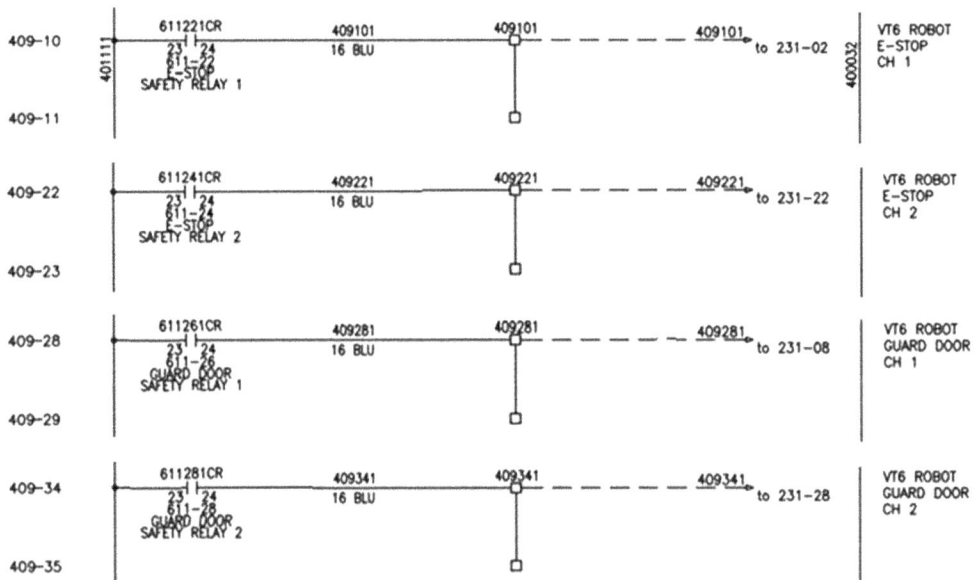

Figure 338 - Robot Safety Connections, Sheet 409

Robot E-Stop and Guard safety connections are shown in Figure 338.

The redundancy is handled within the robot controller. E-Stops and guard doors are treated differently by the robot. Emergency stops halt all motion immediately as usual, while the Safety Door inputs on the robot can be configured to temporarily hold up the program or reduce speed during teaching operations.

The timing on the Emergency Stop and Safety Door inputs is also different. If the two inputs differ by two seconds or more, the system regards it as a critical error. This is a much longer period than a safety relay or safety PLC.

Robot Connections

Details for the robot controller connections are on sheets 230 and 231. The controller is built into the base of the robot in the case of this robot, but controllers are often separate units connected to the robot with various cables.

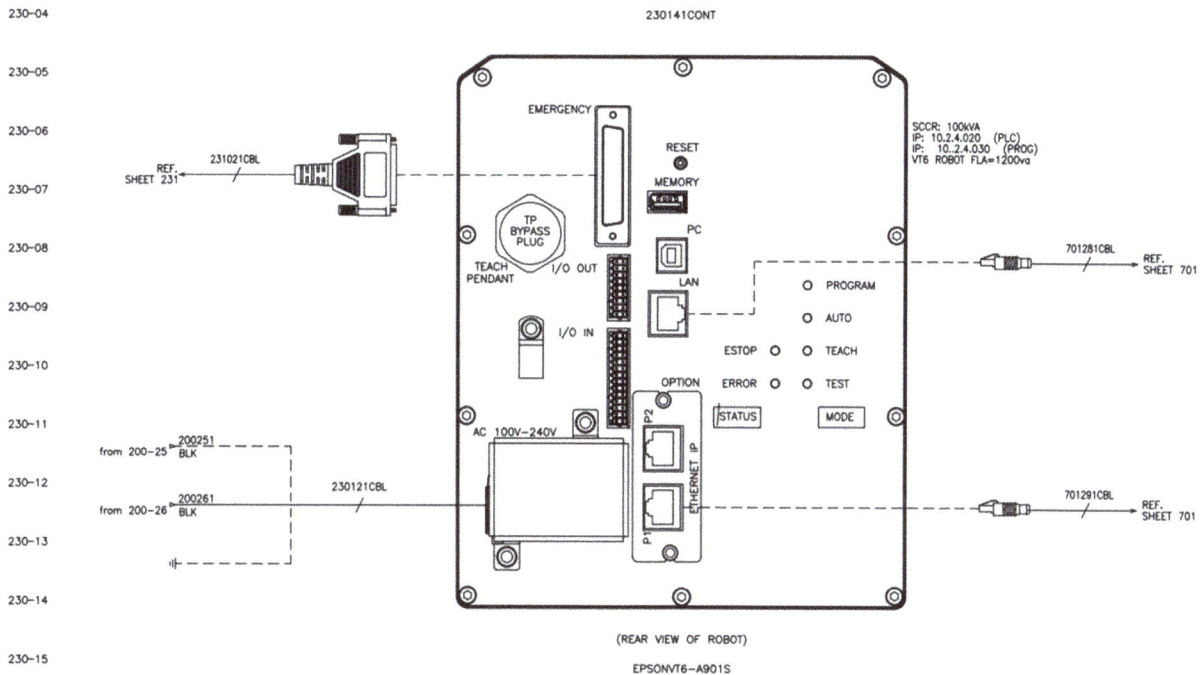

Figure 339 - Robot Controller Connections

Power to the robot is connected from the circuit breaker previously shown in Figure 305. Safety connections are detailed on Sheet 231 and are connected using a DB25 connector, contacts have been detailed in the previous section.

The DB25 is a molded cable that plugs into the port labeled Emergency. The other end of the cable is stripped, and the individual wires were connected to terminal blocks. This stripped end of the cable is known as "**flying leads**".

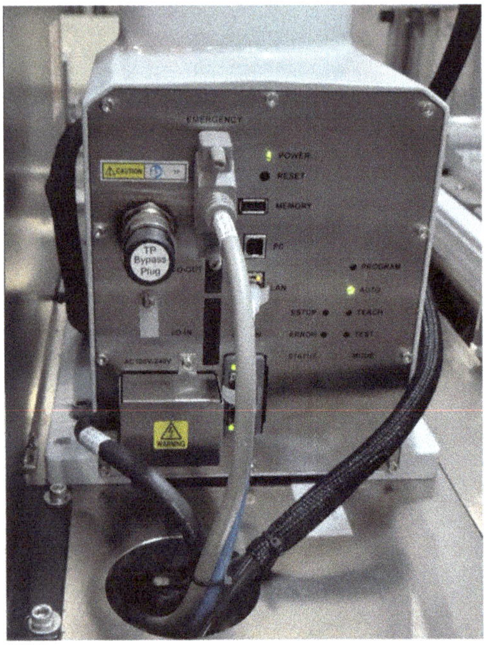

The cable extends through a hole in the base of the machine, along with several other cables. Underneath the base there are a series of cable glands that provide entry to the control enclosure and strain relief to the cables as shown in Figure 341.

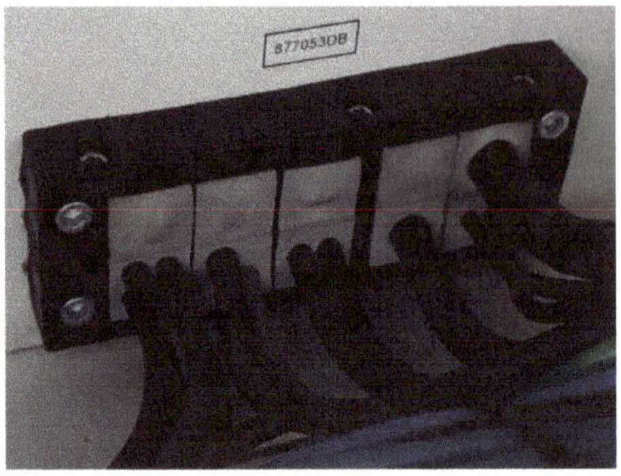

Figure 340 - Robot Base Connections

Figure 341 - Enclosure Cable Entry

After entering the back of the enclosure, wires are connected to terminal blocks as shown in Figure 342.

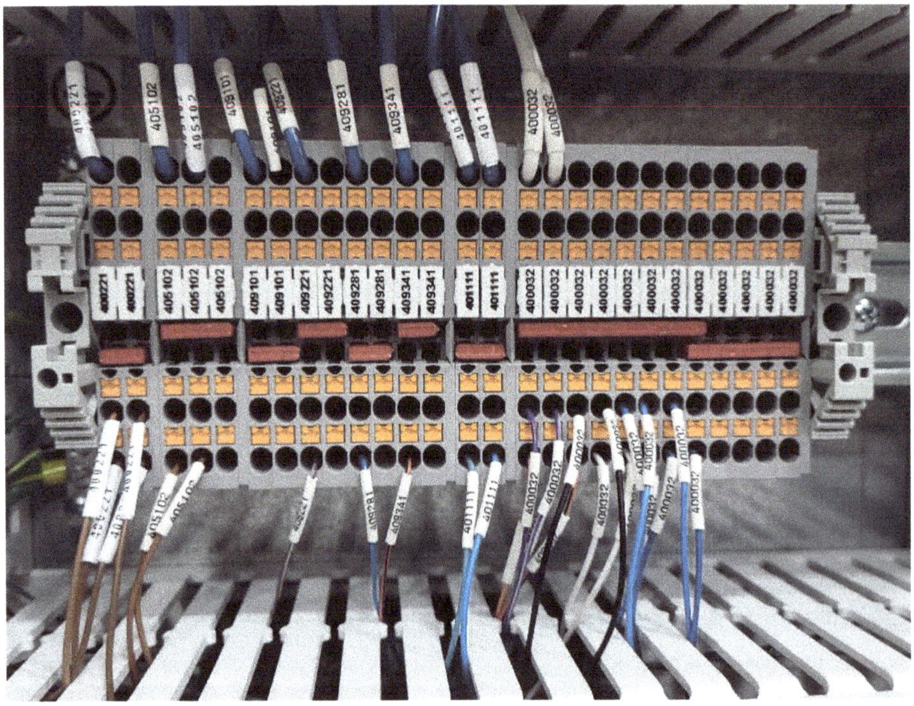

Figure 342 - Terminal Block Connections

The top of the terminal block is the connection to the PLC or control device in the enclosure, and the bottom is "field" wiring. The bottom wire connected to terminal 409341 is the connection for SD21, Safety Door Input 2 in Figure 344. It can be found visually by looking for the Red/Black wire.

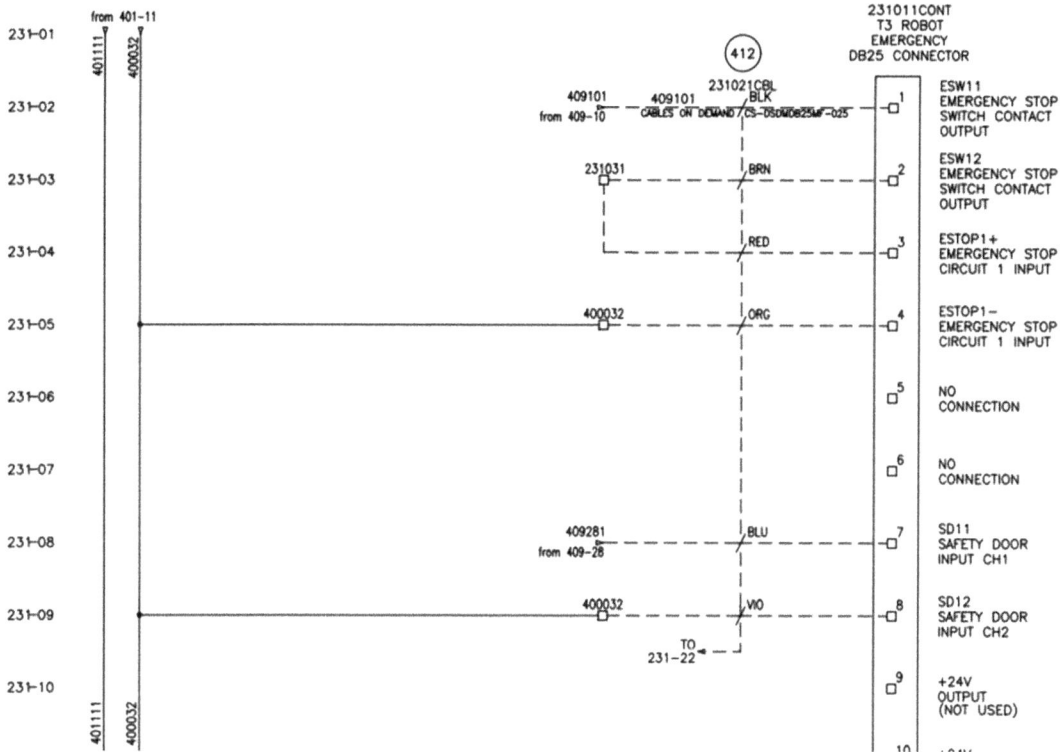

Figure 343 - Robot Safety Pins 1-9

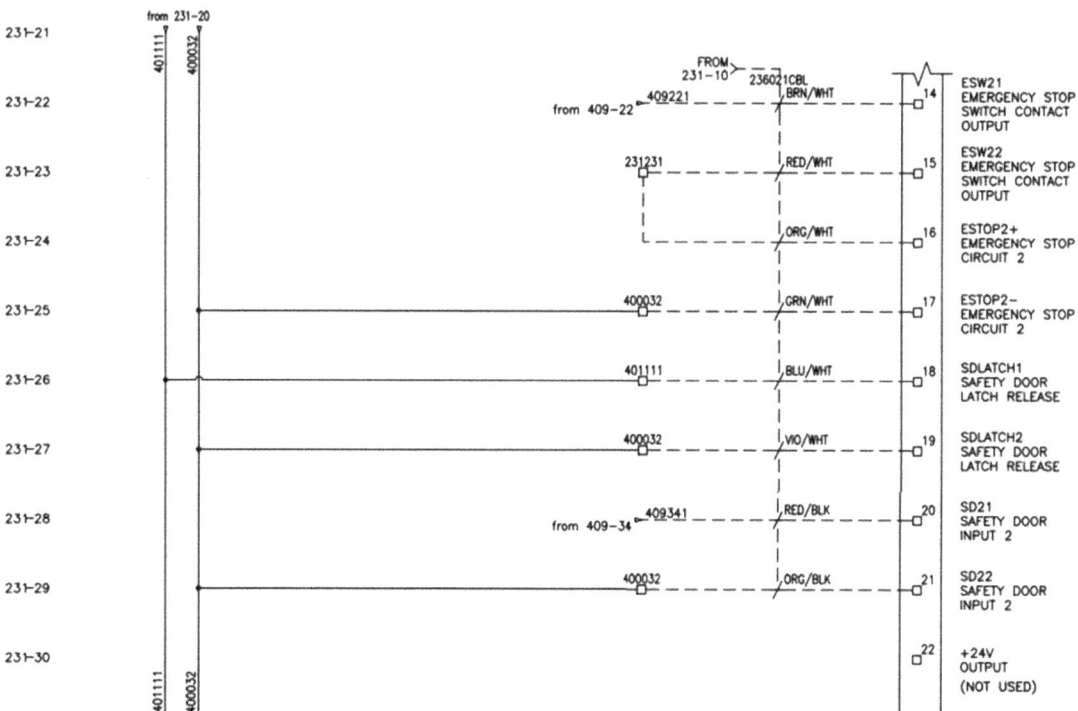

Figure 344 - Robot Safety Pins 14-22

A few more connections to note on the robot controller: The TP Bypass Plug is an optional connection for a Teach Pendant. It is necessary to insert the plug if not using a pendant, since there are E-Stop connections for the pendant. The robot will not run without the plug.

The I/O Out and I/O In connections are 12-24VDC I/O points that are assignable in the robot setup. More information can be found in the manual for the robot and in the software. Inputs can be used for starting and stopping the robot, selecting programs and for user defined tasks. Outputs are used to provide the robot's status, such as ready, paused, running or error. The I/O points are not wired for the Sweet Machine, instead I/O is transferred over an ethernet connection.

The LAN and Ethernet IP RJ45 connections are used to program the robot using Epson's RC+ programming software (LAN) and communicate with the PLC (Ethernet IP). There are cables in the LAN and P1 ports that run to the ethernet switch on Sheet 701, IP address assignments are also shown there. The Ethernet IP ports can also be used to control I/O directly and the port module can be exchanged with other fieldbuses.

The PC connector is a USB "B" style connector that can be used to connect a programming PC to.

Device Communications

In addition to the IO-Link network, there are several Ethernet devices on the Sweet Machine.

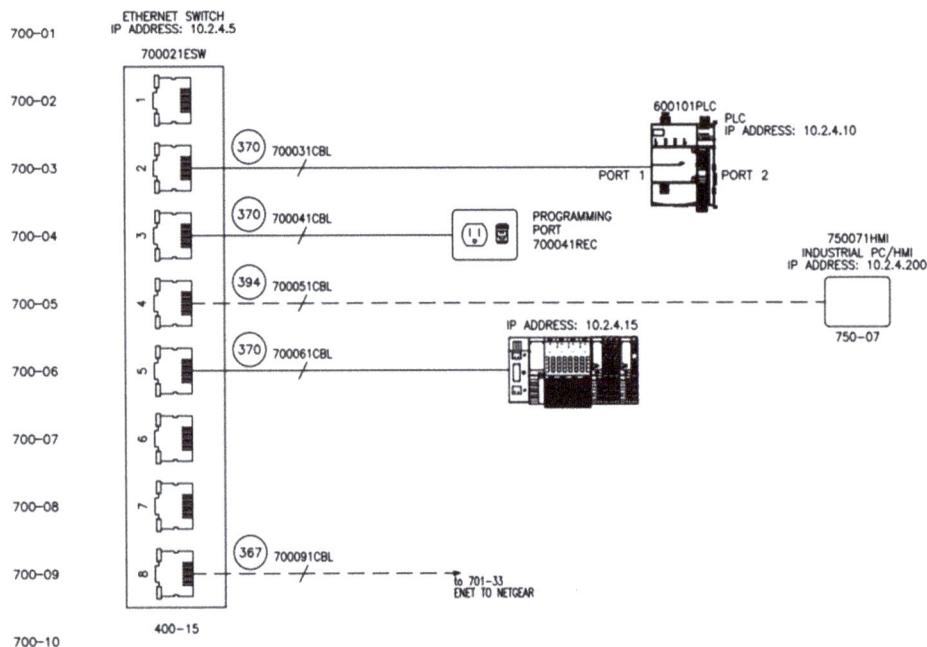

Figure 345 - Main Ethernet Switch

Figure 345 shows the Moxa brand ethernet switch in the main enclosure. It is connected to the PLC processor, the HMI, the Point I/O module, and a programming port on the enclosure door.

It is also connected to an additional 24 port PoE (Power over Ethernet) capable switch underneath the base of the machine. PoE carries both DC power (typically 48VDC) and communications to devices on the network. This NetGear brand switch has 12 PoE ports and 12 regular (communications only) ports.

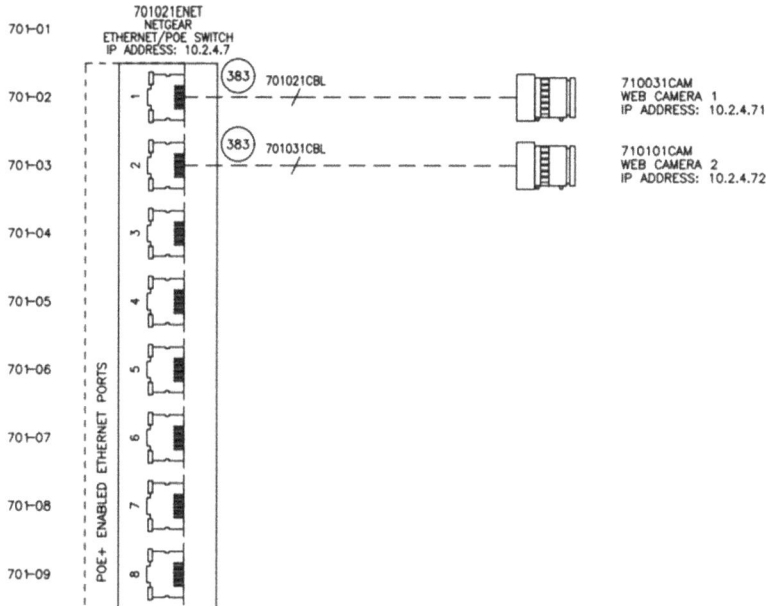

Figure 346 - NetGear 24 Port Managed PoE Switch

Figure 346 shows the first 8 ports on the NetGear switch. The only devices that require power are the two web cameras on the machine.

Note also that the IP address is documented for all devices. On the ethernet switches shown in the schematics, the switches themselves have IP addresses. This is because they are **Managed Switches**. At a basic level, unmanaged switches allow devices to be "plug-and-play", automatically allowing a device to communicate with other devices on the network based on only the address and mask. Since the IP addresses are listed on all the devices, it is important to ensure that whatever programming device (usually a laptop computer) is connected to the programming port (port 3 in figure 345) has a compatible network address that does not duplicate any of the other devices.

Managed switches allow for individual configuration of ports and monitoring settings in the LAN. Among other things, they allow prioritizing certain channels, creating new virtual LANs to keep smaller groups of devices segregated, and better manage traffic. Managed switches offer redundancy features that duplicate and recover data in the event of a device or network failure. They also allow for remote troubleshooting. Both switches are "**Gigabit**" switches, which means they can transfer data at 1 billion bits/second or more.

The web cameras on the Sweet Machine simply feed images to the HMI computer, but in a critical application the video can be recorded and used for machine troubleshooting or process verification. These are not the same as the "Smart Cameras" or machine vision cameras described in this book; they do not process the images for information.

Figure 347 - IP Web Camera

The unpowered side of the NetGear switch is connected to field control devices located outside of the control enclosure. These consist of several pneumatic valve banks, the IO-Link Master, and the Epson robot.

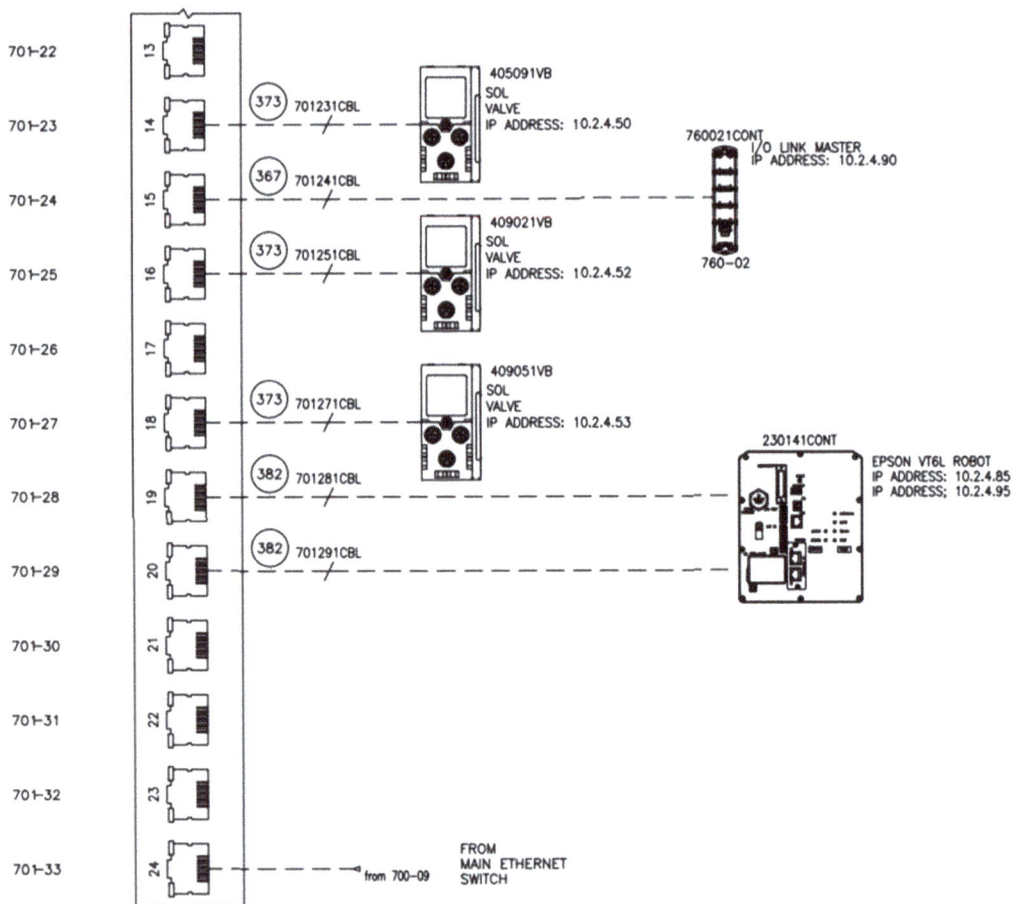

Figure 348 - Unpowered NetGear Switch Ports

Control I/O – Point I/O

There are three different methods of accessing I/O from the PLC in the Sweet Machine. Figure 322 several pages back showed the Safety PLC with a Point I/O module next to it. This module is connected to the main ethernet switch in the control enclosure. It communicates using Ethernet IP.

Also on the Ethernet IP network are the three pneumatic valve banks shown in Figure 338. The IO-Link Master is also connected this way, as is the Epson robot.

Ethernet IP devices are mapped into the PLC using EDS (Electronic Data Sheet) files that describe the functionality of the connected device to the controller. These files are obtained from the vendor of the device rather than from Allen-Bradley, the PLC manufacturer. More on this topic will be covered in the software section of this book.

The first three modules of the Point IO are assigned to safety I/O. These connections are used for emergency stops, safety relays and monitoring, and were discussed extensively in the previous section on safety wiring.

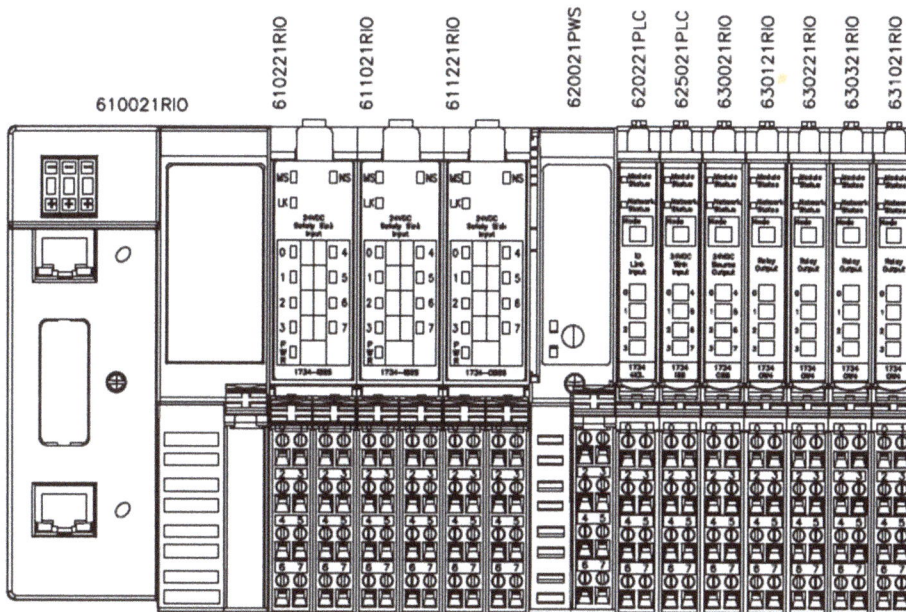

Figure 349 - Point IO Module, Sheet 609

Sheet 609 shows the layout of the point IO module. This module communicates to the PLC over Ethernet/IP through the main ethernet switch in the control enclosure. Listed above the layout are references to the page numbers where the wiring for each module can be found. Note that there are gaps in the page numbers so that modules can be inserted if necessary.

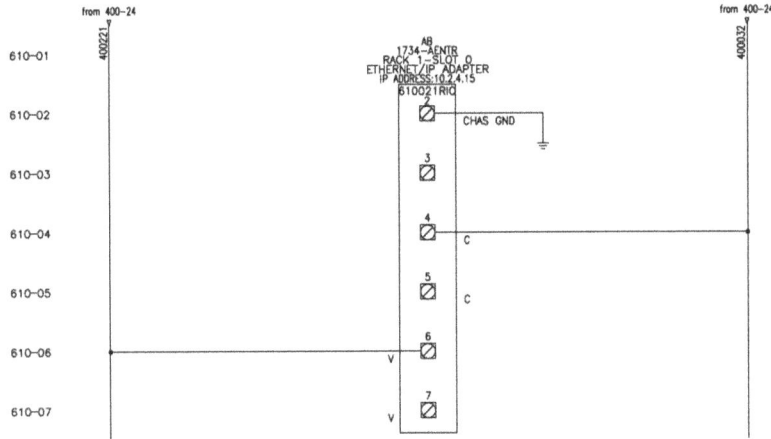

Figure 350 - Point IO Adapter Power, Sheet 610

Power for the Point IO modules is supplied at two locations on the rack. The Ethernet/IP adapter itself receives power as shown on Sheet 610, and an additional power supply adaptor is inserted between slots 3 and 4, shown on Sheet 620. Power for slots 1 through 3 comes from the adapter's power supply.

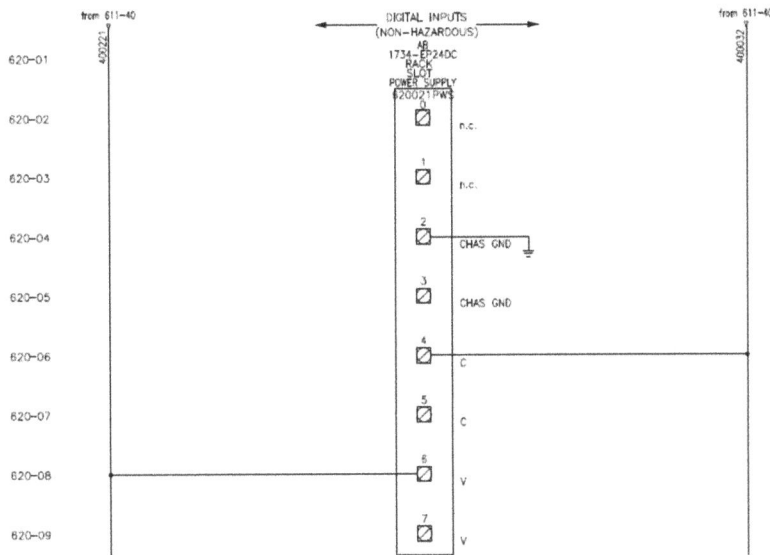

Figure 351 - Point IO Non-Hazardous Module Power, Sheet 620

The adapter on sheet 620 is used to separate the power for the device and other I/O modules. If hazardous power was required for output modules, a module could be inserted between the hazardous and non-hazardous sections.

Slot 4 has already been discussed. It is the IO-Link module used to supply the circuit breaker device. It was shown in Figure 321 and appears on the right side of Sheet 620.

Slot 5 is an 8 point 24vdc digital input module. It appears on Sheet 625 and is shown in Figure 352.

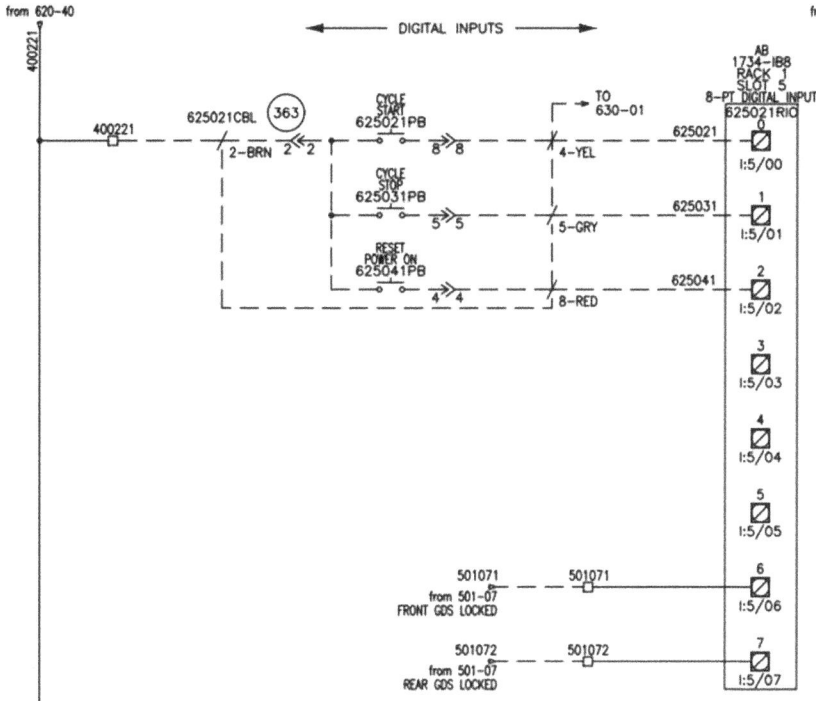

Figure 352 - Input Slot 5, Sheet 625

All the I/O modules to the right of the non-hazardous power module use the applied voltage as a supply. This means the inputs are configured as sinking, with the devices sourcing +DC voltage into the input.

Inputs 0-2 of slot 5 are a Fortress brand self-contained three-pushbutton box with two illuminated green buttons and one non-illuminated red button. The buttons are used to start, stop, and apply control power to the machine.

Inputs 6 and 7 are feedback from the locking guard door switches, on when the doors are locked.

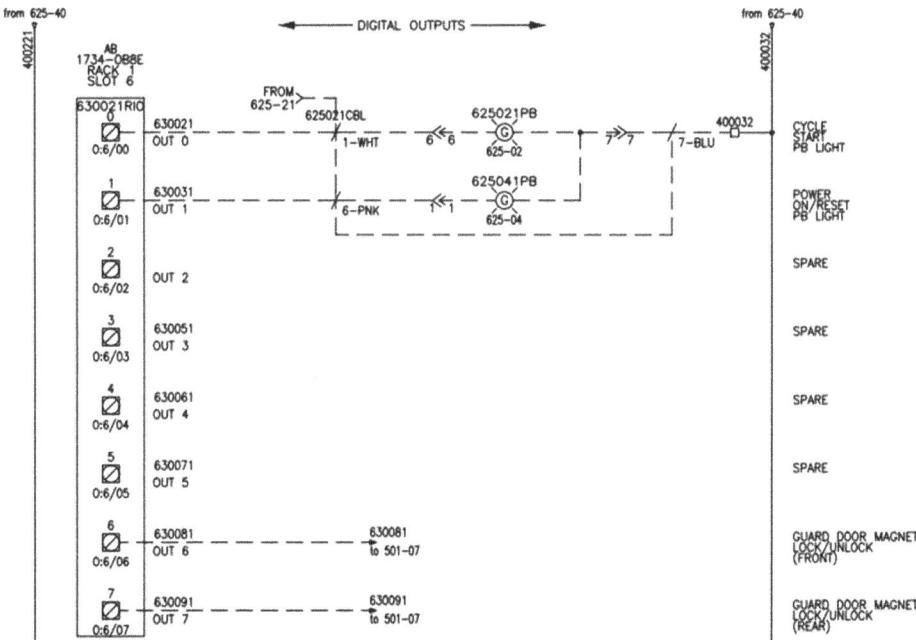

Figure 353 – Output Slot 6, Sheet 630

Slot 6 is an output card that connects to the lights on the pushbutton box (Outputs 0 and 1) and to the locks on the guard switches (Outputs 6 and 7). Since none of these devices are hazardous, the power is fed directly from the non-hazardous power module.

Figure 354 - Cycle Start and Stop, Power On Buttons

The Cycle Start and Cycle Stop buttons are common when operating machines and lines. A typical function would be for a horn or buzzer to sound for several seconds while the start button is pushed, alerting personnel of impending motion. If the button is released before a time elapses, the machine remains stopped. When the Cycle Stop button is pressed, the machine will complete its current operations before stopping. This is all accomplished in the PLC program.

The Reset Power On button is often connected to a safety circuit to energize the Master Control Relay (MCR), provided the E-Stop button is pulled up and guard doors are closed. It is also usually connected to a standard input. When pressed, it is used to programmatically reset faults and warning messages.

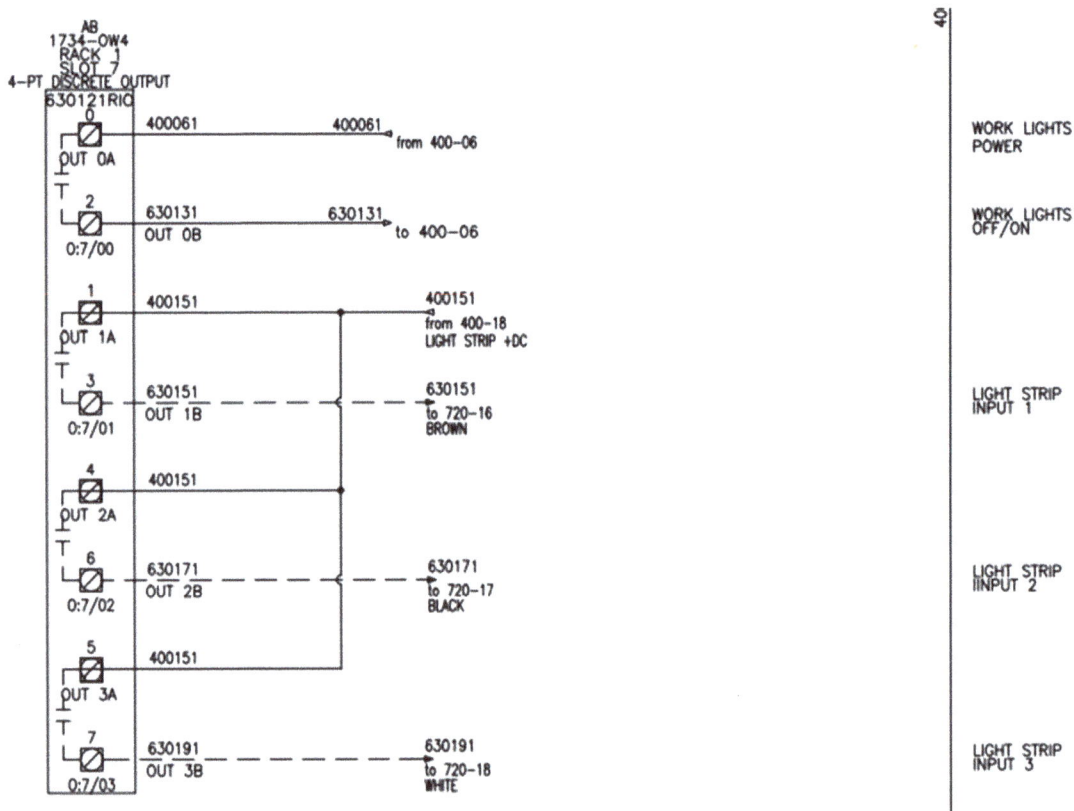

Figure 355 – Output Slot 7, Sheet 630

Slot 7 is also an output card, this time with only 4 points. Because each point is isolated from the others, power can be supplied from different sources.

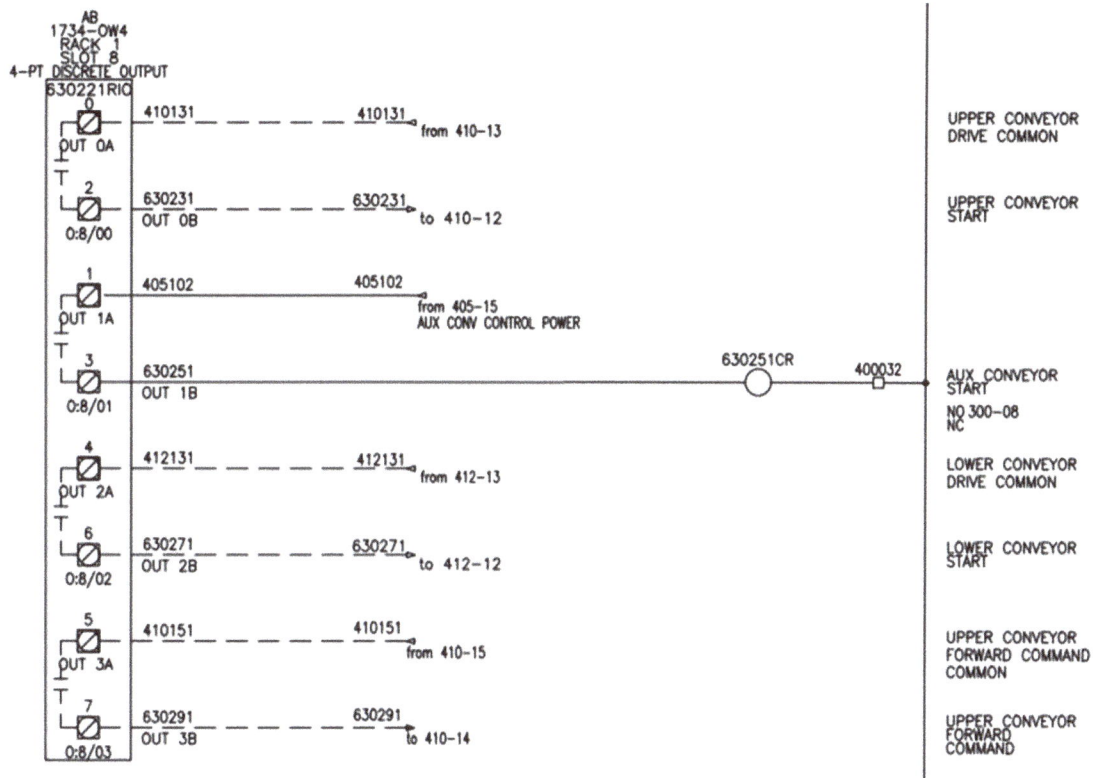

Figure 356 - Output Slot 8, Sheet 630

Slot 8 is also a 4 point isolated card. Note here that the power to the Aux. Conveyor Start relay, 630251CR, comes from the switched power of Sheet 405. This was shown in Figure 337, power is supplied through both the E-Stop Safety Relays and the Guard Door Safety Relays.

There is also a reference to the contacts of the relay in Sheet 300. Figure 313 shows that the power supplied to the Secondary or Aux conveyor is 120VAC and it passes through an overload.

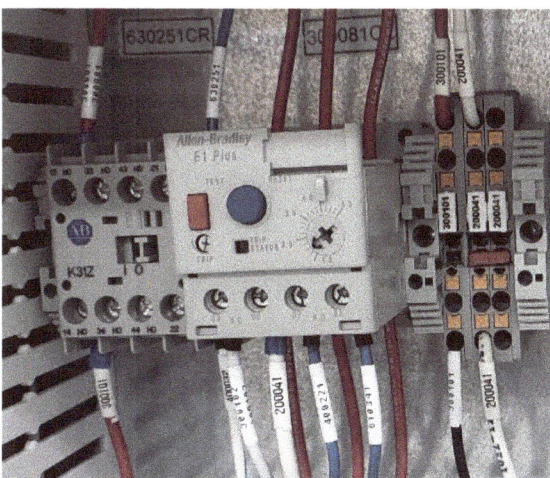

Figure 357 shows the overload and control relay for the motor. When examining all the sheets referenced above (630, 300, 405) there are several places where the relay would be disconnected from power. The safety circuit would disconnect it with an E-Stop or guard door opening, and if the overload tripped the Neutral side of the power to the contactor 630251CR would be opened.

Figure 357 - Aux. Conveyor Overload and Contactor

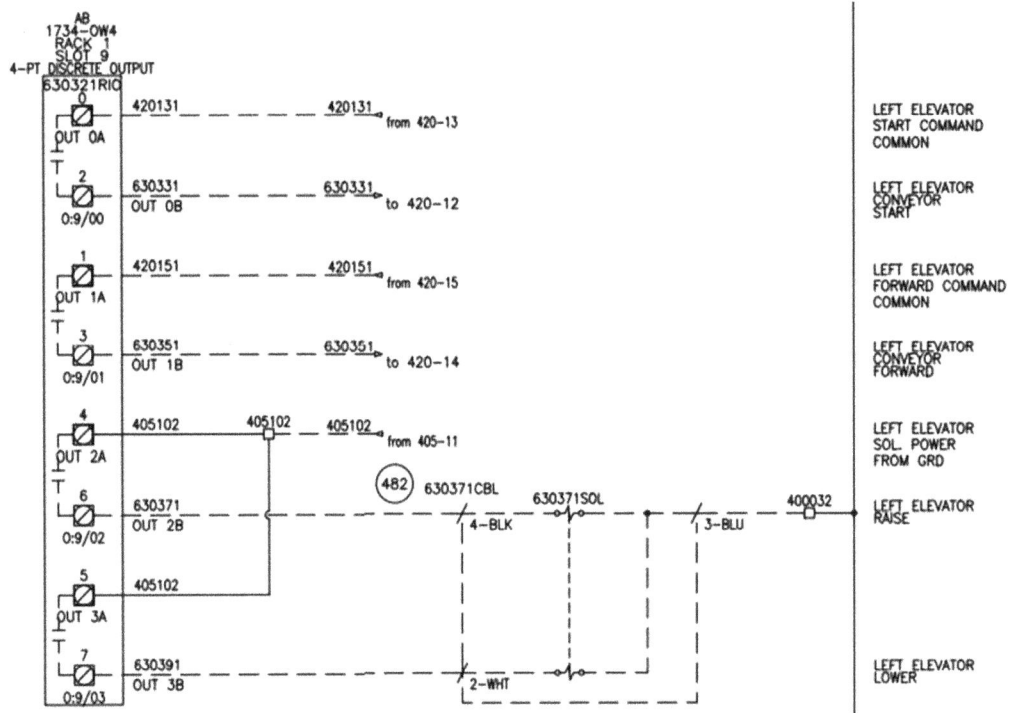

Figure 358 - Output Slot 9, Sheet 630

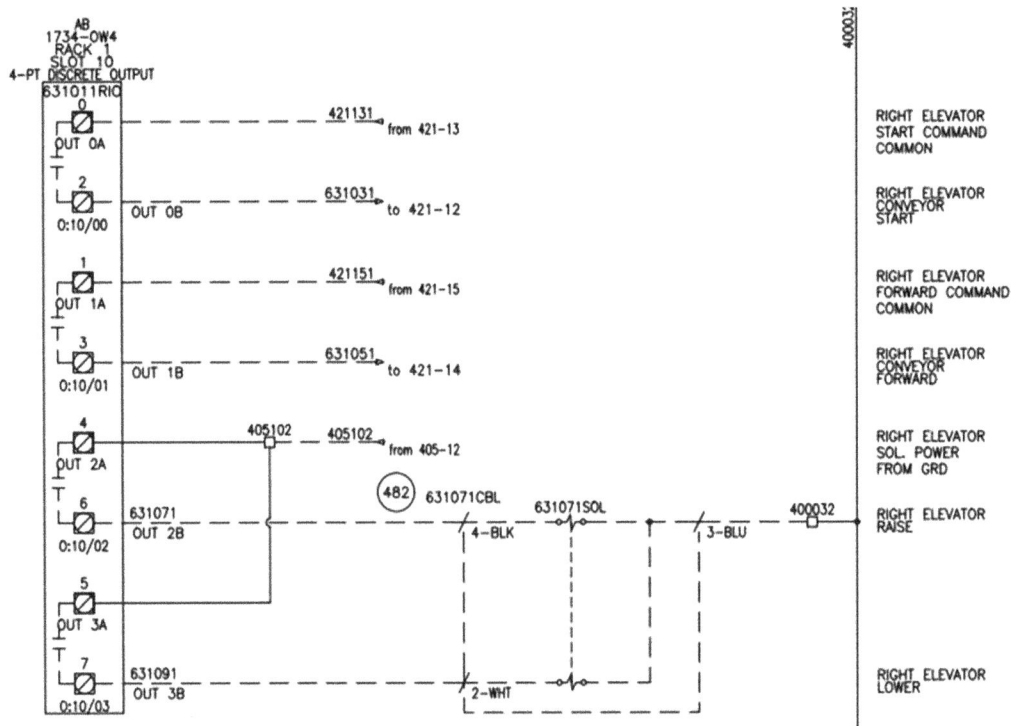

Figure 359 - Output Slot 10, Sheet 631

Slots 8, 9 and 10 show isolated contacts that connect to the 4 DC drives for conveyors. The drive diagrams are detailed on Sheets 410, 412, 420 and 421.

Sweet Machine DC Drives

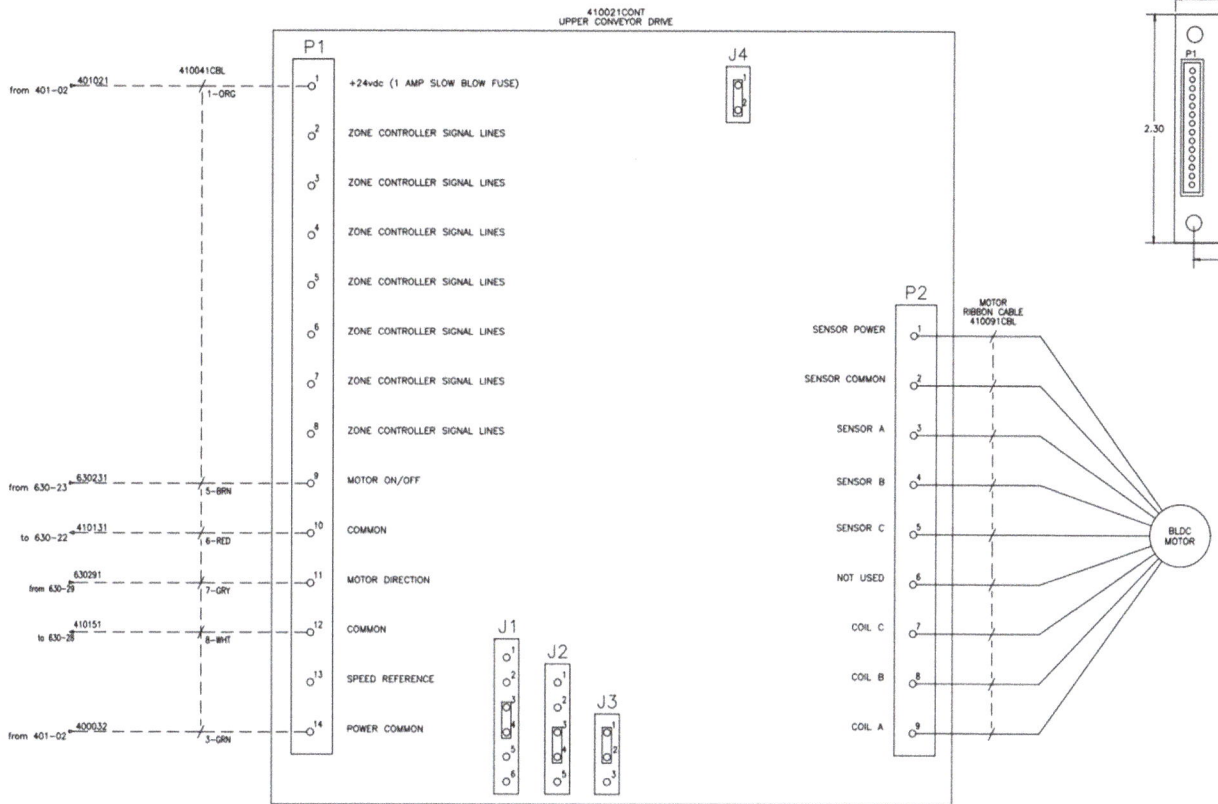

Figure 360 - Upper Conveyor Drive, Sheet 410

Figure 360 shows the connections for the Upper Conveyor drive. All four drives are the same, so the figure is not duplicated for the other drives.

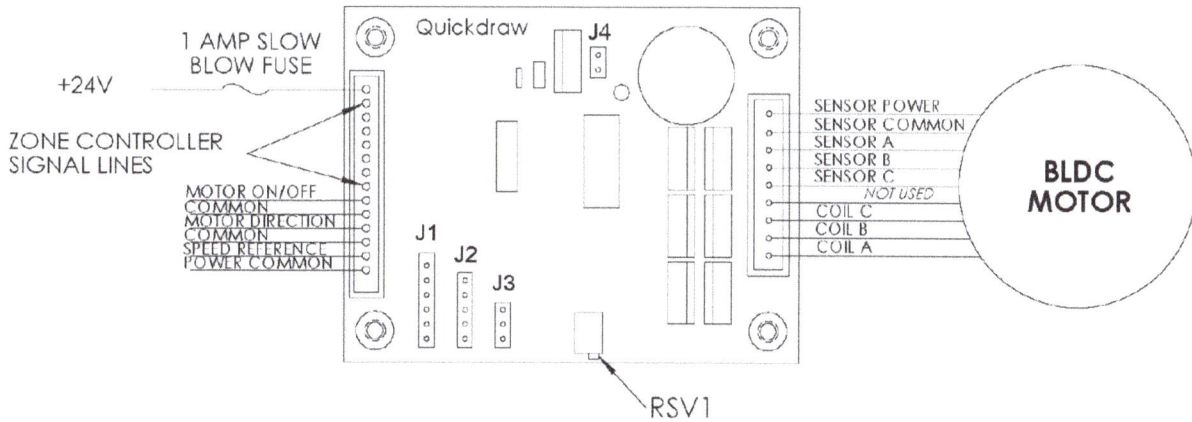

Figure 361 - Quickdraw Integral Motor Controller

The four roller conveyors on the Sweet Machine are used for transporting pallets on the sweet machine. They are made by a company called Quickdraw Systems, and use an integral drive built into the motor assembly to control the conveyor.

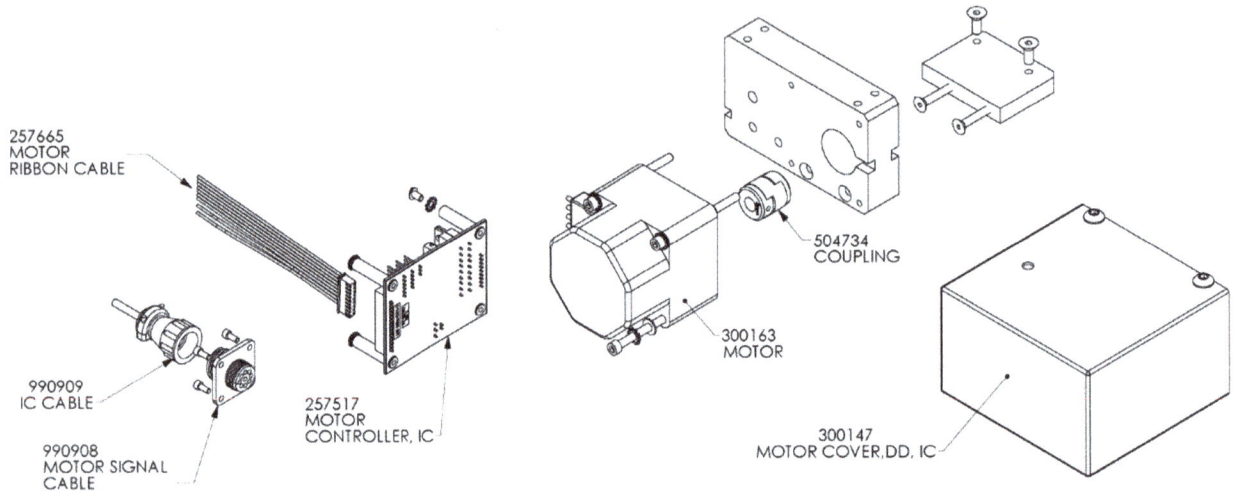

Figure 362 - Motor Assembly Exploded View, Page 6 of Maintenance Guide

Figures 361 and 362 show diagrams from the Quickdraw MR Conveyor Maintenance Manual. Designers of a machine or system rely on manufacturer's documentation for both mechanical and electrical information. These "cut sheets" are also often included in the operator's manual for a machine.

In order to diagnose problems with the conveyor, it would possibly be necessary to download the Maintenance Manual. Although there is a pinout key for setting jumpers on the drives, much more information is presented in the manual.

Cable	Upper 410	Lower 412	L. Elev. 420	R. Elev. 421
1-ORG +24VDC	401021	401021	401021	401021
9-BRN Motor On/Off	630231	630271	630331	631031
10-RED Common	410131	412131	420131	421131
11-GRY Motor Direction	630291		630351	631051
12-WHT Common	410151		420151	421151
14-GRN Power Common	400032	400032	400032	400032

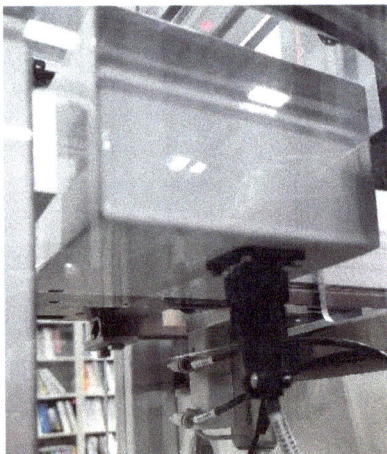

The table above shows the connections to the 3 additional drives not shown.

Figure 363 shows the cable connection to the drive assembly.

Figure 363 - Drive Connection

Pneumatic Elevator Control

Output slots 9 and 10 also showed the connections to the solenoid valves that raise and lower the elevators. There are several valve banks on the Sweet Machine to control actuators, but these two double acting 5/3 valves are separately mounted at the bottom of each elevator.

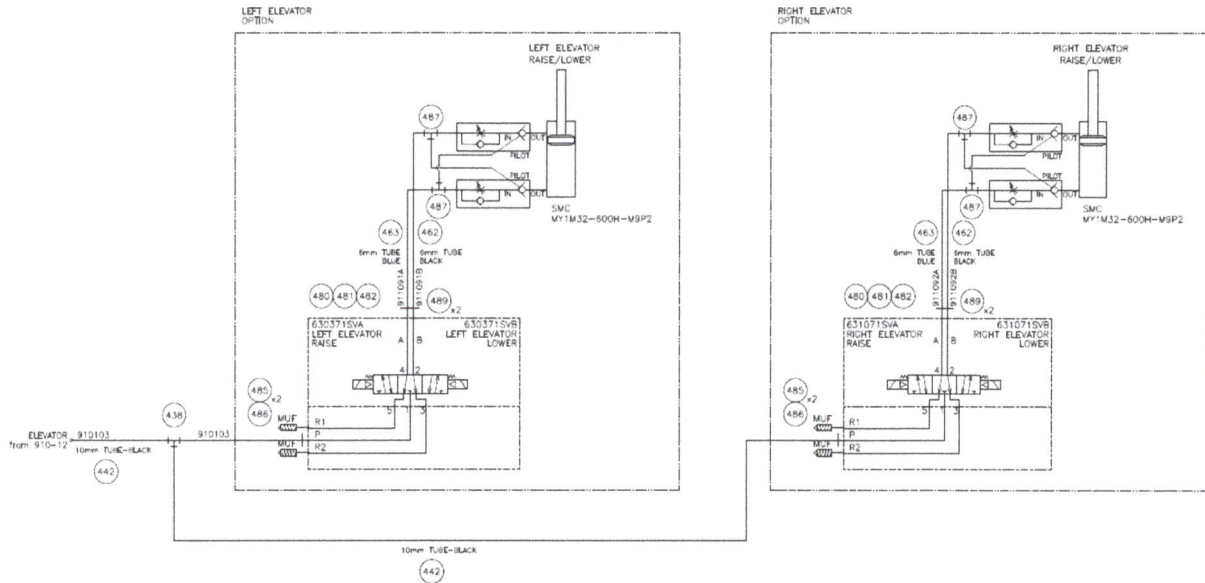

Figure 364 - Elevator Pneumatics, Sheet 917

Though not technically part of the electrical schematics, the pneumatic drawings start at sheet 900 and are also used for troubleshooting.

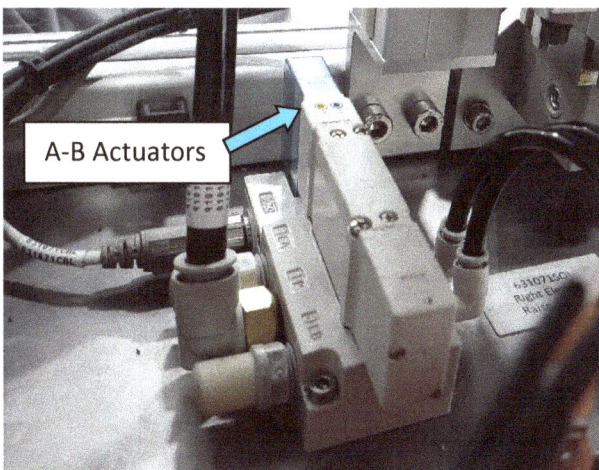

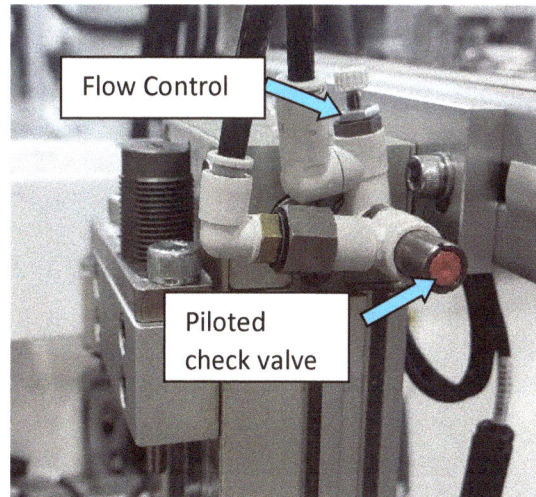

Figure 365 - Elevator Valve and Flow Controls

Figure 365 shows the valve (left) and the piloted flow controls (right) at each end of the elevator cylinder. There are several manually operated actuators that are important for a technician to understand.

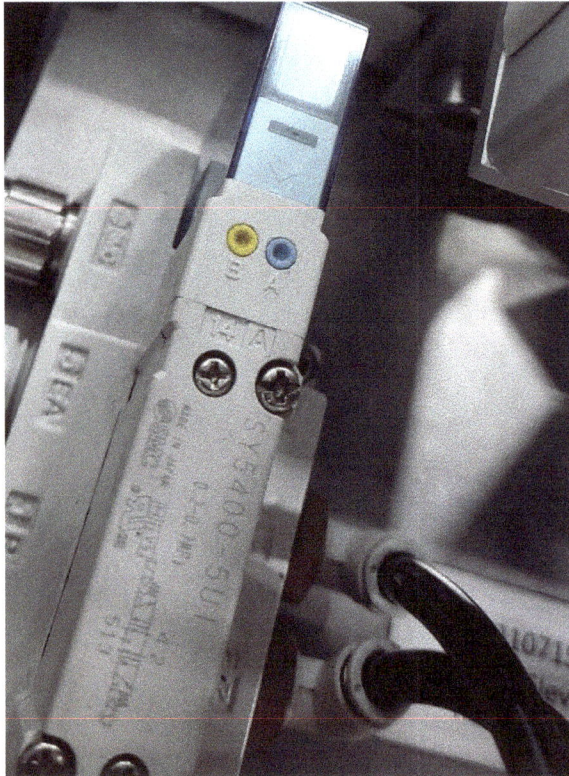

Figure 366 - Valve Manual Actuators A and B

The **flow control** allows for the speed of the elevator to be controlled. As described in the pneumatic section of this book, adjusting the outflow side (meter out) is how the speed is set for a specific direction. To adjust speed in the up direction, the knob at the top would be adjusted, and to adjust the down speed, adjust the bottom flow control. The hex nut can be tightened to lock the position.

The **piloted check valves** are used to hold air inside the cylinder if pressure drops or is turned off. The red button allows air to be released manually to perform maintenance.

The **A and B manual actuation** buttons can be pressed to actuate the valve without an electrical signal. This is usually done with a small screwdriver. A raises the elevator, and B lowers it.

The elevators can also be raised and lowered by using the buttons on the HMI in Manual Mode, actuating the outputs to electrically operate the valve.

Control I/O – Valve Banks

Electrical supply connections to the first valve bank were shown in Figure 337 on Sheet 405. There are three valve banks, and the connections to the other two are shown on Sheet 409.

Unswitched power is used to operate the controller in the valve bank, while switched power operates the solenoids themselves. All three valve banks receive switched power through the E-Stop and Safety Relays on 405102. The valve banks are all located underneath the machine base and are accessible through the lower front doors.

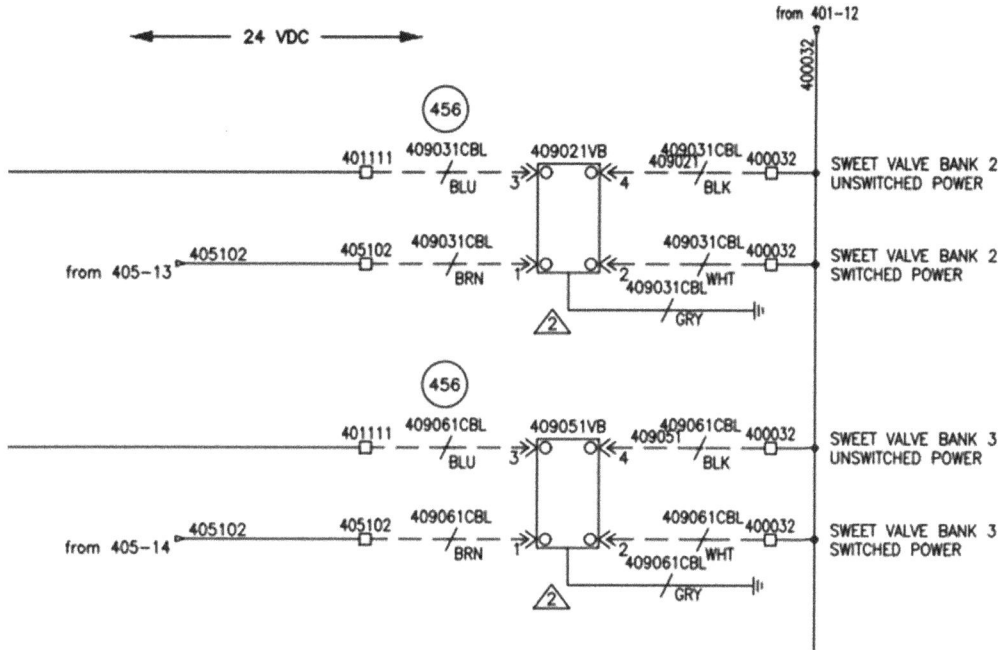

Figure 367 - Valve Bank 2 and 3 Power, Sheet 409

The outputs for the valves themselves are listed on the layout drawings on sheets 780-782, but they are not wired directly into the valves from the PLC. The Ethernet switch shown in Figure 348 showed the three valve banks with their IP addresses. The addresses for the outputs are mapped unto PLC tags as shown in Figure 368:

Each valve bank has the ability to control up to 32 output points. These valve manifolds each hold eight valves, each with two solenoids, so 16 outputs are used in the I/O list. The data is mapped into 4 SINTs or bytes, the first two SINTs are used for the valves.

Figure 368 - Valve Output Mapping

▸ SweetMachine_Valve_Bank_1:I	{…}	{…	_00…	St…
◢ SweetMachine_Valve_Bank_1:O	{…}	{…	_00…	St…
◢ SweetMachine_Valve_Bank_1:O.Data	{…}	{… D	SIN…	St…
◢ SweetMachine_Valve_Bank_1:O.Data[0]	0	D	SINT	St…
SweetMachine_Valve_Bank_1:O.Data[0].0	0	D	BO…	St… Lwr Cnv Stop Lwr
SweetMachine_Valve_Bank_1:O.Data[0].1	0	D	BO…	St… Spare
SweetMachine_Valve_Bank_1:O.Data[0].2	0	D	BO…	St… Upr Cnv Pre-Stop Lwr
SweetMachine_Valve_Bank_1:O.Data[0].3	0	D	BO…	St… Spare
SweetMachine_Valve_Bank_1:O.Data[0].4	0	D	BO…	St… Upr Cnv Stop Lwr
SweetMachine_Valve_Bank_1:O.Data[0].5	0	D	BO…	St… Spare
SweetMachine_Valve_Bank_1:O.Data[0].6	0	D	BO…	St… Spare
SweetMachine_Valve_Bank_1:O.Data[0].7	0	D	BO…	St… Spare
▸ SweetMachine_Valve_Bank_1:O.Data[1]	0	D	SINT	St…
▸ SweetMachine_Valve_Bank_1:O.Data[2]	0	D	SINT	St…
▸ SweetMachine_Valve_Bank_1:O.Data[3]	0	D	SINT	St…
▸ SweetMachine_Valve_Bank_2:I	{…}	{…	_00…	St…
▸ SweetMachine_Valve_Bank_2:O	{…}	{…	_00…	St…
▸ SweetMachine_Valve_Bank_3:I	{…}	{…	_00…	St…
▸ SweetMachine_Valve_Bank_3:O	{…}	{…	_00…	St…

SOL A 0:0.0 LOWER FLIP STA. PALLET STOP LOWER
SOL B 0:0.1 SPARE
SOL A 0:0.2 LOWER FLIP STA. PALLET PRE-STOP LOWER
SOL B 0:0.3 SPARE
SOL A 0:0.4 C1 PICK STA. PALLET STOP LOWER
SOL B 0:0.5 SPARE
SOL A 0:0.6 C1 PICK STA. PALLET PRE-STOP LOWER
SOL B 0:0.7 SPARE
SOL A 0:1.0 C1 PLACE STA. PALLET STOP LOWER
SOL B 0:1.1 SPARE
SOL A 0:1.2 C1 PLACE STA. PALLET PRE-STOP LOWER
SOL B 0:1.3 SPARE
SOL A 0:1.4 L. ELEVATOR LOAD STOP LOWER
SOL B 0:1.5 SPARE
SOL A 0:1.6 R. ELEVATOR LOAD STOP LOWER
SOL B 0:1.7 SPARE

405091VB VALVE BANK

405101CBL POWER CABLE (456)

701231CBL ETHERNET CABLE (455)

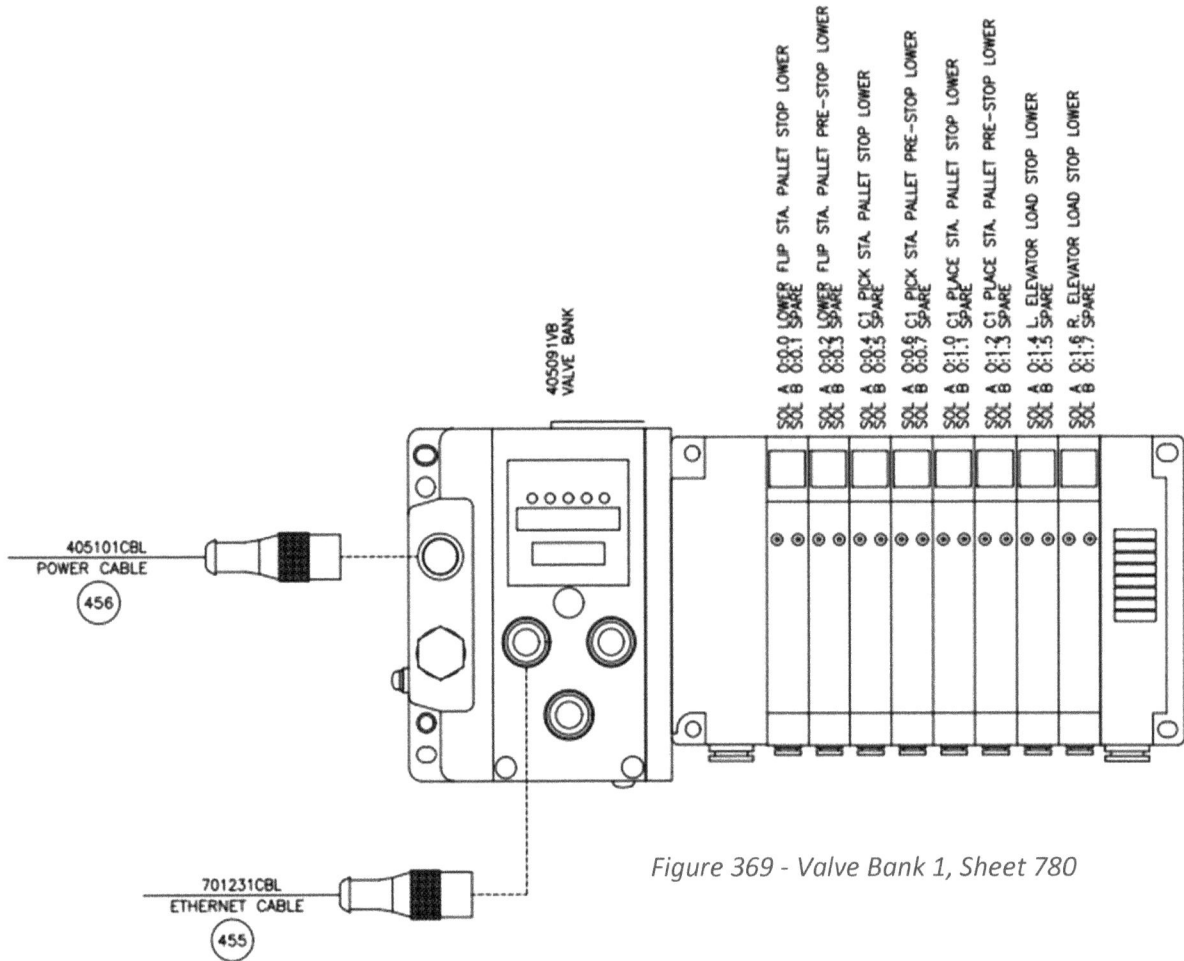

Figure 369 - Valve Bank 1, Sheet 780

The valves on Valve Bank 1 are all single acting valves with a spring return on the other side as illustrated in Figure 370. When the A solenoid is activated, the spool moves to the A side and energizes the port, when turned off the spool moves to the B side.

Figure 370 - Valve Bank 1 Pneumatics, Sheet 911 (First Three Valves)

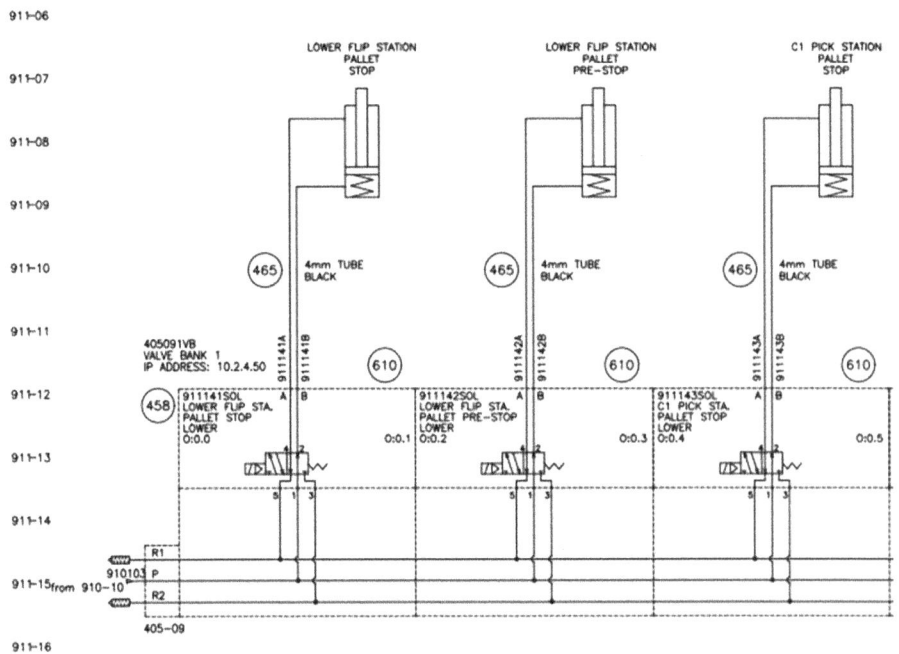

911-06
911-07
911-08
911-09
911-10
911-11
911-12
911-13
911-14
911-15 from 910-10
911-16

LOWER FLIP STATION PALLET STOP

LOWER FLIP STATION PALLET PRE-STOP

C1 PICK STATION PALLET STOP

(465) 4mm TUBE BLACK

(465) 4mm TUBE BLACK

(465) 4mm TUBE BLACK

405091VB VALVE BANK 1 IP ADDRESS: 10.2.4.50

(610)

(610)

(610)

(458) 911141SOL LOWER FLIP STA. PALLET STOP LOWER 0:0.0

911142SOL LOWER FLIP STA. PALLET PRE-STOP LOWER 0:0.2

911143SOL C1 PICK STA. PALLET STOP LOWER 0:0.4

0:0.1

0:0.3

0:0.5

911141A 911141B

911142A 911142B

911143A 911143B

R1
910103 P
R2

405-09

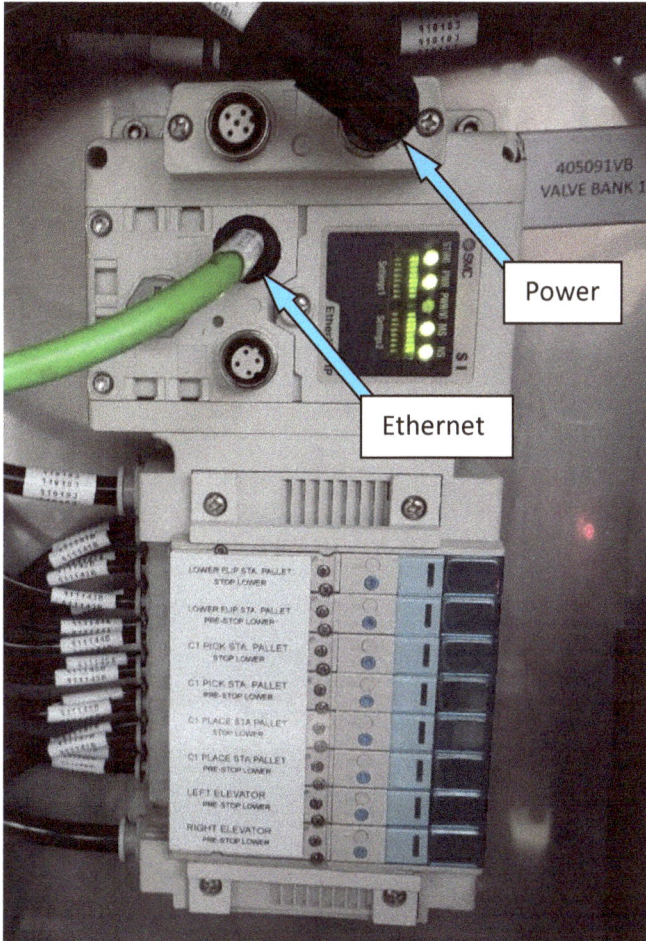

Like the valves on the elevators, there are manual actuators on each valve body. For single acting valves, only the A actuator is used.

Figure 371 shows Valve Bank 1 with the location of the ethernet and power cables. Notice that there are labels on all cables and hoses. There are also engraved labels on the backplane. This is not true in every machine, making it even more important that the schematic pages show the location of physical components.

Valve Banks 2 and 3 are also located underneath the machine base and are labeled similarly.

Air is supplied to the pneumatic components through a Filter Regulator with a manual shut-off valve, and from there through a soft-start valve.

Figure 371 - Valve Bank 1

More On Pneumatics

Air is applied to the machine using a standard bayonet fitting as shown in Figure 372. The fitting is at the top of the machine near the power entry.

Figure 372 - Air and Power Service Area

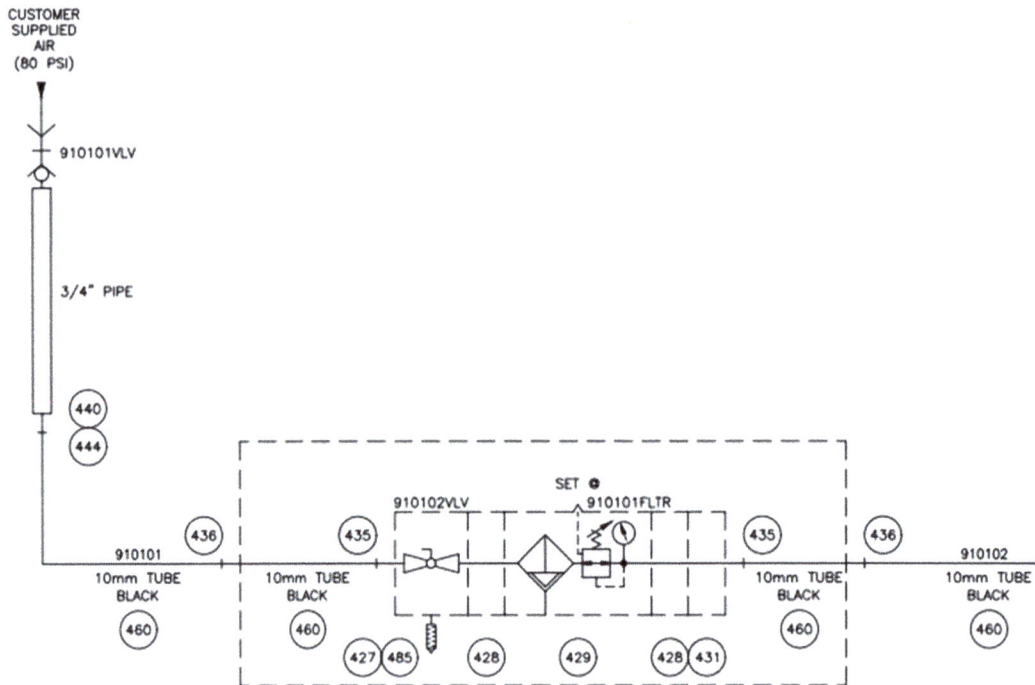

Figure 373 - Air Supply and Filter Regulator, Sheet 910

Figure 373 shows the incoming air to the system with the Manual Shut-Off and Filter Regulator. Like electrical schematics, components and hoses are numbered according to the Sheet and Line number where they appear.

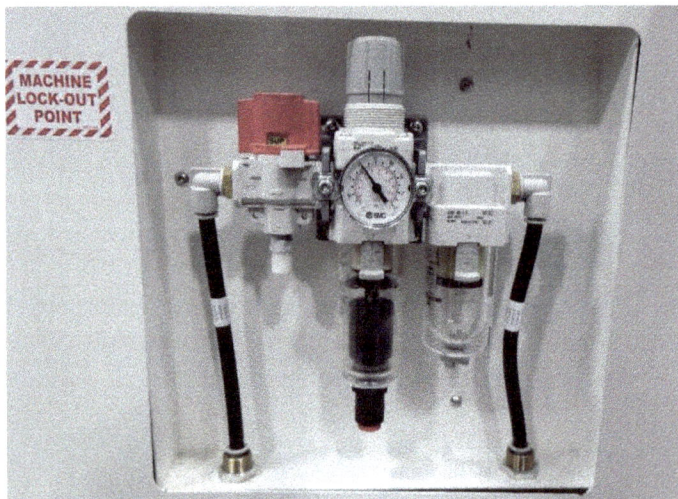

Figure 374 - Filter Regulator and Manual Shut-Off

The Filter Regulator is on the outside of the machine opposite the side where the air is applied. The part number designations in Figure 373 show the various fittings and components required to pipe air through the machine, they are listed on Sheet 980.

The manual shut-off valve shown in red can be locked out for servicing of the machine.

Lockout/Tagout or **LOTO** is used to ensure that dangerous machines are properly shut off and not able to be started again prior to the completion of maintenance or repair work. Hazardous energy sources are "isolated and rendered inoperative" before work is started. The worker (or workers) locks the sources and keeps the key, placing a tag on the lock identifying the worker.

There are various regulations and standards applying to LOTO and they can differ by country and company. Ensure that you understand the applicable requirements for your country and area before attempting to work on equipment.

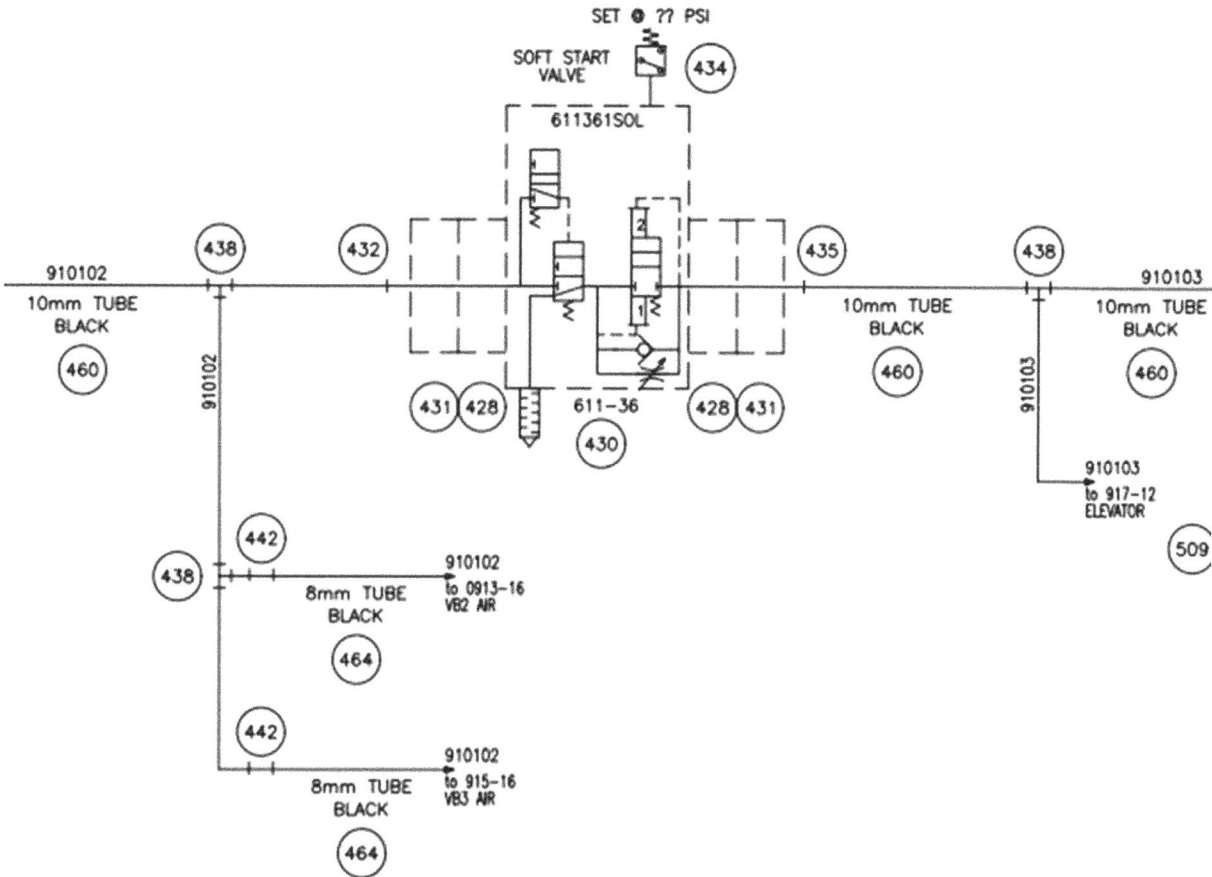

Figure 375 - Air Distribution and Soft Start

Air from the shut-off and filter-regulator on the outside of the machine is fed to the inside of the machine through 10mm black tubing. The 910102 connections to Valve Bank 2 (VB2) and Valve Bank 3 (VB3) connect to grippers on the flip station and the robot, so they are not disconnected by the Soft Start valve.

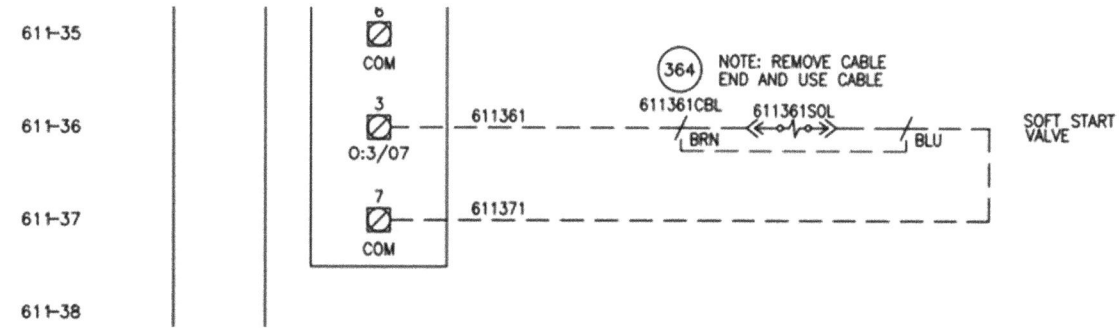

Figure 376 - Soft Start Control, Sheet 611 Lower Right

The **Soft Start** valve serves two purposes. It removes air from the system if the emergency stop circuit or guard doors are opened, and it ensures that air is reapplied slowly to the system to avoid sudden movement or jerking of the actuators if air is applied too quickly. As shown in Figure 376, the Soft Start is controlled by the safety output card of the PLC.

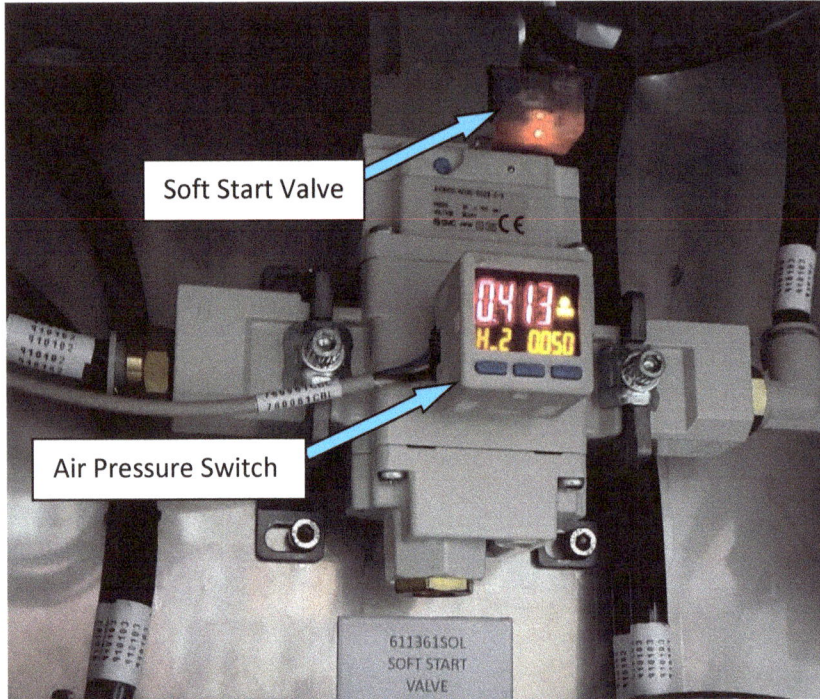

Figure 377 - Soft Start Valve and Pressure Switch

The lighted red connector at the top of the assembly is the connection to the valve itself. This is a "field wireable" connection where there are internal screw terminals inside the connector. It is important to read the documentation for this type "Y" din connector when replacing or servicing it.

The **Air-Pressure Switch** on the front of the valve is numbered 760061 on the connecting cable, which refers to the IO-Link Master sheet 760.

IO-Link devices can transfer much more than simple digital and analog signals. The pressure switch transfers several integers worth of data to the PLC, where an Add-On Instruction (AOI) extracts the air pressure from the device and compares it with low and high limits to determine if the pressure is OK.

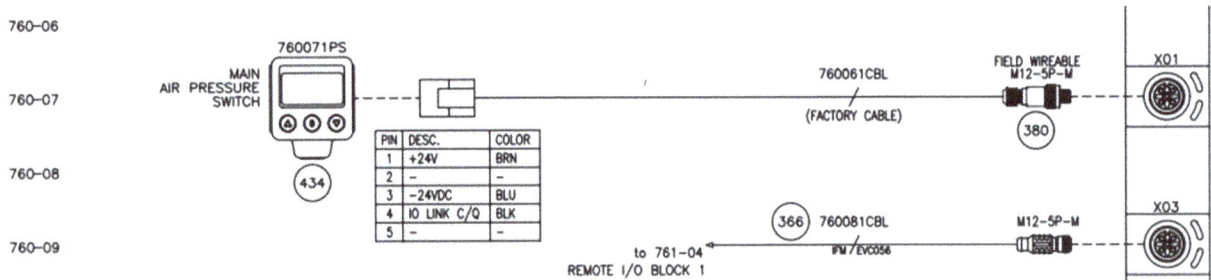

Figure 378 - IO-Link Connection to Pressure Switch

Continuing from the output side of the Soft Start valve air is connected to all of the valve banks.

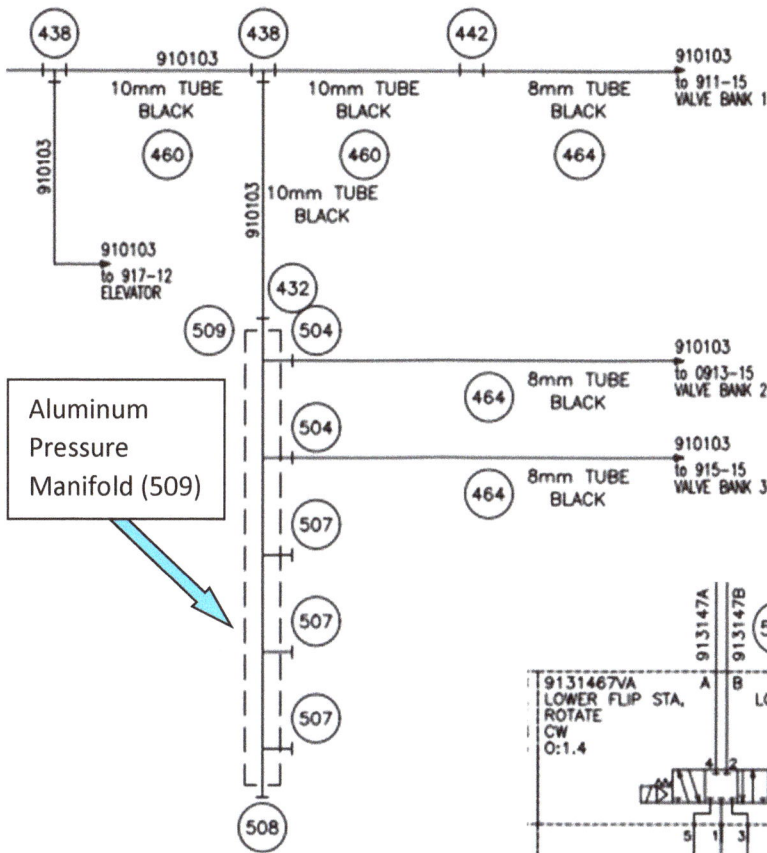

Figure 379 shows the distribution of air to the Valve Banks. As mentioned previously, the air before the Soft Start is also routed to some of the valves in VB2 and VB3. The air feed through the body of the valve manifold applied separately from the ends of the valve banks, and a "blocking disk assembly" is built into the manifold.

Figure 379 - Air Distribution and Blocking Disk

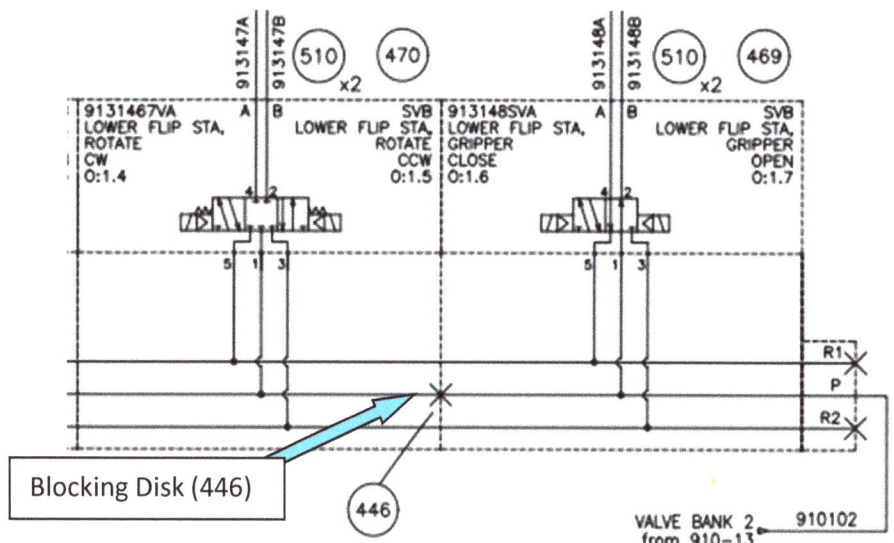

Item 446 in the pneumatic BOM shows a part number for the blocking disk, but the valve banks are generally built at the factory before shipping, with the different valve types, plugs, and fittings specified by the designer.

Item 509 is an aluminum manifold ordered from McMaster-Carr. This is a standard item with the number and sizes of ports specified in the part number; it is once again important to look at the manufacturer's details carefully when ordering replacements or fittings and plugs for the ports. The inlet ports are specified as 3/8" NPT and the outlet ports are specified as ¼"NPT. All of the fittings in the BOM are also detailed with their thread and hole sizes.

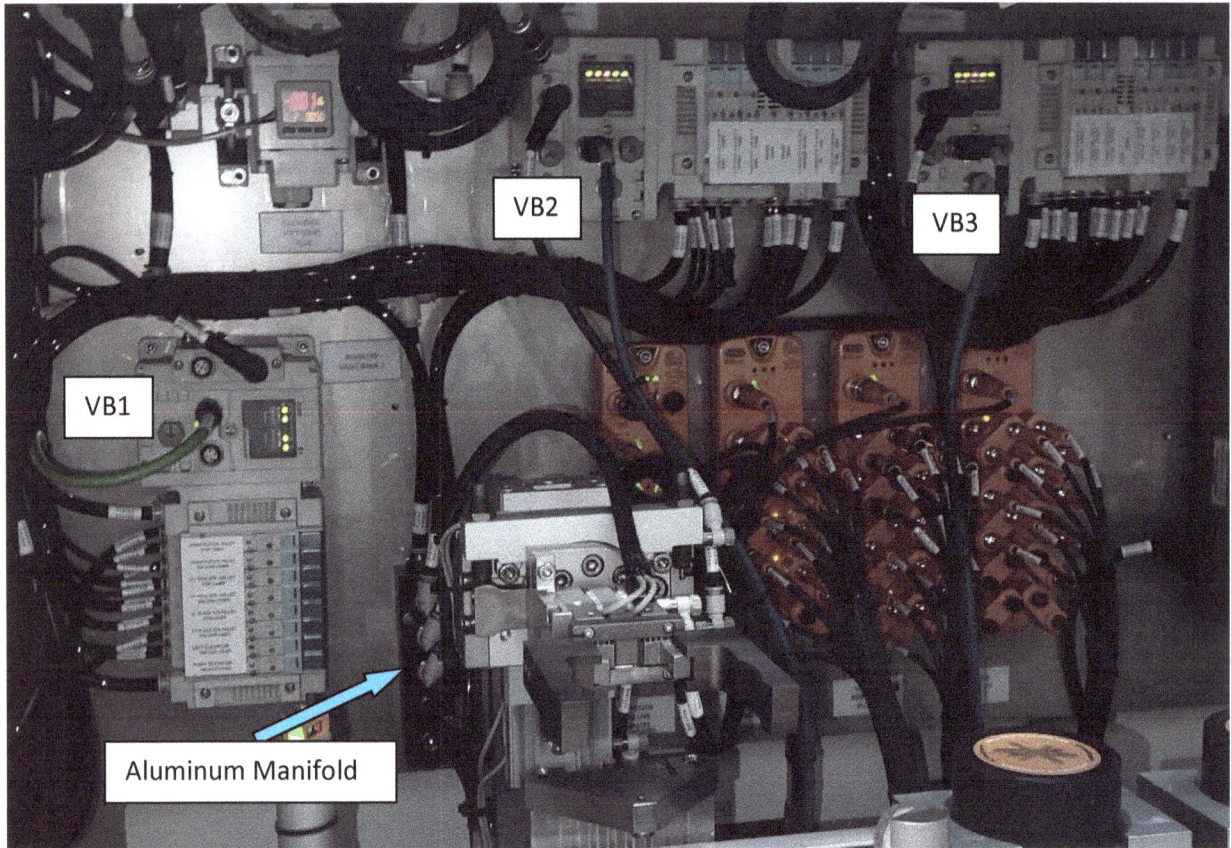

Figure 380 - Valves and Manifold

Figure 380 shows the plumbing to the valve banks and manifold. All hoses and components are labeled with the designations from the drawings.

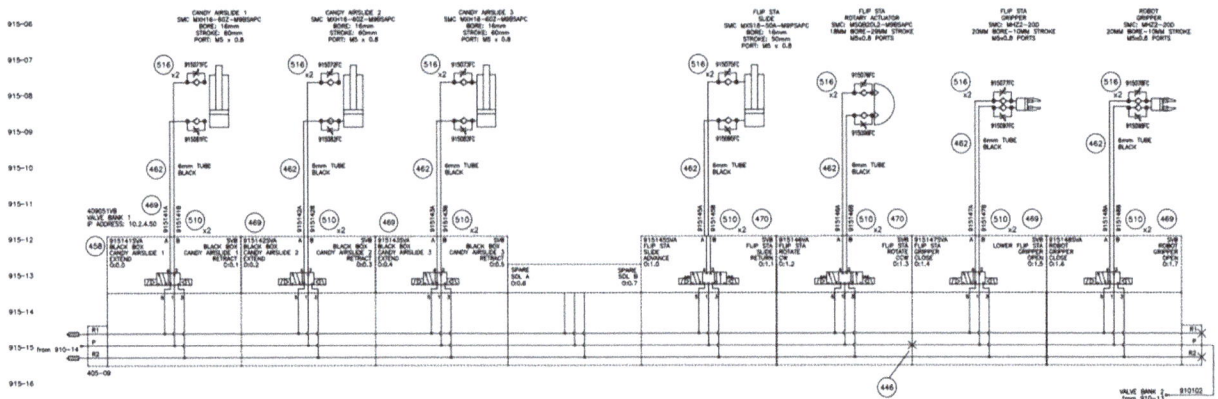

Figure 381 - Valve Bank 3

Figure 381 shows Sheet 915 with Valve Bank 3. While this page is sized for 11" x 17" prints and the text can't be read here, it is useful to see the different shapes of the cylinders for linear cylinders, rotary actuator, and grippers.

IO-Link Input Blocks and Stack Light

In addition to the **IO-Link Master** that was previously discussed, there are 5 IO-Link Remote Input Blocks.

The Master block (Item 281, IFM Part Number AL1122) is connected via Ethernet/IP to the ethernet switch as described previously. It has 8 device ports that can connect to IO-Link devices such as the DC circuit breakers and pressure switch discussed earlier. An additional IO-Link device connected to the master is a Balluff stack light with horn, Item 283 in the BOM, shown in Figure 382.

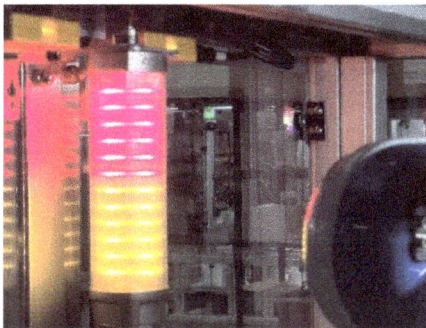

Figure 382 - Sweet Machine Stack Light

Stack lights are used to indicate the status of a machine or system by changing color. They also often contain a horn or buzzer. Typical stack lights are discrete output devices with fixed colors such as red, yellow, and green, but these

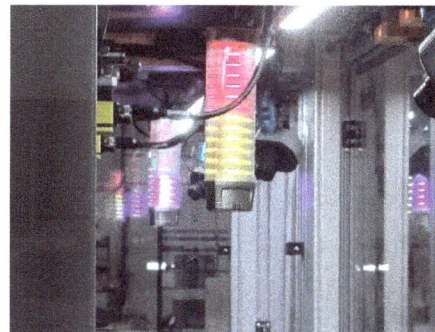

Figure 383 - Balluff SmartLights

"SmartLight" devices can show various additional colors as shown on the lights in Figure 283. Because the IO-Link Master can carry more information than simple on and off signals, these lights can be configured for various display formats, sounds, and modes.

IO-Link devices can transmit and receive multiple bytes of data depending on the configuration of the device, larger frames of data take more time to transmit. The input blocks each have 10 I/O ports, each of which can accommodate two digital signals plus the status of the port and health of each device.

The five input blocks on the Sweet Machine are listed below:

Block	Port	Sheet
Remote Block 1 - Base	X03	761
Remote Block 2 - Base	X04	762
Remote Block 3 - Base	X05	763
Remote Block Left Elevator	X07	765
Remote Block Right Elevator	X08	770

Because each port on the I/O block has two inputs available, a "Y Splitter" connector is used on each port. The port on the input module itself is an M12 connector, while each of the sensor connections is an M8.

Figure 384 - Right Elevator IO-Link Module

Figure 384 shows the input module for the right hand elevator. Notice that there is no room to insert fingers to remove the splitter module, and very little space to access the end of the sensor cable. This can make maintenance tasks difficult: it may be necessary to remove multiple cables to access the one you want.

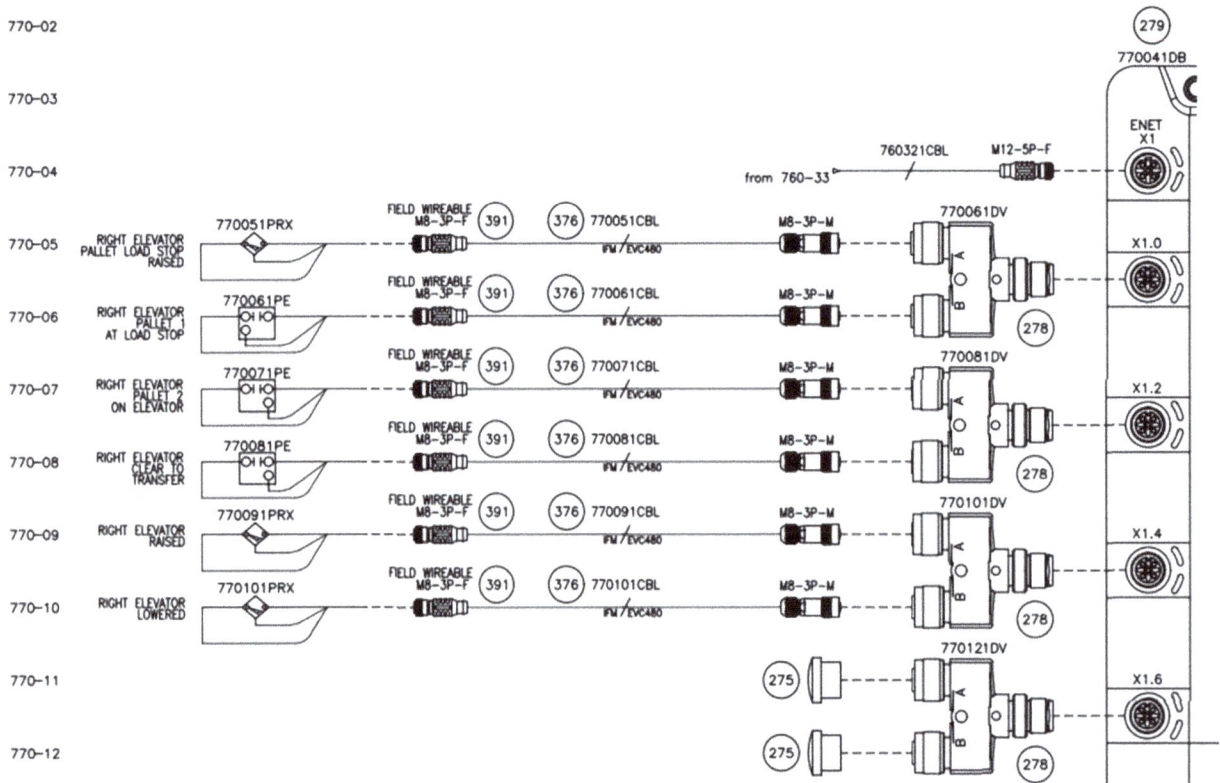

Figure 385 – Right Elevator Block, Sheet 770

Figure 385 shows the connections to the right elevator block. There are field wireable connectors specified for the end connected to the Y splitter and to the end of the cable where the sensor is attached.

Figure 386 - 3 Pin IFM Field Wireable Connectors

Figure 386 shows a 3 pin male field wireable connector before it is assembled. The small plastic pieces at the lower right are where the wires are inserted prior to assembling the connector, there are colored numbers on the pieces that show which wire is to be inserted for each pin. Since these field wireable connectors are also made by IFM, they mate up properly with the Y Splitters and the ports on the input block. The colors brown and blue for + and – DC and black for the signal are universal, but the pin numbers may differ for other manufacturers. It is important to either ensure that the same product is used when replacing the connector, or that the documentation is checked carefully before using a similar product from a different manufacturer.

Figure 387 - IFM "Y Splitter"

The Y splitter units themselves may also need to be replaced, they are not serviceable and cannot be disassembled. They are labeled with an A and B corresponding to the input used on the port.

There are 67 input devices connected to the five input blocks, leaving 33 spare inputs for expansion. All blocks have Y splitters attached to the ports, and all unused connections have plastic caps covering them. This is to ensure that liquid or debris cannot contact the electrical contacts.

As long as connections aren't hit with an object, they should stay intact for a long period of time. Vibration, moisture, heat, and other environmental conditions can affect these so-called "quick disconnect" connectors and it is important that technicians understand how they work.

Additional Schematic Sheets

The Sweet Machine drawings include several additional sheets that are very helpful to a technician. Sheets 800, 820 and 821 show the main enclosure's exterior and interior layout, including BOM reference numbers for components.

Sheets 875 and 876 show the overall cable layout and a detail of the sensor cables connected to the IO-Link input blocks.

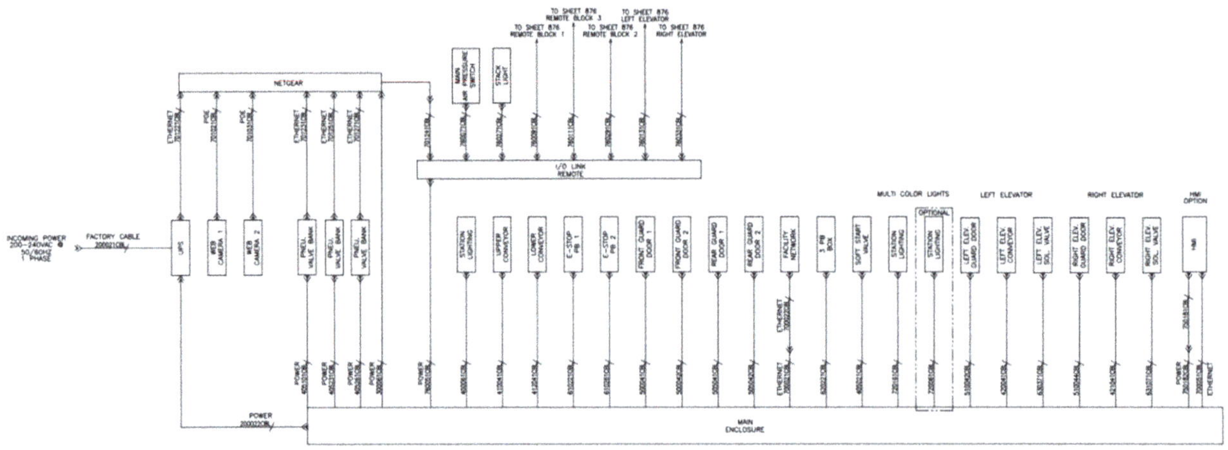

Figure 388 - Overall Cable Layout

Though the items can't be read in the figure above, the diagram indicates the relative positions of the cable entries on the back and side of the enclosure.

Sheet 877 shows details of the Icotek brand grommet frame used for cable entry to the enclosure. Sheet 880 contains the machine tags for nameplates, including the main plate mounted on the front of the enclosure.

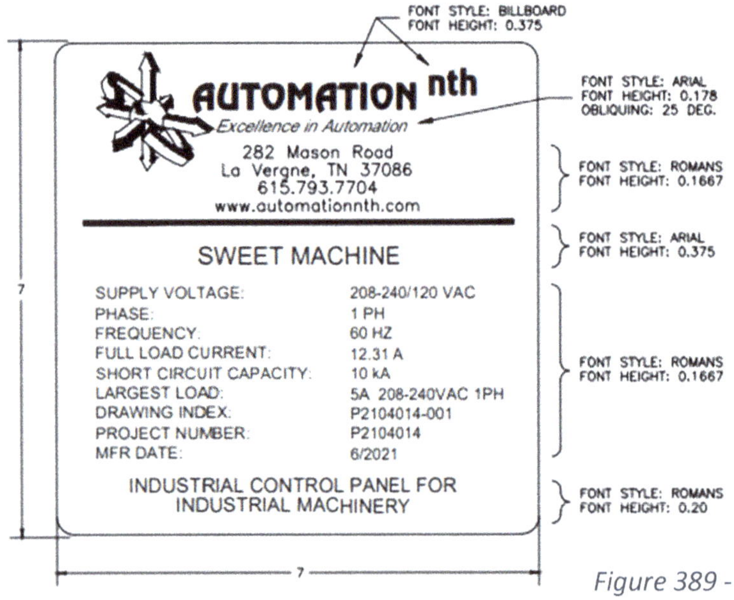

Figure 389 - Machine Nameplate

Sheet 886 shows the field device panel under the front of the machine, and as mentioned previously, sheets 890 and 891 contain the BOM (Bill Of Materials) for the electrical components.

The pneumatic drawings begin at sheet 900. A very useful page show the pneumatic drawing symbols on Sheet 901. The Pneumatic BOM is on Sheet 980.

An important note about schematic drawings: This set of drawings is very detailed in comparison with many of the machines you will find in the field. If information is missing for the machinery in your plant, it may be up to you to document the machine. Even handwritten notes are better than nothing at all, you can save a lot of time with proper documentation.

Software

Microsoft Excel

Excel is one of the most useful software tools available for creating lists and project documentation. PLC and HMI tags can be exported as .csv files and used to generate text descriptions for AutoCAD or other programs.

One of the most important features of spreadsheets like Excel are that formulas can be entered into a cell that references values in other cells. This allows calculations to be performed on values and updated when the referenced numbers change. Columns can also easily be totalized, and the format of cells can be changed to text, currency, or other number formats.

Cells can be highlighted in different colors, and borders can be set up to highlight or separates worksheet areas from each other. This makes Excel one of the most powerful presentation tools for engineering design.

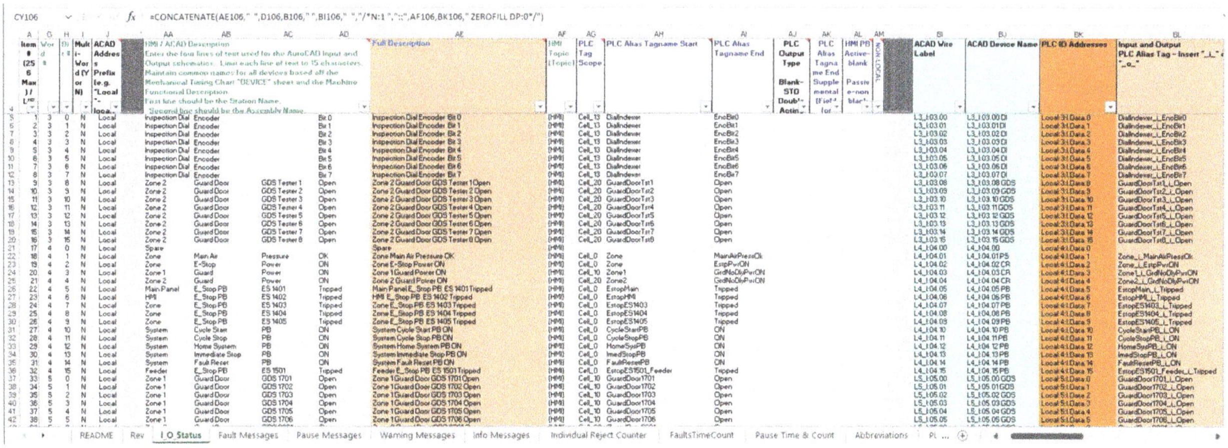

Figure 390 - IO List Data

Figure 390 shows a very large spreadsheet with many different tabs used for PLC I/O data. This tool is used to design very large programs as described previously; it even has macro code that automatically generates rungs for the Allen-Bradley RSLogix 5000 platform. Since all ladder logic has a text-based equivalent (Instruction List), logic can be exported and manipulated within the spreadsheet to create code!

Item	Abbreviation
Input	I
Input Status	Stat
Output	o
HMI Pushbuttons	hPb
HMI Indicators	hInd
Permissives	Perm
Fault	Flt
Auto Sequence Summary	Auto
Home Sequence Summary	Home

AddrForm#	Select Address Format:
2	Allen Bradley CLGX
	Example:
	Local:1:I.Data.0, Local:2:O.Data.0

TagOrder#	Select Tag Order::
1	Device-Assy-Action
	Device-Assy-Action
	Assy-Device-Action
	Assy-Action-Device
	Action-Assy-Device

Component	Abbreviation
Proximity Switch	PX
Limit Switch	LS
Reed Switch	RS
Pushbutton	PB
Pressure Switch	PS
Pressure Indicator	PI
Master Control Relay	MCR
Emergency Stop	ESTP
Guard Door Switch	GSW
Spare IO (Not Used)	Spare
Solenoid Valve	SV
Pilot Light	PL
Buzzer	BZ
Motor Starter	MS

Slot	Description or Part #	Type	Prefix	Inputs	Outputs	SlotType#
0		Processor	Local	0	0	2
1		Digital In	Local	16	0	3
2		Digital In	Local	16	0	3
3		Digital In	Local	32	0	3
4		Digital Out	Local	0	16	4
5		Digital Out	Local	0	16	4
6		Analog In	Local	4	0	5
7		Analog Out	Local	0	4	6
8		Communications Serial	Local	0	0	11
9				0	0	0
10				0	0	0
11				0	0	0
12				0	0	0
13				0	0	0
14				0	0	0
15				0	0	0
16				0	0	0
17				0	0	0
18				0	0	0
19				0	0	0

Program Heirarchy	Label
Level 1 (Top)	Zone
Level 2 (Mid)	Station
Level 3 (Lowest)	Cell

Figure 391 - IO Configuration

Figure 391 shows a configuration screen that converts cells in the IO List Data sheet to the desired format.

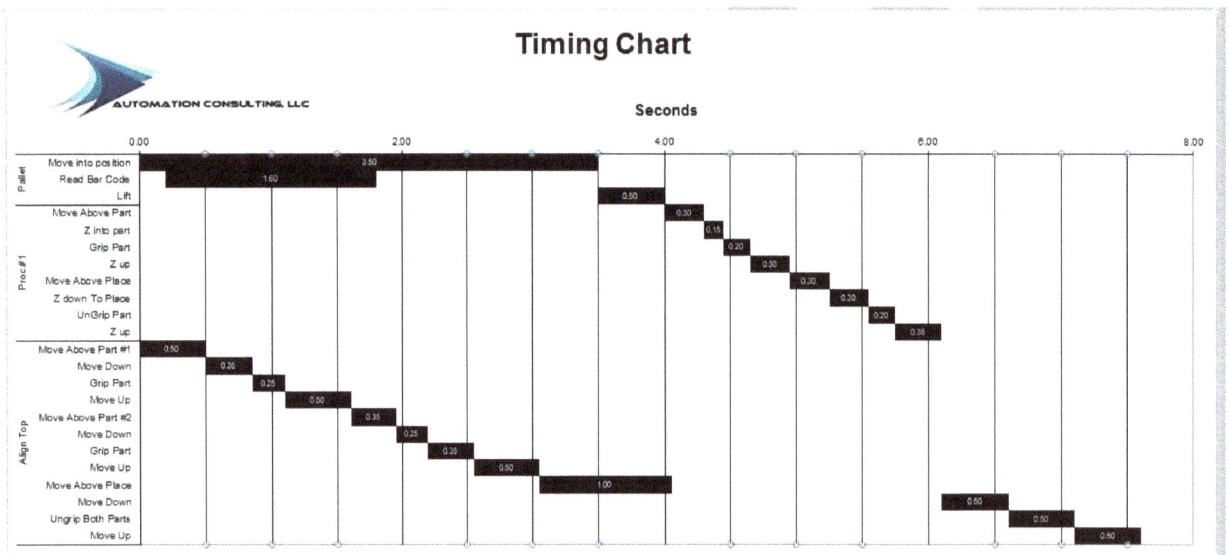

Figure 392 - Timing Chart

The ability to create graphics like pie charts and bar graphs based on data makes Excel useful for presentation and analysis. The timing chart above was created by entering data into the form in Figure 393.

Category	Action	Device Name	Absolute	Start	Duration	Elapsed	Start	Duration
Pallet	Move into position	Conveyor	Absolute	0.00	3.50	3.50	0.00	3.50
	Read Bar Code	Conveyor	Start of 3:Move into position+0.2	0.20	1.60	1.80	0.20	1.60
	Lift	Conveyor	Finish of 3:Move into position	3.50	0.50	4.00	3.50	0.50
Proc #1	Move Above Part	Picker #1	Finish of 5:Lift	4.00	0.30	4.30	4.00	0.30
	Z into part	Picker #1	Finish of 6:Move Above Part	4.30	0.15	4.45	4.30	0.15
	Grip Part	Conveyor	Finish of 7:Z into part	4.45	0.20	4.65	4.45	0.20
	Z up	Picker #1	Finish of 8:Grip Part	4.65	0.30	4.95	4.65	0.30
	Move Above Place	Conveyor	Finish of 9:Z up	4.95	0.30	5.25	4.95	0.30
	Z down To Place	Picker #1	Finish of 10:Move Above Place	5.25	0.30	5.55	5.25	0.30
	UnGrip Part	Picker #1	Finish of 11:Z down To Place	5.55	0.20	5.75	5.55	0.20
	Z up	Picker #1	Finish of 12:UnGrip Part	5.75	0.35	6.10	5.75	0.35
Align Top	Move Above Part #1	Robot	Absolute	0.00	0.50	0.50	0.00	0.50
	Move Down	Robot	Finish of 14:Move Above Part #1	0.50	0.35	0.85	0.50	0.35
	Grip Part	Robot	Finish of 15:Move Down	0.85	0.25	1.10	0.85	0.25
	Move Up	Robot	Finish of 16:Grip Part	1.10	0.50	1.60	1.10	0.50
	Move Above Part #2	Robot	Finish of 17:Move Up	1.60	0.35	1.95	1.60	0.35
	Move Down	Robot	Finish of 18:Move Above Part #2	1.95	0.25	2.20	1.95	0.25
	Grip Part	Robot	Finish of 19:Move Down	2.20	0.35	2.55	2.20	0.35
	Move Up	Robot	Finish of 20:Grip Part	2.55	0.50	3.05	2.55	0.50
	Move Above Place	Robot	Finish of 21:Move Up	3.05	1.00	4.05	3.05	1.00
	Move Down	Robot	Finish of 13:Z up	6.10	0.50	6.60	6.10	0.50
	Ungrip Both Parts	Robot	Finish of 23:Move Down	6.60	0.50	7.10	6.60	0.50
	Move Up	Robot	Finish of 24:Ungrip Both Parts	7.10	0.50	7.60	7.10	0.50
			Finish of 24:Ungrip Both Parts					

Figure 393 - Sequence List for Timing Chart

Timing charts are an important design and analysis tool for mechanical designers.

Databases and Sequential Query Language (SQL)

Closely related to Excel are **Databases**. Information is stored in rows and columns like Excel, but multiple tables can be built and related to each other. By using **Sequential Query Language (SQL)**, tables can be searched for information and easily updated by users.

A database is simply an organized collection of data stored and accessed electronically. There are many different database models, and vendors often make proprietary commercial versions as part of their software platform. Graphical front-ends are used to make data easily obtainable and help avoid mistakes from users not knowing SQL.

A **Database Management System (DBMS)** is the software that interacts with end users, applications, and the database itself to capture and analyze data. It also includes the core tools required to administer the database.

Management of databases usually includes the following four functions:

Data Definition – Creation, modification and removal of definitions that define data organization.

Update – Insertion, modification, and deletion of the actual data.

Retrieval – Providing information in a form usable to an operator or for processing by another application. This often involves combining information from the database.

Administration – Registering and monitoring users, enforcing data security, monitoring performance, maintaining data integrity, and recovering information that has been corrupted by unexpected system failures.

CAD and Drawing Software

Most professional CAD and drawing packages require a license and can be expensive. Examples of electrical CAD packages are AutoCAD, ETAP and EPlan. There are free or inexpensive software packages available also. Most CAD packages are dxf (Drawing Exchange Format) compatible, but less expensive CAD packages can be less than user-friendly.

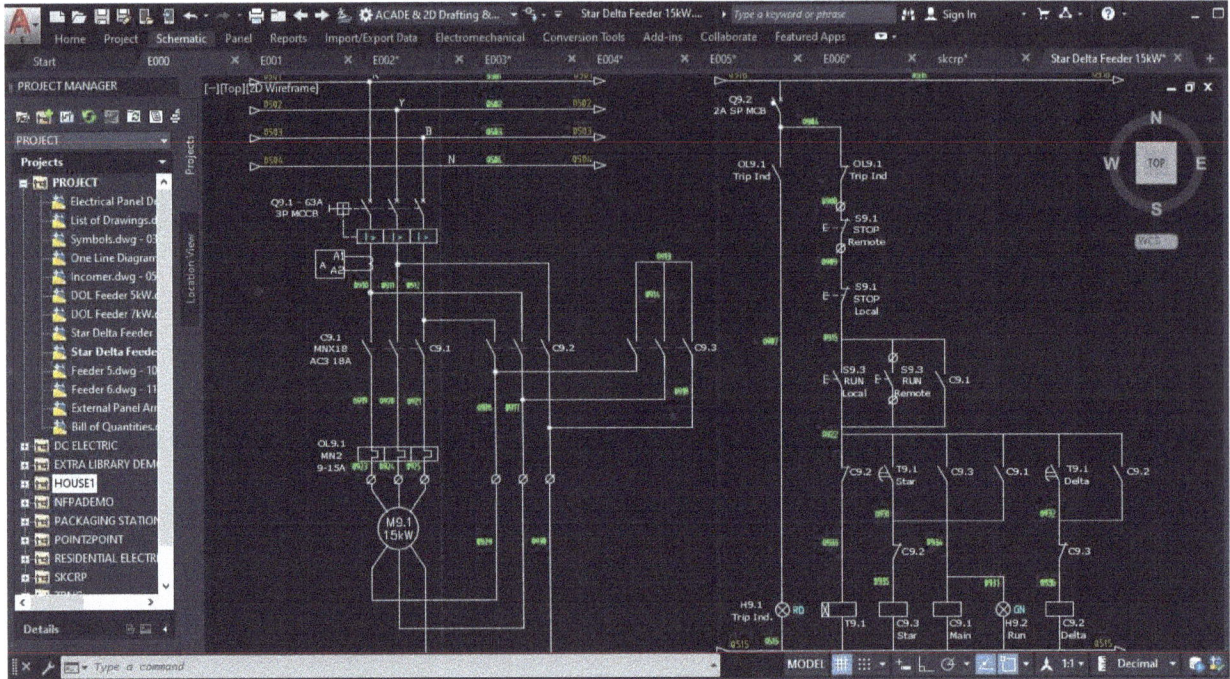

Figure 394 - AutoCAD Electrical

Professional mechanical drawings are often done using 3D modeling software. SolidWorks, AutoCAD Inventor, Fusion 360 and Pro/ENGINEER are examples. 3D software can also be used to create models for 3D printers. Some of these 3D modeling packages such as Blender, Sketchup, or FreeCAD are inexpensive or free!

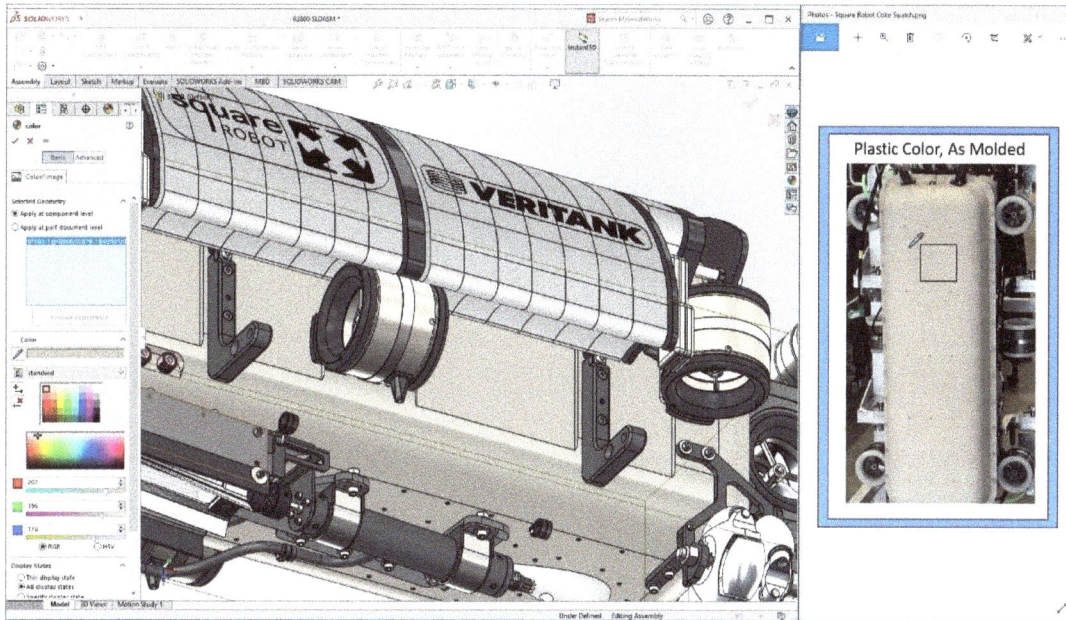

Figure 395 - SolidWorks 3D Modeling Software

Other software packages like Microsoft's Word, Excel, and Visio have drawing utilities that can be useful for creating basic sketches while documenting projects or machinery. MS Paint is also useful for both drawing and modifying or annotating pictures.

PLC Software

PLC software is dependent on the platform or brand being used. Most of the major manufacturers require a license for each instance of software, and the license can be very expensive. These companies also generally have less expensive platforms where the software costs less or is even free, but functionality will be reduced.

There may be versions of the software that only allow viewing of the logic, but not editing. Since there are 5 IEC PLC programming languages, the basic software package may not include all of the languages and require a separate add-on module.

Allen-Bradley

Allen-Bradley or Rockwell Software has several PLC software platforms including RSLogix 500, FactoryTalk Studio 5000, and Control Components Workbench, or CCW. In addition, RSLinx software is required for communications configuration for all PLC and HMI platforms.

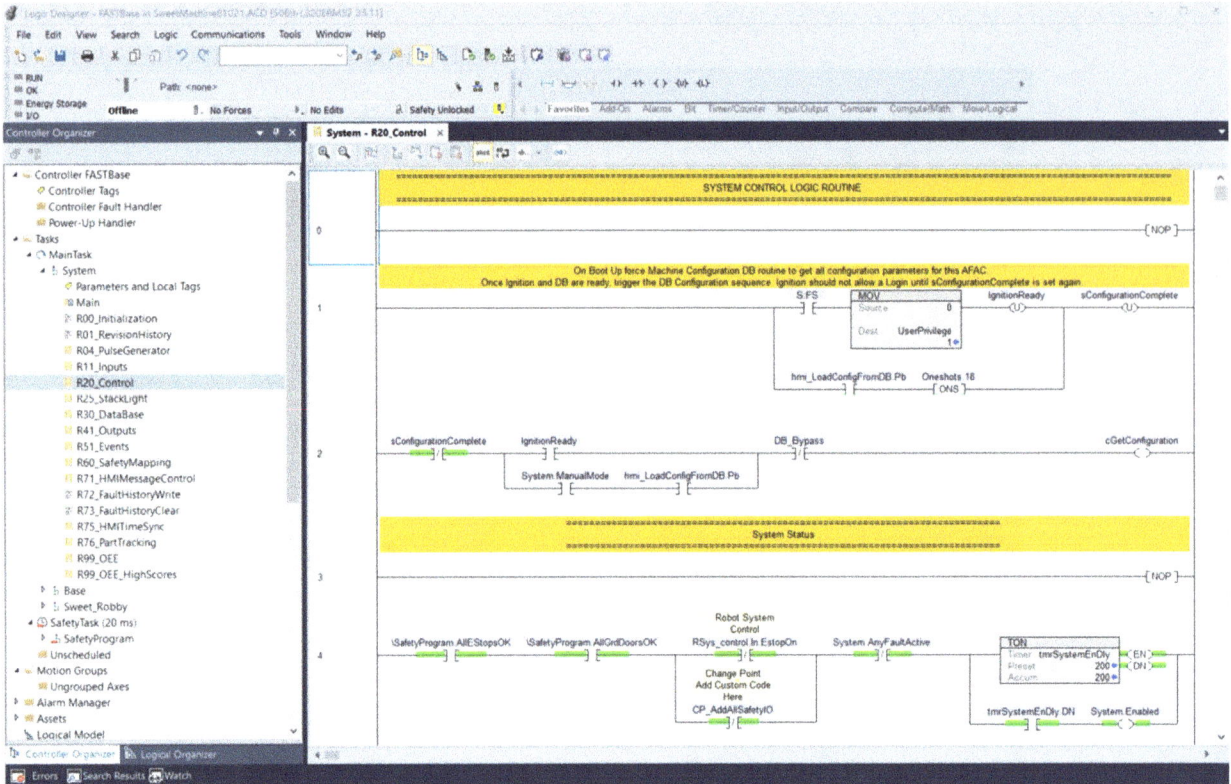

Figure 396 - RSLogix Studio 5000 PLC Software

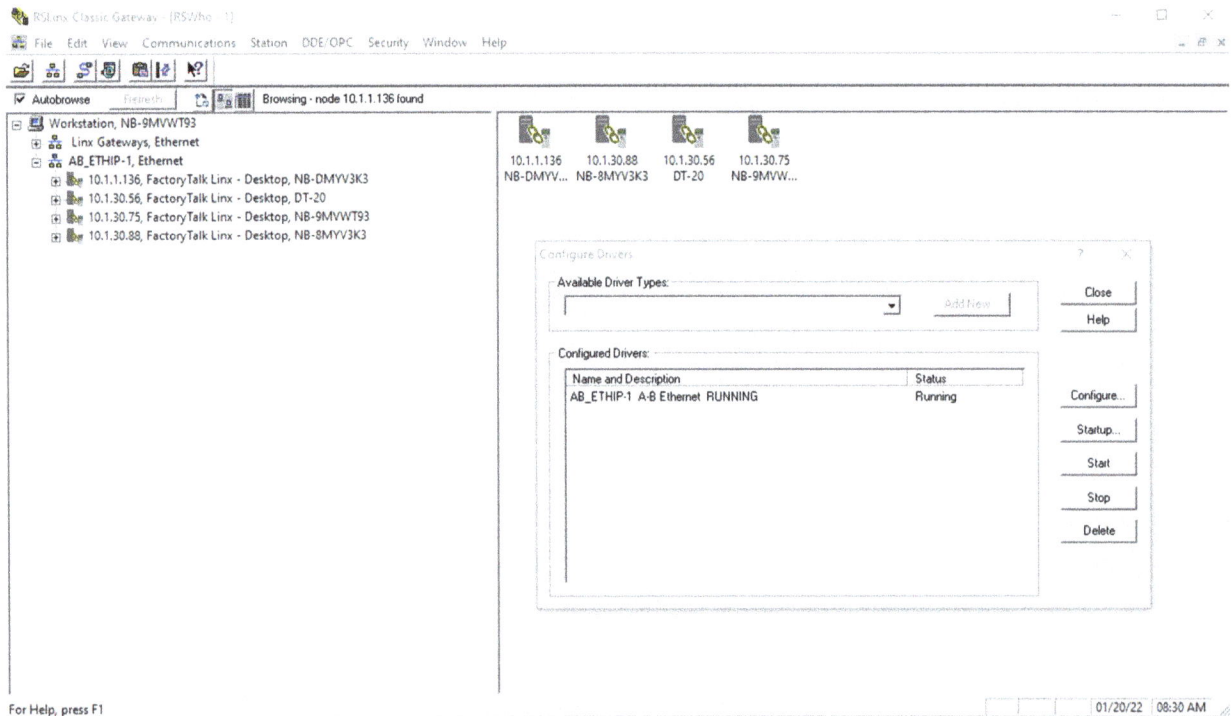

Figure 397 - RSLinx Communication Driver Software

Siemens

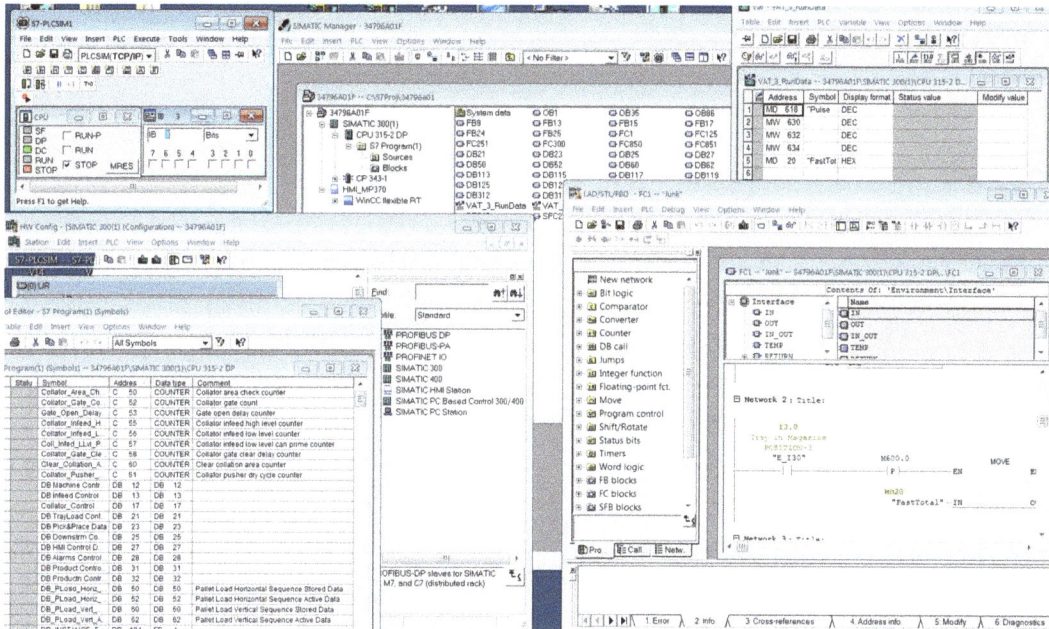

Figure 398 - Siemens Step 7 Software

Figure 398 shows a screenshot of Siemens Step7 software, used for the 300 and 400 controllers. This software is composed of many separate programs and can be difficult to manage.

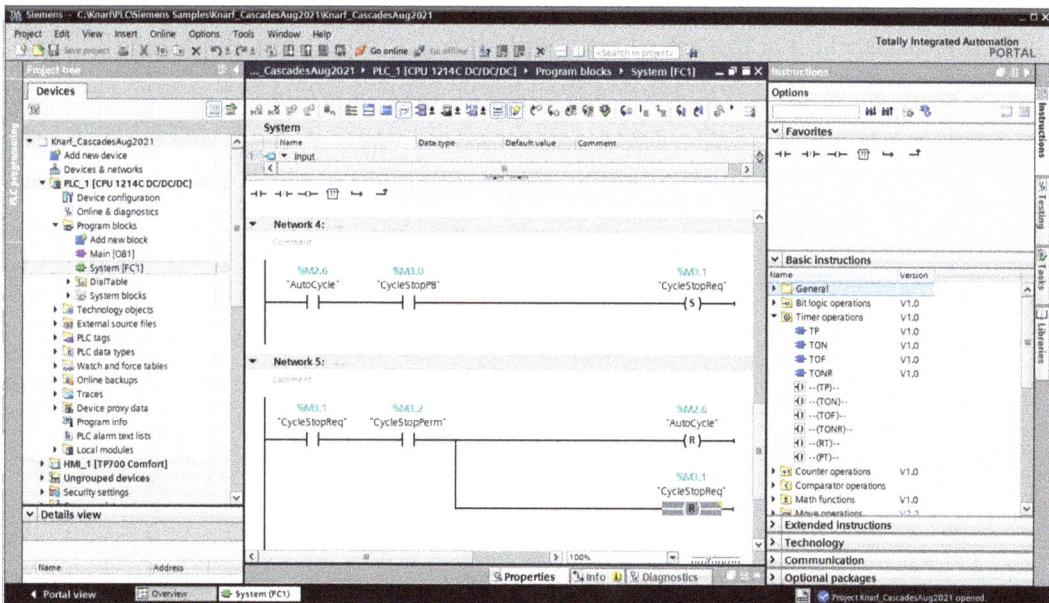

Figure 399 - Siemens TIA Portal

Siemens TIA Portal, used for newer tag-based systems, is an integrated environment that is more user friendly.

Simulation and Emulators

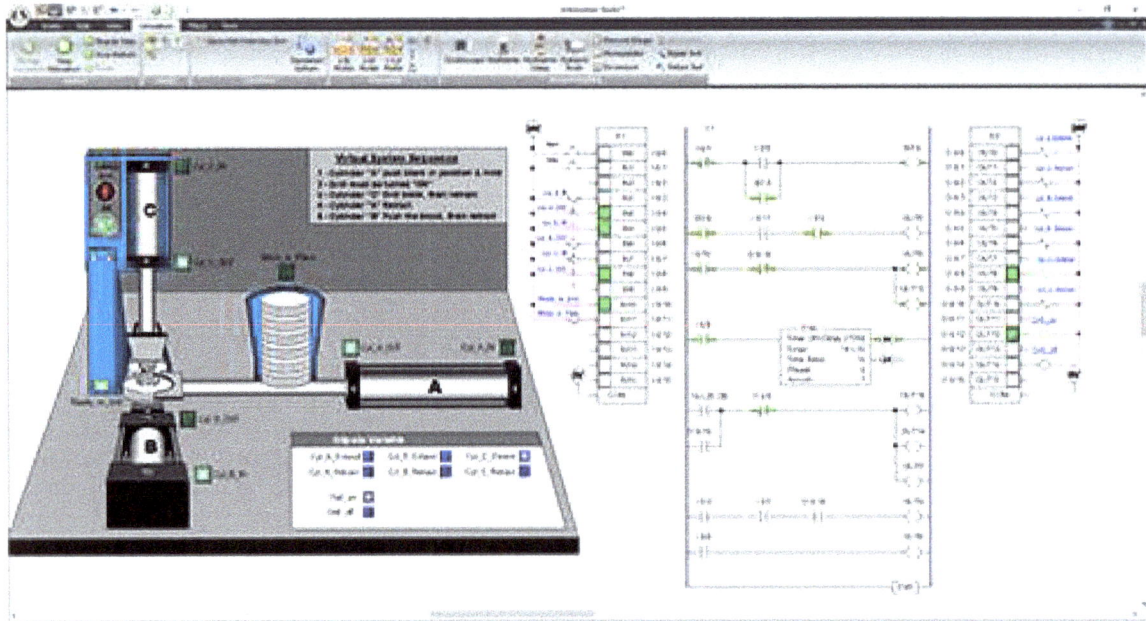

Figure 400 - Famic Automation Studio

Figure 400 shows a PLC simulation package that allows students to practice programming with a mock-up. This is great for students but can also be used to test code. Many of the major manufacturers such as Allen-Bradley and Siemens have emulation software like this.

Of course, you must also have the appropriate software for the brand and model of PLC in your plant!

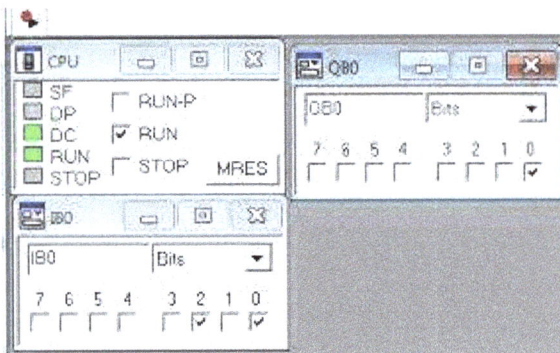

Siemens PLCSIM software is built into the Step 7 Professional package and makes it easy to test programs. Allen Bradley's RSEmulator program must be purchased separately.

Most HMI software allows screens to be connected to a PLC or a simulation to see how the screens and objects look also.

Figure 401 - Siemens PLCSIM

HMI Software

HMIs (Human-Machine Interfaces) differ from SCADA in that they usually run on a dedicated interface or touchscreen. As with PLCs, the software is dependent on the platform or brand. It is very common for the HMI to be the same brand as the PLC, since the communication driver is common to both.

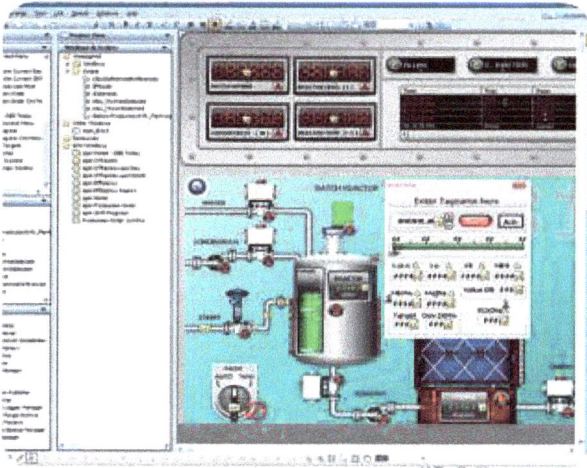

Often the HMI software will be built into the overall programming package, as with Siemens WinCC. There also brands that are independent of the PLC manufacturer such as ProFace or Weintek/Weinview. HMI software is graphically driven, where objects are drawn on the screen and linked to an address in the controller.

Figure 402 - HMI Development Software

SCADA Software

SCADA is an acronym for "Supervisory Control and Data Acquisition". It is used for overall monitoring and control of plant and machine functions. It generally includes all the features of an HMI, but also adds computer-based features such as Data Historians and the ability to use SQL.

SCADA development software is usually quite a bit more expensive than HMI software but has much more capability. It is sometimes priced by the number of "tags" or data points exchanged with devices.

Ignition

Ignition by Inductive Automation is a widely used SCADA software that is used on many of Automation NTH's machines and production lines instead of an HMI. It is also used for plant-wide data acquisition and monitoring at American Beverage Depot, described in the Systems part of this book on beverage processing.

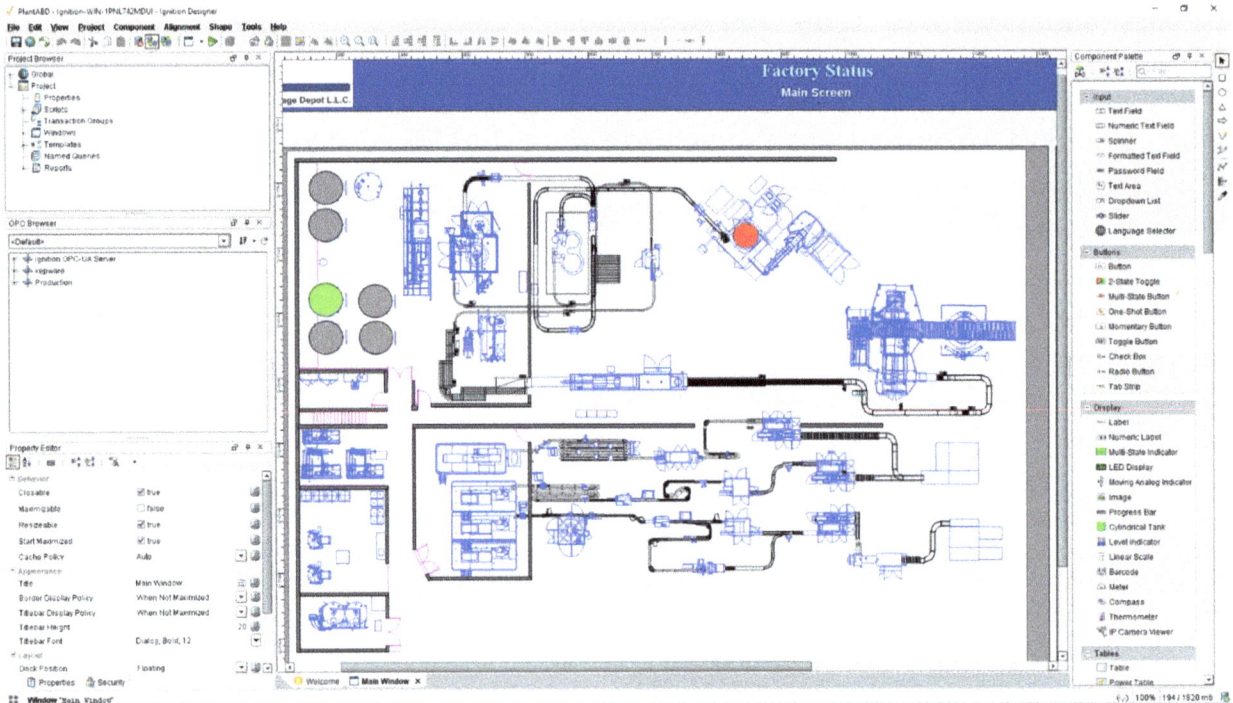

Figure 403 - Ignition Development Screen – American Beverage Depot

In the picture above, a layout of the plant provides a place to show the presence of trouble spots and machine status, and by clicking on an area can be used for screen navigation.

Figure 404 shows an overall status of production machinery by line, including a list of alarms on the right side. The alarms currently active are at the top in red.

Figure 404 - Status Screen – ABD Lines

Sweet Machine HMI

The HMI on the previously described Sweet Machine also uses Ignition. Complex machines may use a computer as an operator interface so that data can be stored, and auxiliary software can run on it.

Below are some of the screens used on the Sweet Machine:

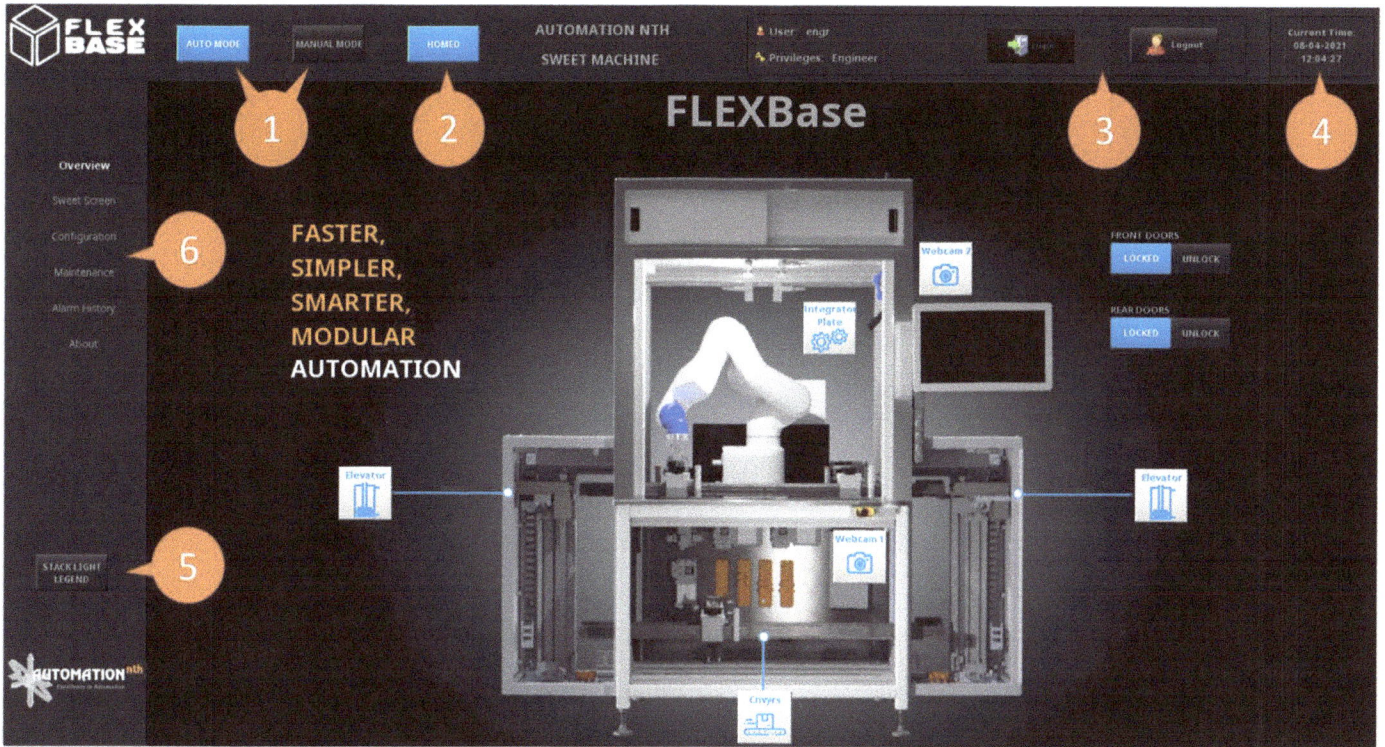

Figure 405 - Sweet Machine Overview Screen

Number	Name	Description
1	Auto/Manual Pushbuttons	Switch between Auto and Manual Operation Modes
2	Home All Pushbutton	Homes all stations of the system. It shows Homed when complete
3	User Access	Login or logout to gain access to more privileges
4	Time	Displays the current Time and Date
5	Stack Light Legend	Opens a pop-up window depicting what the stack light is indicating
6	Navigation Bar	Navigates to other screens of the HMI

Notice that 3D models of the machine and components are used to improve operator interaction.

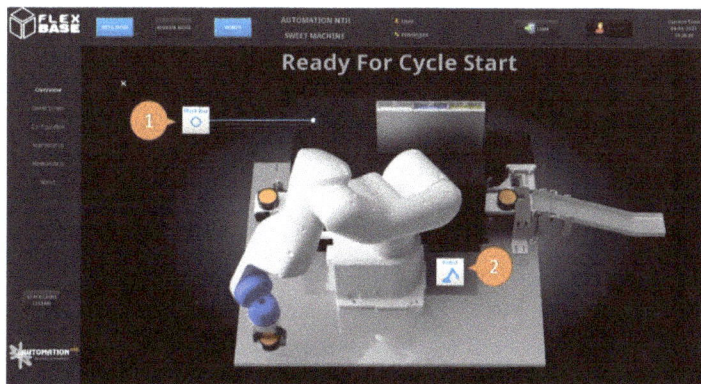

Figure 406 - Robot Main Screen

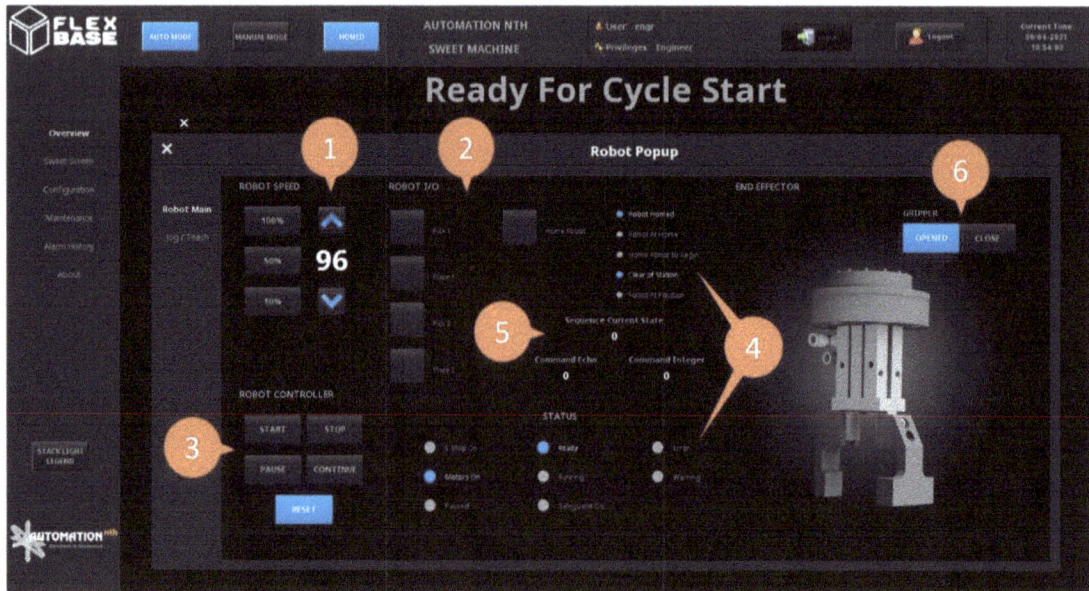

Figure 407 - Robot Teach/Jog Screen

Number	Name	Description
1	Robot Speed	Used to change the robot speed
2	Robot IO	In manual mode, move the robot to any station
3	Robot Controller	Control the robot, stop the program, pause and continue and clear faults
4	Robot Status	Monitor the status of the Robot Controller
5	Commands	Monitor the commands and that go to the robot
6	Gripper	Open and close the gripper in manual mode

The Epson robot on the Sweet Machine does not use a teach pendant, but instead can be controlled and taught from the HMI.

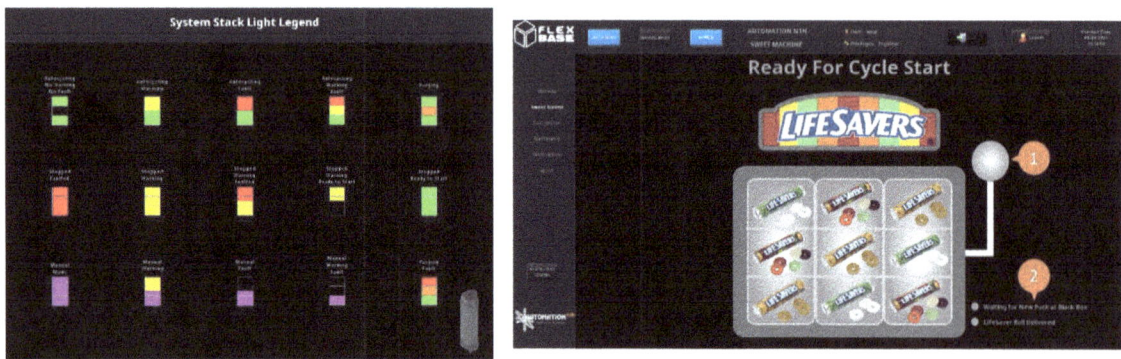

Figure 408 - Light Stack Legend (Left) and Sweet Screen (Right)

Colorful screens can enhance the appeal of any machine!

Device and Configuration Software

There are many utility software packages that are used to configure specific devices, set up communications, or assist in selecting and sizing components. These packages are usually provided by the manufacturer of the equipment being configured, but sometimes can be used for other companies' equipment, such as with BootP servers.

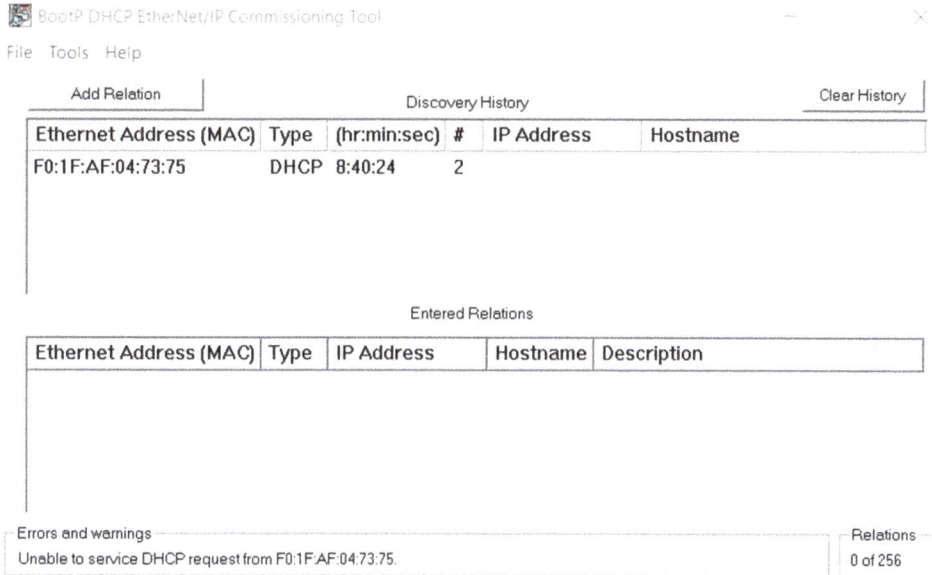

Figure 409 - BootP Server Commissioning Tool

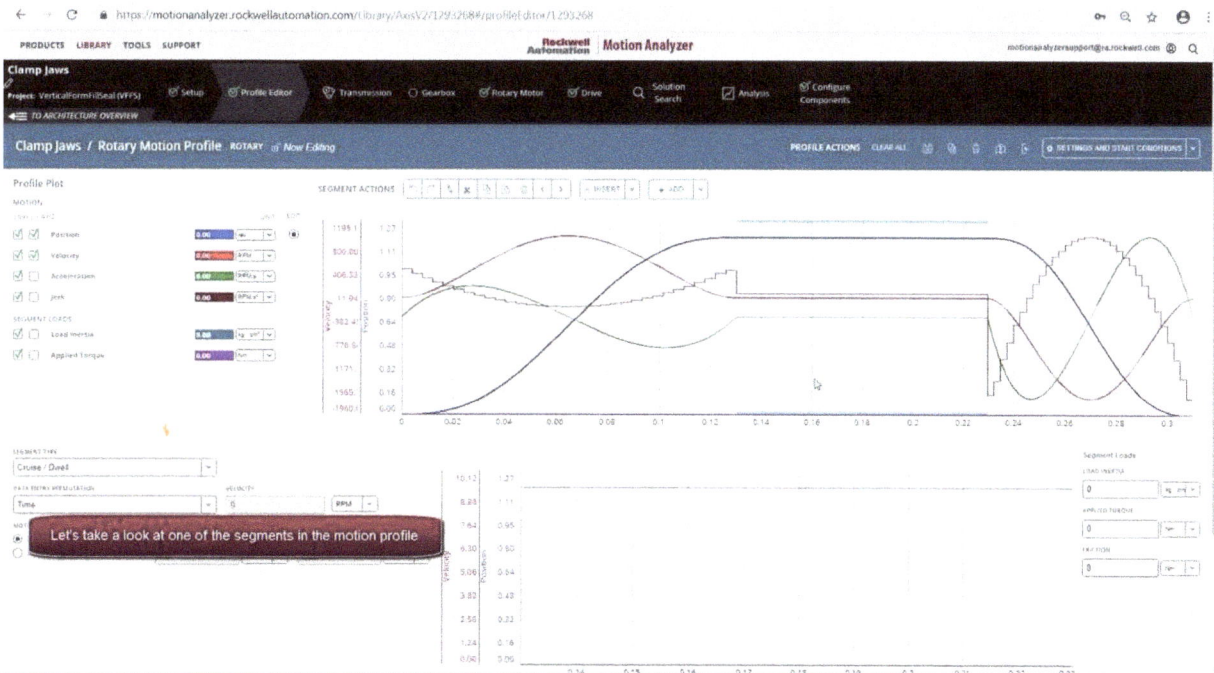

Figure 410 - Rockwell Motion Analyzer

The OEE Optimizer

An interesting tool developed by Automation NTH is the **OEE Optimizer**. This software can be installed on a machine to help improve the operation of a machine by analyzing the causes of downtime.

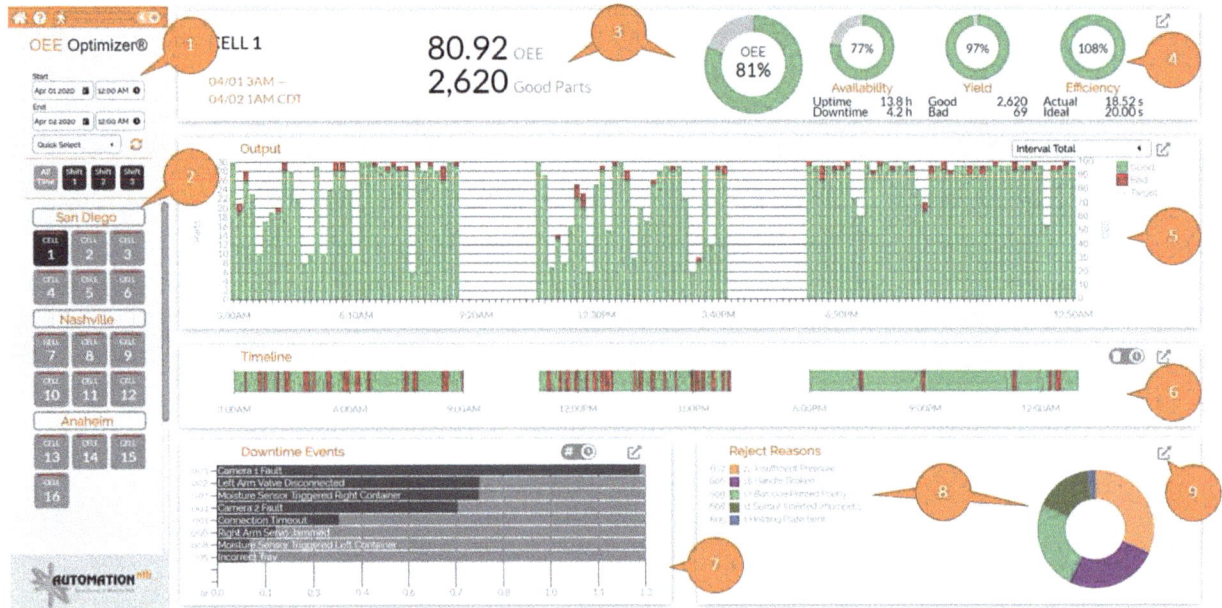

Figure 411 - OEE Optimizer Webpage Screen

KEY		
#	**WIDGET**	**DESCRIPTION**
1	Date/Time Picker	Select date and time range for visible data
2	Site Picker	Select location from which to pull OEE data
3	Overall OEE Score	Resulting OEE product
4	Individual OEE Metrics	Factors which when multiplied yield OEE product
5	Yield	Ratio of good to total parts
6	Availability	Ratio of uptime to planned total uptime
7	Efficiency	Ratio of actual cycle time to a designed ideal cycle time
8	Reject Reasons	Cause for machine to reject a part
9	Export	Exports data from a given section

The OEE optimizer is a web-based visualization tool for real-time, in-process data and analytics. It includes a range of features that guides an operator or technician to target items that are most impacting OEE score.

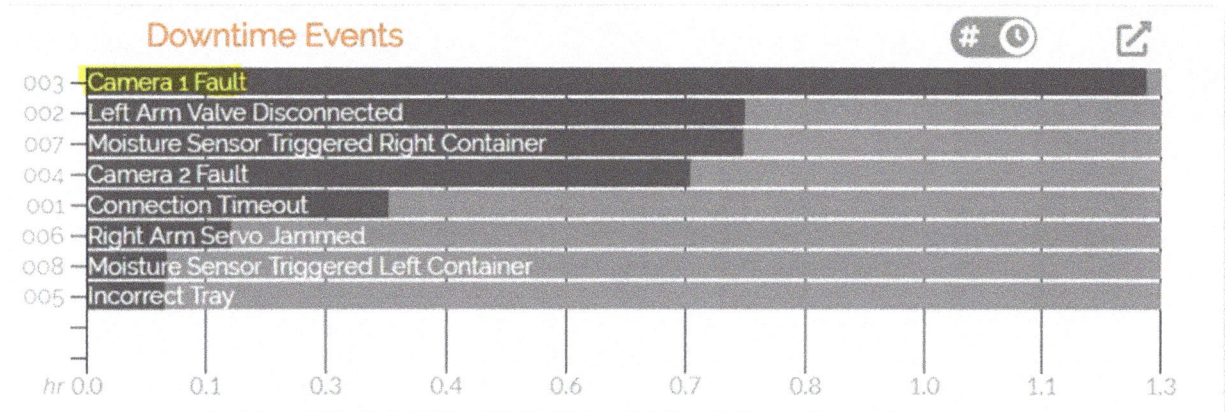

Figure 412 - Downtime Events

The usefulness of the OEE Optimizer is in its ability to highlight the biggest contributors to events that cause a reduction in OEE Score. For example, in Figure 412, you can easily see that the Downtime Event that occurred for the most cumulative time (not to be confused with the event that happened most frequently) is "Camera 1 Fault". Remedying the "Camera 1 Fault" would result in the greatest improvement in Availability and increase the overall OEE Score accordingly.

Being web based the software can also be accessed from various device including phones and tablets.

A Troubleshooting Journey

This final section uses the knowledge and techniques discussed in this book to take the reader through a series of steps to troubleshoot a complex piece of equipment.

The machine used in this exercise is a custom unit built by Automation NTH in Lavergne, Tennessee, near Nashville. It is used primarily as a trade show demo that "magically" converts pucks into Life Saver candies.

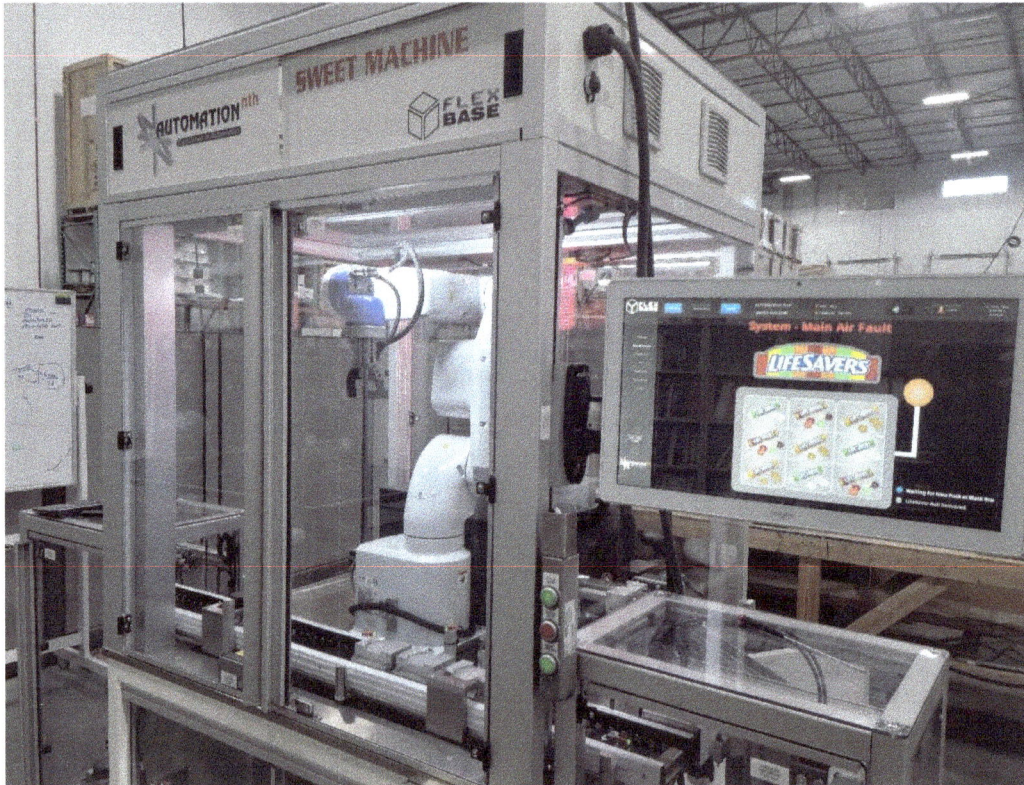

Figure 413 - The Sweet Machine

Much of the electrical and pneumatic system for the Sweet Machine has been described in the sections on reading schematics. The HMI was also discussed in the previous section.

Control System Fundamentals

At a basic level, automated control systems use input information from sensors and operator interfaces to control the movements and signals of machinery. The movements are a result of output signals that activate actuators, which may be in the form of air cylinders, motors, or even complex systems like robots.

The movements and signals are a result of logic in some type of controller, in the case of the Sweet Machine the main control system is a Safety PLC.

PLC code does not "break". If the code worked previously for months or years and suddenly the operation of the machinery changes, it is NOT a result of the logic changing. Sensors can become misaligned or malfunction, wires can become disconnected, actuators can break, and parts can get stuck, but the code is going to remain the same. PLCs either work or they don't.

Fortunately, there is usually a great deal of logic dedicated to helping a technician diagnose problems. Faults and alarms determine whether actuators and processed parts move where they are supposed to, and the HMI displays the resulting fault message. This in turn leads the technician to investigate the source of the problem.

Understanding the Process

As mentioned in the System/Theory chapter at the beginning of this book, the best way to learn how a machine runs is to observe it in operation. There is also often some kind of operations manual that explains how the machine is supposed to work.

As a last resort, the PLC program can be examined to determine exactly how sections of the machine work, but there are thousands of lines of code, and it can be tedious to read through all the different routines and take every effect into account. However, understanding the overall layout of the program is often useful before having to use it in troubleshooting tasks.

An important element of the PLC program is the addressing of the inputs and outputs. The electrical and pneumatic schematic section of this book mentions how the various signals are connected, but there is more to understand when finding how the I/O points are located in the PLC program.

Sweet Machine PLC I/O

There are several ways in which inputs and outputs are connected to the Sweet Machine as described in the schematics:

1. Point I/O Module (Ethernet/IP)

2. Valve Banks (Ethernet/IP)

3. IO-Link Master (Ethernet/IP)

4. IO-Link Input Modules (IO-Link)

5. Epson Robot (Ethernet/IP)

Ethernet/IP modules are placed into the I/O Configuration of the PLC program by selecting them from available EDC (Electronic Data Sheet) files when writing the program. Allen-Bradley devices are already included in the Logix Designer software, but third part devices like the robot, IO-Link modules and SMC valve banks are imported from files obtained from the manufacturer.

▲ I/O Configuration
    ▲ 5069 Backplane
        [0] 5069-L320ERMS2 FASTBase
    ▲ A1/A2, Ethernet
        5069-L320ERMS2 FASTBase
        ▲ 1734-AENTR/B PointIO
            ▲ PointIO 11 Slot Chassis
                [0] 1734-AENTR/B PointIO
                [1] 1734-IB8S/B Slot1
                [2] 1734-IB8S/B Slot2
                [3] 1734-OB8S/B Slot3
                [4] 1734-4IOL/A AB_IO_Link
                [5] 1734-IB8/C Slot5
                [6] 1734-OB8E/C Slot6
                [7] 1734-OW4/C Slot7
                [8] 1734-OW4/C Slot8
                [9] 1734-OW4/C Slot9
                [10] 1734-OW4/C Slot10
            EX600-SEN3/4 SweetMachine_Valve_Bank_1
            EX600-SEN3/4 SweetMachine_Valve_Bank_2
            EX600-SEN3/4 SweetMachine_Valve_Bank_3
            AL1122 IOLink_Master
            ETHERNET-MODULE Base_Robot_Cont

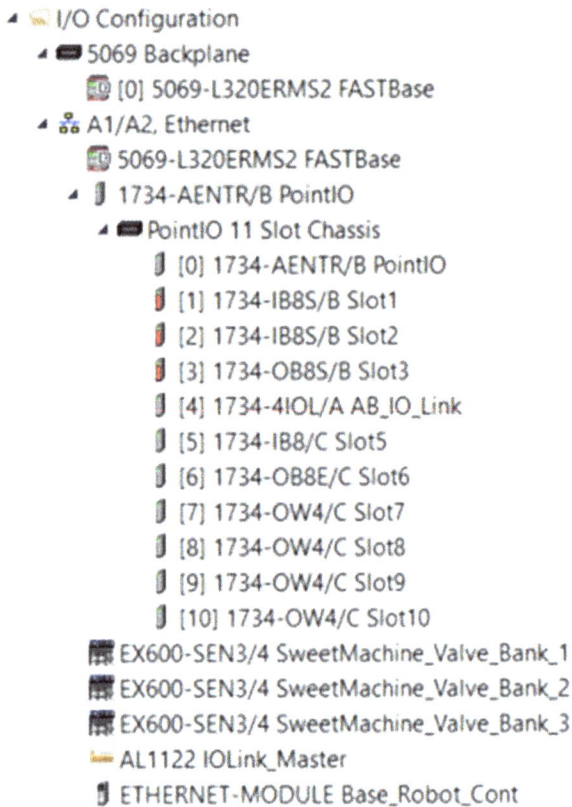

Figure 414 - PLC I/O Configuration Tree

Figure 414 to the left shows the I/O configuration tree in the PLC software. When devices are inserted into the tree under the Ethernet connections, names are assigned to the devices along with their ethernet addresses. These names can then be found by looking in the Controller Tags of the PLC, where they are listed alphabetically.

The tags fall into three basic categories: I (Input), O (Output), and C (Control). The data types in the tags are structured as UDTs (User Defined Tags) as explained in the PLC section of this book. Depending on the manufacturer's control structure, these may be further divided into Bytes, Ints, DINTs, Reals or Bools.

The Base_Robot_Cont tag shown below in Figure 415 illustrates how Epson (the robot manufacturer) has structured their I/O for data exchange with the PLC.

▲ Base_Robot_Cont:C	AB:ETHERNET_MODULE:C:0	Standard
▶ Base_Robot_Cont:C.D...	SINT[400]	Standard
▶ Base_Robot_Cont:I	AB:ETHERNET_MODULE_INT_128Bytes:I:0	Standard
▶ Base_Robot_Cont:O	AB:ETHERNET_MODULE_INT_128Bytes:O:0	Standard

Figure 415 - Base_Robot_Cont Tag Structure

Note that the robot data structure is configured as bytes and SINTs. A SINT or Single Integer is Allen-Bradley's term for a byte. The I data is information sent from the robot to the PLC, and the O data is commands sent from the PLC to the robot. C data is generally configuration data telling the device how to behave.

Point I/O data is structured in a similar way; however, the data is also mapped to additional tags by **aliasing**. This technique was discussed in the PLC section of this book; the idea is that any tag or structure can be linked to an equivalent structure in the tag database. This allows more descriptive names to be connected to I/O tags or array elements.

▸ PointIO:9:C			AB:1734_DO4:C:0
▸ PointIO:9:O	PointIO:O.Data[9]	PointIO:O.Data[9]	SINT
▸ PointIO:10:C			AB:1734_DO4:C:0
▸ PointIO:10:O	PointIO:O.Data[10]	PointIO:O.Data[10]	SINT
◢ PointIO:I			AB:1734_11SLOT:I:0
▸ PointIO:I.SlotStatusBits0_31			DINT
▸ PointIO:I.SlotStatusBits32_63			DINT
▸ PointIO:I.Data			SINT[11]
◢ PointIO:O			AB:1734_11SLOT:O:0
▸ PointIO:O.Data			SINT[11]

Figure 416 - Point IO Tags and Aliasing

Figure 416 above shows that individual PointIO:X:O output tags are aliased to individual array SINTs as PointIO:O.Data[X] tags. PointIO:X:I input tags are aliased in the same way.

Name	≡≡|▲ Usage	Alias For	Base Tag	Data Type
o_LeftElevCnvRunningRev	Local			BOOL
o_LeftElevLwr	Local	PointIO:9:O.3(C)	PointIO:O.Data[9].3(C)	BOOL
o_LeftElevRse	Local	PointIO:9:O.2(C)	PointIO:O.Data[9].2(C)	BOOL
o_LockBackDoor	Local	PointIO:6:O.7(C)	PointIO:O.Data[6].7(C)	BOOL
o_LockFrontDoor	Local	PointIO:6:O.6(C)	PointIO:O.Data[6].6(C)	BOOL

Figure 417 - Descriptive Tag Aliasing

Figure 417 shows where descriptive tags representing the Elevator Raise and Lower tags have been further aliased to the Point I/O tags. The descriptive tag is then used in the logic to control the output. The (C) in the alias field indicates that the tag can be found in the Controller (Global) tags, the o_LeftElevRse tag is in a Local tag directory "Base".

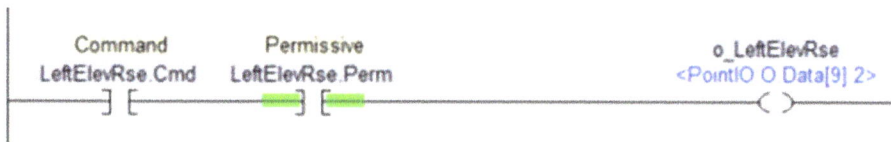

Figure 418 - Left Elevator Control

Cross referencing the tags as described in the PLC chapter allows a technician to find these tags and their control points.

The valve banks are aliased in a similar way, with output points accessed through a SINT array.

▸ SweetMachine_Valve_Bank_1:I	_0007:EX600_SEN34_538A8EA9:I:0
◢ SweetMachine_Valve_Bank_1:O	_0007:EX600_SEN34_AAB94180:O:0
◢ SweetMachine_Valve_Bank_1:O.Data	SINT[4]
▸ SweetMachine_Valve_Bank_1:O.Data[0]	SINT
▸ SweetMachine_Valve_Bank_1:O.Data[1]	SINT
▸ SweetMachine_Valve_Bank_1:O.Data[2]	SINT
▸ SweetMachine_Valve_Bank_1:O.Data[3]	SINT

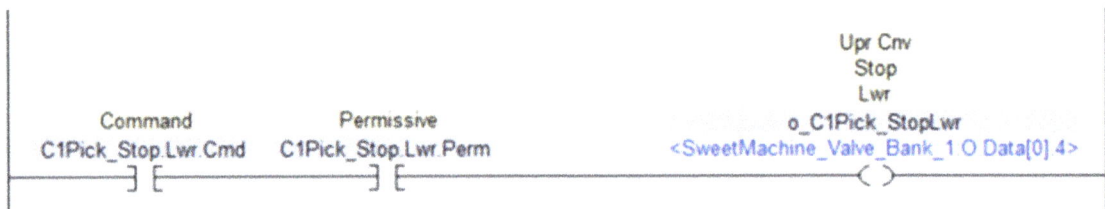

Figure 419 - Valve Bank Structure and Aliased Tag

The IO-Link Master and its I/O input blocks are configured as a Control tag and two large arrays of INTs for the I and O structures.

▸ IOLink_Master:C	_0142:AL1122_4FBFE91A:C:0
◢ IOLink_Master:I1	_0142:AL1122_04852EE3:I:0
IOLink_Master:I1.ConnectionFaulted	BOOL
▸ IOLink_Master:I1.Data	INT[223]
◢ IOLink_Master:O1	_0142:AL1122_23FE281C:O:0
▸ IOLink_Master:O1.Data	INT[151]

Figure 420 – IO-Link Master Data Structure

The IOLink_Master:C controller tag contains a great deal of configuration and interface information for both the master and the connected blocks. Most of the content is set up using the LR Device software described in the IO-Link section of the schematics chapter. The data from the controller tag is not accessed when cross referencing it in the software, but it could be.

The IO blocks themselves on the other hand do have setup information that accesses the IOLink Master data. An Add-On Instruction (AOI) from IFM is used in the program to set up data for the block.

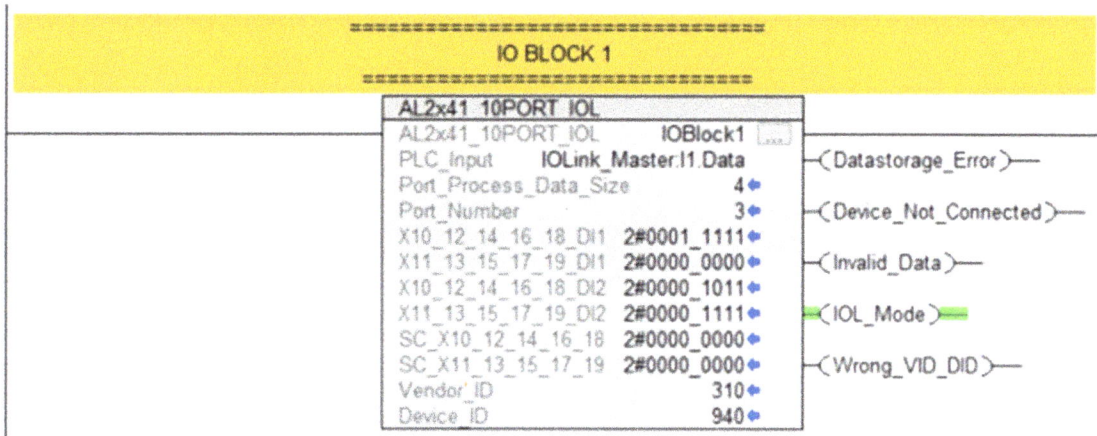

Figure 421 - IO-Link Add-On Instruction

Add-On Instructions are not generally meant to be modified by technicians; in fact they are often locked so that the logic cannot be examined.

Similar AOIs are used to access the more complex data for devices attached to the IO-Link Master directly. The Air Pressure switch, Stack Light and DC Circuit breakers all have associated AOIs. The circuit breaker block in particular has quite a bit of logic associated with the setup of the device.

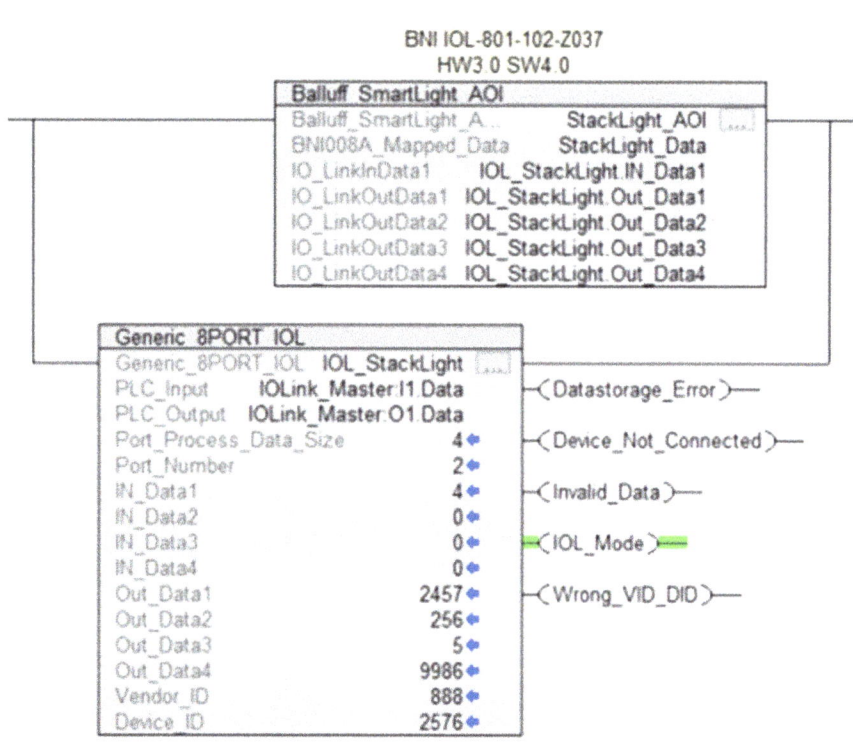

Figure 422 - Stack Light AOI

⊿ IOBlock1	Local	AL2x41_10PORT_IOL
IOBlock1.EnableIn		BOOL
IOBlock1.EnableOut		BOOL
IOBlock1.Datastorage_Error		BOOL
IOBlock1.Device_Not_Connected		BOOL
IOBlock1.Invalid_Data		BOOL
IOBlock1.IOL_Mode		BOOL
▸ IOBlock1.Port_Process_Data_Size		INT
▸ IOBlock1.Port_Number		INT
▸ IOBlock1.Process_Data		INT
IOBlock1.Wrong_VID_DID		BOOL
▸ IOBlock1.X10_12_14_16_18_DI1		SINT
▸ IOBlock1.X11_13_15_17_19_DI1		SINT
▸ IOBlock1.X10_12_14_16_18_DI2		SINT
▸ IOBlock1.X11_13_15_17_19_DI2		SINT
▸ IOBlock1.SC_X10_12_14_16_18		SINT
▸ IOBlock1.SC_X11_13_15_17_19		SINT
▸ IOBlock1.Vendor_ID		INT
▸ IOBlock1.Device_ID		INT
▸ IOBlock2	Local	AL2x41_10PORT_IOL
▸ IOBlock3	Local	AL2x41_10PORT_IOL

Figure 9 - IO_Block1 Tags (Local, Base Program)

Input signals to the IO-Link input modules are mapped to SINTs within the IOBlockX tag structure. These are further exposed in the AOI shown back in Figure xxx. Each individual bit of the SINT is then entered into another AOI called a "DI_Handler". These are instructions that condition the inputs by allowing an on or off-delay timer to be applied or for the signal to be artificially "forced" on or off, or to be inverted.

All of the input bits applied to the DI_Handler AOIs start with the DI prefix and are local in scope. The output variable from the DI_Handler uses the .Value variable to represent the tag in logic.

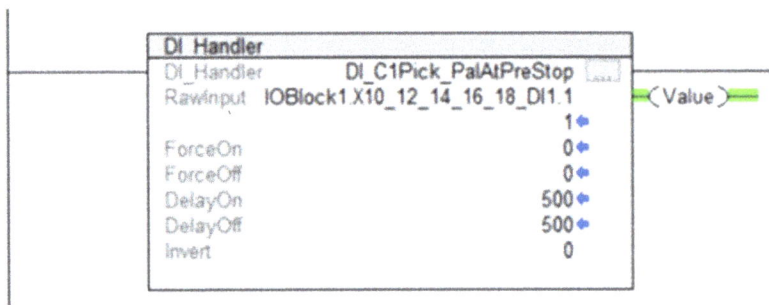

Figure 423 - DI Handler

The DI_Handler instructions can all be found in the Inputs routine of each program. It is important to examine these blocks to see whether a time value (DelayOn, DelayOff), Inversion or Force has been applied to the input.

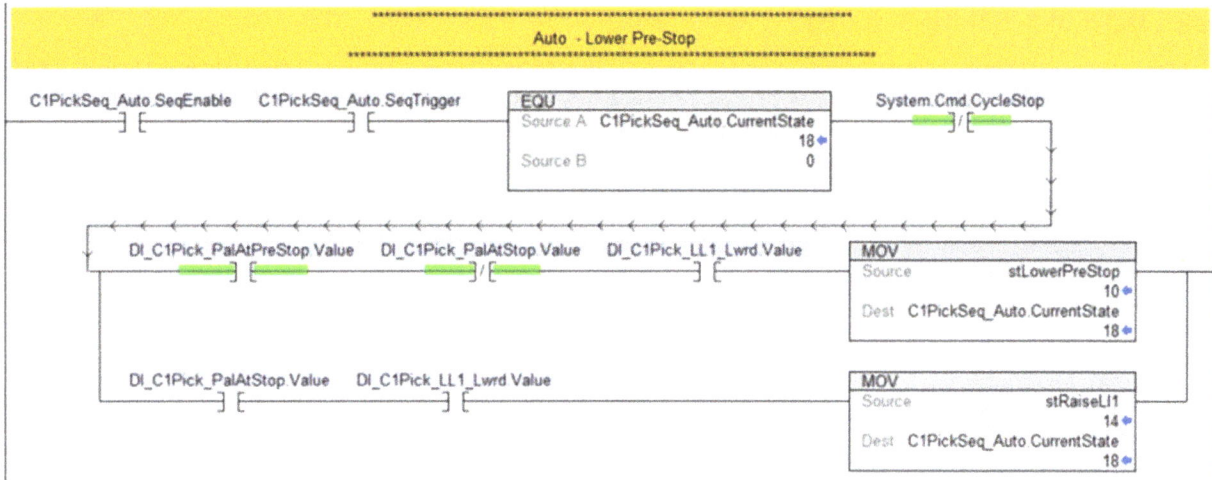

Figure 424 - DI_Handler Value Tags in Logic

Figure 424 shows several different ".Value" tags used in a section of sequence logic.

IO Lists

Following is a list of Inputs and Outputs from the Sweet Machine:

Point I/O Slot 1 (Safety Inputs)

Address	Tag	Description
PointIO:1:I.Pt00Data	si_EStopPb1Ch1	E-Stop Pushbutton 1
PointIO:1:I.Pt01Data	si_EStopPb1Ch2	E-Stop Pushbutton 1
PointIO:1:I.Pt02Data	si_EStopPb2Ch1	E-Stop Pushbutton 2
PointIO:1:I.Pt03Data	si_EStopPb2Ch2	E-Stop Pushbutton 2
PointIO:1:I.Pt04Data	si_LwrGrdCh1	Lower Guard Door Switch
PointIO:1:I.Pt05Data	si_LwrGrdCh2	Lower Guard Door Switch
PointIO:1:I.Pt06Data		Spare
PointIO:1:I.Pt07Data		Spare

Point I/O Slot 2 (Safety Inputs)

Address	Tag	Description
PointIO:2:I.Pt00Data	si_FrontGrdCh1	Front Guard Door Switch
PointIO:2:I.Pt01Data	si_FrontGrdCh2	Front Guard Door Switch
PointIO:2:I.Pt02Data	si_RearGrdCh1	Rear Guard Door Switch
PointIO:2:I.Pt03Data	si_RearGrdCh2	Rear Guard Door Switch
PointIO:2:I.Pt04Data	si_ElevGrdCh1	Elevator Guard Door Switch
PointIO:2:I.Pt05Data	si_ElevGrdCh2	Elevator Guard Door Switch
PointIO:2:I.Pt06Data	si_EStopSafetyRelayFdbk	E-Stop Safety Relays Monitoring
PointIO:2:I.Pt07Data	si_GrdSafetyRelayFdbk	Guard Door Safety Relay Monitoring

Point I/O Slot 3 (Safety Outputs)

Address	Tag	Description
PointIO:3:O.Pt00Data	so_EStopSafetyRelay1	E-Stop Safety Relay 1
PointIO:3:O.Pt01Data	so_EStopSafetyRelay2	E-Stop Safety Relay 2
PointIO:3:O.Pt02Data	so_GrdSafetyRelay1	Guard Safety Relay 1
PointIO:3:O.Pt03Data	so_GrdSafetyRelay2	Guard Safety Relay 2
PointIO:3:O.Pt04Data		Spare
PointIO:3:O.Pt05Data		Spare
PointIO:3:O.Pt06Data		Spare
PointIO:3:O.Pt07Data	so_main_air_sol	Main Air Soft Start SV

Point I/O Slot 4 (IO-Link Inputs)

Address	Tag	Description
PointIO:4:I.Ch0Data	ETA_MessageDataRead	DC Circuit Breaker Status
PointIO:4:O.Ch0Data	ETA_MessageDataWrite	DC Circuit Breaker Setup
PointIO:4:I.Ch1Data		Spare
PointIO:4:O.Ch1Data		Spare
PointIO:4:I.Ch2Data		Spare
PointIO:4:O.Ch2Data		Spare
PointIO:4:I.Ch3Data		Spare
PointIO:4:O.Ch3Data		Spare

Point I/O Slot 5 (Inputs)

Address	Tag	Description
PointIO:5:I.Pt00Data	DI_StartPB	Cycle Start Pushbutton
PointIO:5:I.Pt01Data	DI_StopPB	Cycle Stop Pushbutton
PointIO:5:I.Pt02Data	I_PwrOnResetPB	Reset Power On Pushbutton
PointIO:5:I.Pt03Data		
PointIO:5:I.Pt04Data		
PointIO:5:I.Pt05Data		
PointIO:5:I.Pt06Data	DI_FrontDoorClosed	Front Locking Guard Switch Locked
PointIO:5:I.Pt07Data	DI_RearDoorClosed	Rear Locking Guard Switch Locked

Point I/O Slot 6 (Outputs)

Address	Tag	Description
PointIO:6:O.Pt00Data	o_CycleStartPBLT	Cycle Start PB Light
PointIO:6:O.Pt01Data	o_PwrOnResetPBLT	Power On/Reset PB Light
PointIO:6:O.Pt02Data	so_GrdSafetyRelay1	Guard Safety Relay 1
PointIO:6:O.Pt03Data	so_GrdSafetyRelay2	Guard Safety Relay 2
PointIO:6:O.Pt04Data		Spare
PointIO:6:O.Pt05Data		Spare
PointIO:6:O.Pt06Data	o_LockFrontDoor	Front Guard Door Magnet Lock
PointIO:6:O.Pt07Data	o_LockBackDoor	Rear Guard Door Magnet Lock

Point I/O Slot 7 (Relay Outputs)

Address	Tag	Description
PointIO:7:O.Pt00Data	o_OverheadWorkLT	Work Lights Power
PointIO:7:O.Pt01Data	o_SL1_Color1	Light Strip Input 1
PointIO:7:O.Pt02Data	o_SL1_Color2	Light Strip Input 2
PointIO:7:O.Pt03Data	o_SL1_Color3	Light Strip Input 3

Point I/O Slot 8 (Relay Outputs)

Address	Tag	Description
PointIO:8:O.Pt00Data	o_uprCnv1Run	Upper Conveyor 1 Run
PointIO:8:O.Pt01Data	o_uprCnv2Run	Upper Conveyor 2 Run
PointIO:8:O.Pt02Data	o_LwrCnv1Run	Lower Conveyor 2 Run
PointIO:8:O.Pt03Data	o_uprCnv1Fwd	Upper Conveyor 1 Forward CMD

Point I/O Slot 9 (Relay Outputs)

Address	Tag	Description
PointIO:9:O.Pt00Data	o_LeftElevCnvRun	Left Elevator Conveyor Run
PointIO:9:O.Pt01Data	o_LeftElevCnvFwd	Left Elevator Conveyor Forward CMD
PointIO:9:O.Pt02Data	o_LeftElevRse	Left Elevator Raise SV
PointIO:9:O.Pt03Data	o_LeftElevLwr	Left Elevator Lower SV

Point I/O Slot 10 (Relay Outputs)

Address	Tag	Description
PointIO:10:O.Pt00Data	o_RightElevCnvRun	Right Elevator Conveyor Run
PointIO:10:O.Pt01Data	o_RightElevCnvFwd	Right Elevator Conveyor Forward CMD
PointIO:10:O.Pt02Data	o_RightElevRse	Right Elevator Raise SV
PointIO:10:O.Pt03Data	o_RightElevLwr	Right Elevator Lower SV

IO Link Master

Address	Tag	Description
X01 (Port 1)	IOL_MainAirPressure	Main Air Pressure PSI
X02 (Port 2)	StackLight_Data	Stack Light Control
X03 (Port 3)	IOBlock1	IO Block 1 Inputs
X04 (Port 4)	IOBlock2	IO Block 2 Inputs
X05 (Port 5)	IOBlock3	IO Block 3 Inputs
X06 (Port 6)		Spare
X07 (Port 7)	LeftElev_IOBlock	Left Elevator Inputs
X08 (Port 8)	RightElev_IOBlock	Right Elevator Inputs

Remote IO Block 1

Address (Base Program, Local)	Tag (Base Program, Local)	Description
IOBlock1.X10_12_14_16_18_DI1.0	DI_LwrFlipStopRsd	Lower Flip Station Pallet Stop Raised PRX
IOBlock1.X10_12_14_16_18_DI2.0	DI_LwrFlip_PalAtStop	Lower Flip Station Pallet at Stop PE
IOBlock1.X11_13_15_17_19_DI1.0	DI_LwrCnvDownstreamClear	Lower Conveyor Downstream Clear PE
IOBlock1.X11_13_15_17_19_DI2.0	DI_C1Pick_PreStopRsd	C1 Pick Station PreStop Raised PRX
IOBlock1.X10_12_14_16_18_DI1.1	DI_C1Pick_PalAtPreStop	C1 Pick Station Pallet at PreStop PE
IOBlock1.X10_12_14_16_18_DI2.1	DI_C1Pick_StopRsd	C1 Pick Station Stop Raised PRX
IOBlock1.X11_13_15_17_19_DI1.1	DI_C1Pick_PalAtStop	C1 Pick Station Pallet at Stop PE
IOBlock1.X11_13_15_17_19_DI2.1	DI_UprCnvDownstreamClear	Upper Conveyor Downstream Clear PE
IOBlock1.X10_12_14_16_18_DI1.2	DI_C1Pick_LL1_Rsd	C1 Pick Station Lift and Locate Raised PRX
IOBlock1.X10_12_14_16_18_DI2.2	DI_C1Pick_LL1_Lwrd	C1 Pick Station Lift and Locate Lowered PRX
IOBlock1.X11_13_15_17_19_DI1.2	DI_C1Place_PalAtStop	C1 Place Station Pallet at Stop PE
IOBlock1.X11_13_15_17_19_DI2.2	DI_C1Place_Stop_Rsd	C1 Place Station Stop Raised PRX
IOBlock1.X10_12_14_16_18_DI1.3	DI_C1Place_PalAtPreStop	C1 Place Station Pallet at PreStop PE
IOBlock1.X10_12_14_16_18_DI2.3	DI_C1Place_PreStopRsd	C1 Place Station PreStop Raised PRX
IOBlock1.X11_13_15_17_19_DI1.3	DI_C1Place_LL2_Rsd	C1 Place Station Lift and Locate Raised PRX
IOBlock1.X11_13_15_17_19_DI2.3	DI_C1Place_LL2_Lwrd	C1 Place Station Lift and Locate Lowered PRX
IOBlock1.X10_12_14_16_18_DI1.4	DI_C1Pick_PuckPres	C1 Pick Station Puck Present at Lift and Locate 1 PE
IOBlock1.X10_12_14_16_18_DI2.4	DI_C1Place_PuckPres	C1 Place Station Puck Present at Lift and Locate 2 PE
IOBlock1.X11_13_15_17_19_DI1.4		Spare
IOBlock1.X11_13_15_17_19_DI2.4		Spare

Remote IO Block 2

Address (Base Program, Local)	Tag (Base Program, Local)	Description
IOBlock2.X10_12_14_16_18_DI1.0	DI_CoinSlot1_CandyPresent	Candy Present in Slot 1 PE
IOBlock2.X10_12_14_16_18_DI2.0	DI_CoinSlot2_CandyPresent	Candy Present in Slot 2 PE
IOBlock2.X11_13_15_17_19_DI1.0	DI_CoinSlot3_CandyPresent	Candy Present in Slot 3 PE
IOBlock2.X11_13_15_17_19_DI2.0	DI_C2_PuckAtPlace	Secondary Conveyor Puck Present at C2 Place Station PE
IOBlock2.X10_12_14_16_18_DI1.1	DI_C2_PuckAtPick	Secondary Conveyor Puck Present at C2 Pick Station PE
IOBlock2.X10_12_14_16_18_DI2.1	DI_C2_CandyAtChute	Secondary Conveyor Candy Present at Candy Chute PE
IOBlock2.X11_13_15_17_19_DI1.1	DI_Robot_Grip_ON	Robot Gripper Closed PRX
IOBlock2.X11_13_15_17_19_DI2.1	DI_Robot_Grip_OFF	Robot Gripper Opened PRX
IOBlock2.X10_12_14_16_18_DI1.2	DI_C2Flip_Grip_Closed	Upper Flip Station Gripper Closed PRX
IOBlock2.X10_12_14_16_18_DI2.2	DI_C2Flip_Grip_Opened	Upper Flip Station Gripper Opened PRX
IOBlock2.X11_13_15_17_19_DI1.2	DI_C2Flip_Rot_Flipped	Upper Flip Station Rotary Actuator Flipped PRX
IOBlock2.X11_13_15_17_19_DI2.2	DI_C2Flip_Rot_Homed	Upper Flip Station Rotary Actuator Home PRX
IOBlock2.X10_12_14_16_18_DI1.3	DI_C2Flip_Slide_Rsd	Upper Flip Station Airslide Raised PRX
IOBlock2.X10_12_14_16_18_DI2.3	DI_C2Flip_Slide_Lwrd	Upper Flip Station Airslide Lowered PRX
IOBlock2.X11_13_15_17_19_DI1.3	DI_CandySlide1_Ext	Secondary Conveyor Candy Air Slide 1 Extended PRX
IOBlock2.X11_13_15_17_19_DI2.3	DI_CandySlide1_Ret	Secondary Conveyor Candy Air Slide 1 Retracted PRX
IOBlock2.X10_12_14_16_18_DI1.4	DI_CandySlide2_Ext	Secondary Conveyor Candy Air Slide 2 Extended PRX
IOBlock2.X10_12_14_16_18_DI2.4	DI_CandySlide2_Ret	Secondary Conveyor Candy Air Slide 2 Retracted PRX
IOBlock2.X11_13_15_17_19_DI1.4	DI_CandySlide3_Ext	Secondary Conveyor Candy Air Slide 3 Extended PRX
IOBlock2.X11_13_15_17_19_DI2.4	DI_CandySlide3_Ret	Secondary Conveyor Candy Air Slide 3 Retracted PRX

Remote IO Block 3

Address (Base Program, Local)	Tag (Base Program, Local)	Description
IOBlock3.X10_12_14_16_18_DI1.0	DI_LwrFlip_PreStop_Rsd	Lower Flip Station Pallet PreStop Cylinder Raised PRX
IOBlock3.X10_12_14_16_18_DI2.0	DI_LwrFlip_PalAtPreStop	Lower Flip Station Pallet at PreStop PE
IOBlock3.X11_13_15_17_19_DI1.0	DI_LwrFlip_LL3_Rsd	Lower Flip Station Lift and Locate 3 Pallet Lifted PRX
IOBlock3.X11_13_15_17_19_DI2.0	DI_LwrFlip_LL3_Lwrd	Lower Flip Station Lift and Locate 3 Pallet Lowered PRX
IOBlock3.X10_12_14_16_18_DI1.1	DI_LwrFlip_PuckPres	Lower Flip Station Puck Present at Lift and Locate 3 PE
IOBlock3.X10_12_14_16_18_DI2.1		Spare
IOBlock3.X11_13_15_17_19_DI1.1	DI_LwrFlip_Slide_Rsd	Lower Flip Station Airslide Raised PRX
IOBlock3.X11_13_15_17_19_DI2.1	DI_LwrFlip_Slide_Lwrd	Lower Flip Station Airslide Lowered PRX
IOBlock3.X10_12_14_16_18_DI1.2	DI_LwrFlip_Grip_Opened	Lower Flip Station Gripper Opened PRX
IOBlock3.X10_12_14_16_18_DI2.2	DI_LwrFlip_Grip_Closed	Lower Flip Station Gripper Closed PRX
IOBlock3.X11_13_15_17_19_DI1.2	DI_LwrFlip_Rot_Homed	Lower Flip Station Rotary Actuator Homed PRX
IOBlock3.X11_13_15_17_19_DI2.2	DI_LwrFlip_Rot_Flipped	Lower Flip Station Rotary Actuator Flipped PRX
IOBlock3.X10_12_14_16_18_DI1.3	DI_C2Flip_PuckAtGrip	Conveyor 2 Upper Flip Station Puck at Gripper PE
IOBlock3.X10_12_14_16_18_DI2.3		Spare
IOBlock3.X11_13_15_17_19_DI1.3		Spare
IOBlock3.X11_13_15_17_19_DI2.3		Spare
IOBlock3.X10_12_14_16_18_DI1.4		Spare
IOBlock3.X10_12_14_16_18_DI2.4		Spare
IOBlock3.X11_13_15_17_19_DI1.4		Spare
IOBlock3.X11_13_15_17_19_DI2.4		Spare

Left Elevator IO Block

Address (Base Program, Local)	Tag (Base Program, Local)	Description
LeftElev_IOBlock.X10_12_14_16_18_DI1.0	DI_LeftElevLoadStopRsd	Left Elevator Pallet Load Stop Raised PRX
LeftElev_IOBlock.X10_12_14_16_18_DI2.0	DI_LeftElevPal1AtLoadStop	Left Elevator Pallet 1 at Load Stop PE
LeftElev_IOBlock.X11_13_15_17_19_DI1.0	DI_LeftElevPal2AtLoadStop	Left Elevator Pallet 2 at Load Stop PE
LeftElev_IOBlock.X11_13_15_17_19_DI2.0	DI_LeftElevPal1OnElev	Left Elevator Pallet 1 On Elevator PE
LeftElev_IOBlock.X10_12_14_16_18_DI1.1	DI_LeftElevPal2OnElev	Left Elevator Pallet 2 On Elevator PE
LeftElev_IOBlock.X10_12_14_16_18_DI2.1	DI_LeftElevClearToXfer	Left Elevator Clear to Transfer PE
LeftElev_IOBlock.X11_13_15_17_19_DI1.1	DI_LeftElevDownstreamClear	Left Elevator Downstream Clear PE
LeftElev_IOBlock.X11_13_15_17_19_DI2.1		Spare
LeftElev_IOBlock.X10_12_14_16_18_DI1.2	DI_LeftElevRsd	Left Elevator Raised PRX
LeftElev_IOBlock.X10_12_14_16_18_DI2.2	DI_LeftElevLwrd	Left Elevator Lowered PRX
LeftElev_IOBlock.X11_13_15_17_19_DI1.2		Spare
LeftElev_IOBlock.X11_13_15_17_19_DI2.2		Spare
LeftElev_IOBlock.X10_12_14_16_18_DI1.3		Spare
LeftElev_IOBlock.X10_12_14_16_18_DI2.3		Spare
LeftElev_IOBlock.X11_13_15_17_19_DI1.3		Spare
LeftElev_IOBlock.X11_13_15_17_19_DI2.3		Spare
LeftElev_IOBlock.X10_12_14_16_18_DI1.4		Spare
LeftElev_IOBlock.X10_12_14_16_18_DI2.4		Spare
LeftElev_IOBlock.X11_13_15_17_19_DI1.4		Spare
LeftElev_IOBlock.X11_13_15_17_19_DI2.4		Spare

Right Elevator IO Block

Address (Base Program, Local)	Tag (Base Program, Local)	Description
RightElev_IOBlock.X10_12_14_16_18_DI1.0	DI_RightElevLoadStopRsd	Right Elevator Pallet Load Stop Raised PRX
RightElev_IOBlock.X10_12_14_16_18_DI2.0	DI_RightElevPal1AtLoadStop	Right Elevator Pallet 1 at Load Stop PE
RightElev_IOBlock.X11_13_15_17_19_DI1.0	DI_RighttElevPal2AtLoadStop	Right Elevator Pallet 2 at Load Stop PE
RightElev_IOBlock.X11_13_15_17_19_DI2.0	DI_RightElevPal1OnElev	Right Elevator Pallet 1 On Elevator PE
RightElev_IOBlock.X10_12_14_16_18_DI1.1	DI_RightElevPal2OnElev	Right Elevator Pallet 2 On Elevator PE
RightElev_IOBlock.X10_12_14_16_18_DI2.1	DI_RightElevClearToXfer	Right Elevator Clear to Transfer PE
RightElev_IOBlock.X11_13_15_17_19_DI1.1	DI_RightElevDownstreamClear	Right Elevator Downstream Clear PE
RightElev_IOBlock.X11_13_15_17_19_DI2.1		Spare
RightElev_IOBlock.X10_12_14_16_18_DI1.2	DI_RightElevRsd	Right Elevator Raised PRX
RightElev_IOBlock.X10_12_14_16_18_DI2.2	DI_RightElevLwrd	Right Elevator Lowered PRX
RightElev_IOBlock.X11_13_15_17_19_DI1.2		Spare
RightElev_IOBlock.X11_13_15_17_19_DI2.2		Spare
RightElev_IOBlock.X10_12_14_16_18_DI1.3		Spare
RightElev_IOBlock.X10_12_14_16_18_DI2.3		Spare
RightElev_IOBlock.X11_13_15_17_19_DI1.3		Spare
RightElev_IOBlock.X11_13_15_17_19_DI2.3		Spare
RightElev_IOBlock.X10_12_14_16_18_DI1.4		Spare
RightElev_IOBlock.X10_12_14_16_18_DI2.4		Spare
RightElev_IOBlock.X11_13_15_17_19_DI1.4		Spare
RightElev_IOBlock.X11_13_15_17_19_DI2.4		Spare

Valve Bank 1

Address (Controller Tags)	Tag (Base Program, Local)	Description
Sweet_Machine_Valve_Bank_1:O.Data[0].0	o_LwrFlip_StopLwr	Lower Flip Station Pallet Stop Lower SVA
Sweet_Machine_Valve_Bank_1:O.Data[0].1		Spare SVB
Sweet_Machine_Valve_Bank_1:O.Data[0].2	o_LwrFlip_PreStopLwr	Lower Flip Station Pallet PreStop Lower SVA
Sweet_Machine_Valve_Bank_1:O.Data[0].3		Spare SVB
Sweet_Machine_Valve_Bank_1:O.Data[0].4	o_C1Pick_StopLwr	C1 Pick Station Stop Lower SVA
Sweet_Machine_Valve_Bank_1:O.Data[0].5		Spare SVB
Sweet_Machine_Valve_Bank_1:O.Data[0].6	o_C1Pick_PreStopLwr	C1 Pick Station PreStop Lower SVA
Sweet_Machine_Valve_Bank_1:O.Data[0].7		Spare SVB
Sweet_Machine_Valve_Bank_1:O.Data[1].0	o_C1Place_StopLwr	C1 Place Station Stop Lower SVA
Sweet_Machine_Valve_Bank_1:O.Data[1].1		Spare SVB
Sweet_Machine_Valve_Bank_1:O.Data[1].2	o_C1Place_PreStopLwr	C1 Place Station PreStop Lower SVA
Sweet_Machine_Valve_Bank_1:O.Data[1].3		Spare SVB
Sweet_Machine_Valve_Bank_1:O.Data[1].4	o_LeftElev_LoadStopLwr	Left Elevator Load Stop Lower SVA
Sweet_Machine_Valve_Bank_1:O.Data[1].5		Spare SVB
Sweet_Machine_Valve_Bank_1:O.Data[1].6	o_RightElev_LoadStopLwr	Right Elevator Load Stop Lower SVA
Sweet_Machine_Valve_Bank_1:O.Data[1].7		Spare SVB

Valve Bank 2

Address (Controller Tags)	Tag (Base Program, Local)	Description
Sweet_Machine_Valve_Bank_2:O.Data[0].0	o_C1Pick_LL1Rse	C1 Pick Station Lift and Locate 1 Raise SVA
Sweet_Machine_Valve_Bank_2:O.Data[0].1	o_C1Pick_LL1Lwr	C1 Pick Station Lift and Locate 1 Lower SVB
Sweet_Machine_Valve_Bank_2:O.Data[0].2	o_C1Place_LL2Rse	C1 Place Station Lift and Locate 2 Raise SVA
Sweet_Machine_Valve_Bank_2:O.Data[0].3	o_C1Place_LL2Lwr	C1 Place Station Lift and Locate 2 Lower SVB
Sweet_Machine_Valve_Bank_2:O.Data[0].4	o_LwrFlip_LL3Rse	Lower Flip Station Lift and Locate 3 Raise SVA
Sweet_Machine_Valve_Bank_2:O.Data[0].5	o_LwrFlip_LL3Lwr	Lower Flip Station Lift and Locate 3 Lower SVB
Sweet_Machine_Valve_Bank_2:O.Data[0].6		Spare SVA
Sweet_Machine_Valve_Bank_2:O.Data[0].7		Spare SVB
Sweet_Machine_Valve_Bank_2:O.Data[1].0		Spare SVA
Sweet_Machine_Valve_Bank_2:O.Data[1].1		Spare SVB
Sweet_Machine_Valve_Bank_2:O.Data[1].2	o_LwrFlip_Airslide_Rse	Lower Flip Station Airslide Raise SVA
Sweet_Machine_Valve_Bank_2:O.Data[1].3	o_LwrFlip_Airslide_Lwr	Lower Flip Station Airslide Lower SVB
Sweet_Machine_Valve_Bank_2:O.Data[1].4	o_LwrFlip_RotAct_Home	Lower Flip Station Rotary Actuator Home SVA
Sweet_Machine_Valve_Bank_2:O.Data[1].5	o_LwrFlip_RotAct_Flip	Lower Flip Station Rotary Actuator Flip SVB
Sweet_Machine_Valve_Bank_2:O.Data[1].6	0_LwrFlip_Grip_On	Lower Flip Station Gripper Close SVA
Sweet_Machine_Valve_Bank_2:O.Data[1].7	0_LwrFlip_Grip_Off	Lower Flip Station Gripper Open SVB

Valve Bank 3

Address (Controller Tags)	Tag (Base Program, Local)	Description
Sweet_Machine_Valve_Bank_3:O.Data[0].0	o_Candy1Pusher_Ext	Black Box Candy Airslide 1 Extend SVA
Sweet_Machine_Valve_Bank_3:O.Data[0].1	o_Candy1Pusher_Ret	Black Box Candy Airslide 1 Retract SVB
Sweet_Machine_Valve_Bank_3:O.Data[0].2	o_Candy2Pusher_Ext	Black Box Candy Airslide 2 Extend SVA
Sweet_Machine_Valve_Bank_3:O.Data[0].3	o_Candy2Pusher_Ret	Black Box Candy Airslide 2 Retract SVB
Sweet_Machine_Valve_Bank_3:O.Data[0].4	o_Candy3Pusher_Ext	Black Box Candy Airslide 3 Extend SVA
Sweet_Machine_Valve_Bank_3:O.Data[0].5	o_Candy3Pusher_Ret	Black Box Candy Airslide 3 Retract SVB
Sweet_Machine_Valve_Bank_3:O.Data[0].6		Spare SVA
Sweet_Machine_Valve_Bank_3:O.Data[0].7		Spare SVB
Sweet_Machine_Valve_Bank_3:O.Data[1].0	o_C2Flip_Airslide_Rse	Upper Flip Station Airslide Raise SVA
Sweet_Machine_Valve_Bank_3:O.Data[1].1	o_C2Flip_Airslide_Lwr	Upper Flip Station Airslide Lower SVB
Sweet_Machine_Valve_Bank_3:O.Data[1].2	o_C2Flip_RotAct_Home	Upper Flip Station Rotate Actuator CW SVA
Sweet_Machine_Valve_Bank_3:O.Data[1].3	o_C2Flip_RotAct_Flip	Upper Flip Station Rotate Actuator CCW SVB
Sweet_Machine_Valve_Bank_3:O.Data[1].4	o_C2Flip_Grip_On	Upper Flip Station Gripper Open SVA
Sweet_Machine_Valve_Bank_3:O.Data[1].5	o_C2Flip_Grip_Off	Upper Flip Station Gripper Close SVB
Sweet_Machine_Valve_Bank_3:O.Data[1].6	o_RobotGripper_ON	Robot Gripper Close SVA
Sweet_Machine_Valve_Bank_3:O.Data[1].7	o_RobotGripper_OFF	Robot Gripper Open SVB

Robot I/O

Address (Controller Tags)	Tag (Sweet_Robby Program, Local)	Description
Base_Robot_Cont:C.Data (SINT[400])	None	Not Used
Base_Robot_Cont:I.Data (INT[64])	Robot_Cont.In (UDT)	Robot Status & Position
Base_Robot_Cont:O.Data (INT[64])	Robot_Cont.Out (UDT)	Robot Commands

The robot UDTs are mapped to various parameters for control.

How Does It Work?

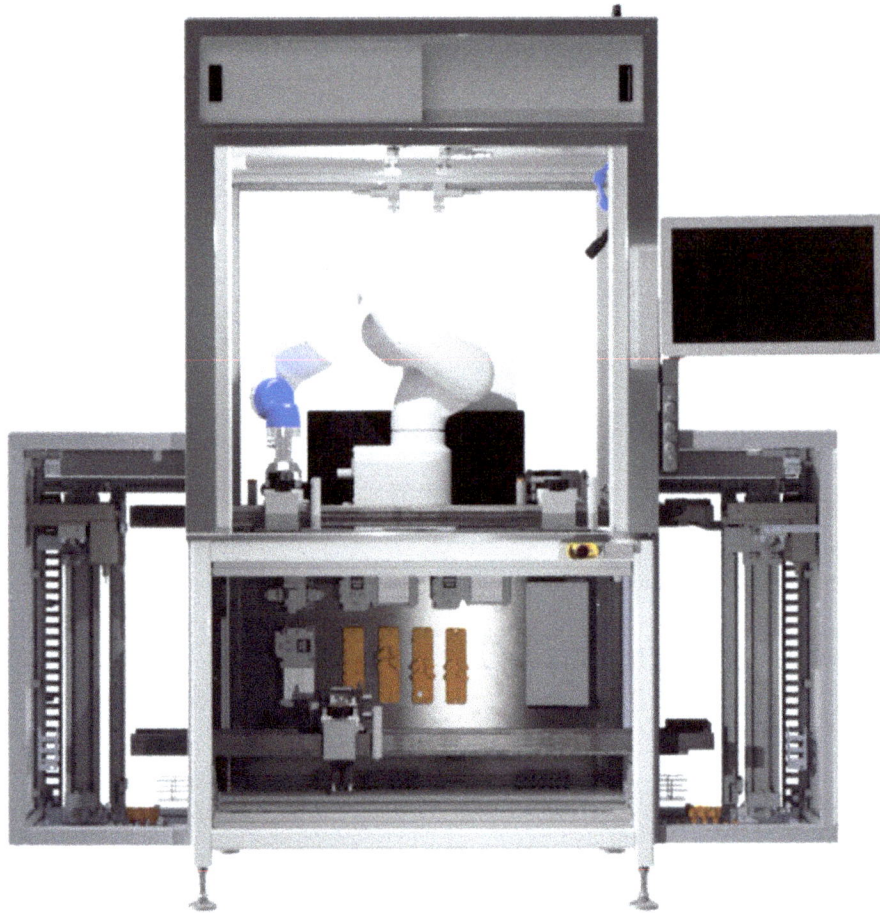

Figure 425 - Sweet Machine Layout

The Sweet Machine is a complex trade show demo that transfers pallets loaded with pucks to demonstrate the technology of a "Flex Base" with two added elevators. Much of the technology is used in Automation NTH projects for customers, Flex Bases can be combined to form manufacturing production lines.

Process Description

In "Normal" mode pallets are moved through the different stations within the machine continuously, with the Robot separating a puck from a pallet and passing it down a secondary conveyor covered with a "Black Box". After being flipped inside of the box, the puck is set back onto the pallet where it makes its way down an elevator, under the base and back up an elevator on the other side. If a button is pressed on the "Sweet" screen, the puck is held inside the Black Box and a candy is dispensed out of the side of the machine with flashing lights and sounds to represent a "magical" transformation effect.

Starting the Machine:

Cycle Start (top) must be held for 3 seconds to begin auto-sequence.

Cycle Stop (middle) halts machine function when it is safe to do so

Power On Reset Pushbutton (bottom) clears fault codes, as well as re-enabling the safety circuits after an Emergency Stop or opening a guard door.

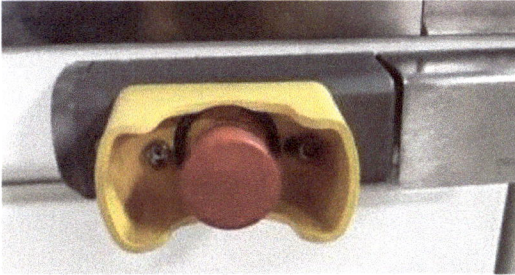

Figure 426 - Operator Start/Stop Buttons

Figure 427 - Emergency Stop Pushbutton

There are Emergency Stop pushbuttons on the front and back of the machine. When pressed the machine comes to an immediate stop and removes hazardous power from all components.

HMI Screens were shown in the previous section. The machine is placed into Auto or Manual mode from the HMI, the machine must be in Auto with no faults present to perform a Cycle Start. Once it is started, the Cycle Stop button is used to make a request to stop. When all stations are in convenient home positions, the machine is automatically taken out of Auto Cycle.

Normal Mode

Pallets with pucks are loaded onto the Left Elevator

- The elevator raises
- The elevator unloads the pallets
- One pallet enters the first Lift and Locate Station
- It lifts and waits for the Robot to grab the puck
- The Robot picks the puck and places it on the Second conveyor.
- The pallet lowers and moves to a second Lift and Locate station and the puck is placed back on it by the Robot.
- The puck that was placed on the second conveyor enters the black box
- It is lifted, flipped, and lowered back onto the conveyor
- It comes out the other side of the Black Box orange.
- The Robot picks it up and places it back onto the pallet at the secondary Lift station.
- The puck and pallet then travel onto the Right Elevator.
- The right elevator lowers and unloads two pucks with pallets onto the lower conveyor.
- One pallet enters the Lower Flip station.
- A pneumatic flipper lifts the puck off the pallet and flips it back to the gray side.
- The puck is placed back onto the pallet and waits to be loaded onto the left elevator.
- The cycle loops from here

Sweet Mode

- The user presses on the Sweet Screen of the HMI
- A puck enters the black box and held
- The selected candy is isolated in the escapement and dropped onto the secondary conveyor
- It travels out of the Black Box on the secondary conveyor.
- It slides out of the enclosure and into a chute where the user can collect their prize

Figure 428 - Pucks and Robot Gripper

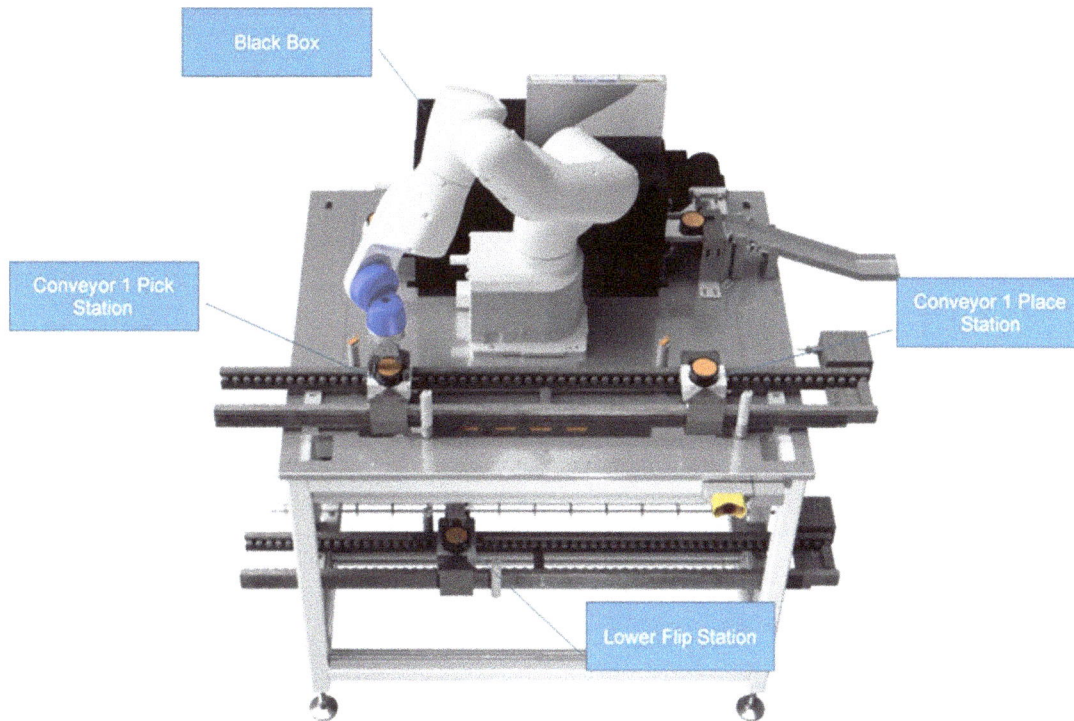

Figure 429 - Base Stations

Stations

The stations for handling pucks and pallets within the base are shown in Figure 429. There are also elevators with small conveyors on the lifts at the left and right sides of the base.

The Black Box behind the robot also contains additional mechanisms as shown below in Figure 430:

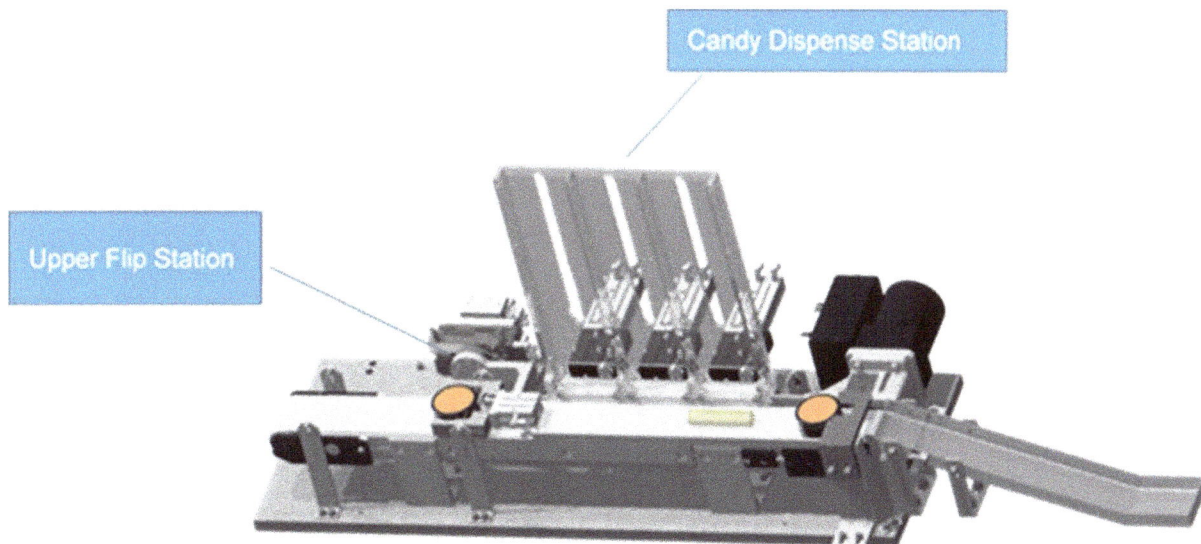

Figure 430 - "Black Box" Stations

There are also pop-up Pre-Stops and Stops before each of the Lift and Locate stations. While the Lift and Locate Stations are handling pallets, the PreStop holds additional pallets back. There are photoeyes at various points along the path to detect pallets and pucks as listed in the I/O List.

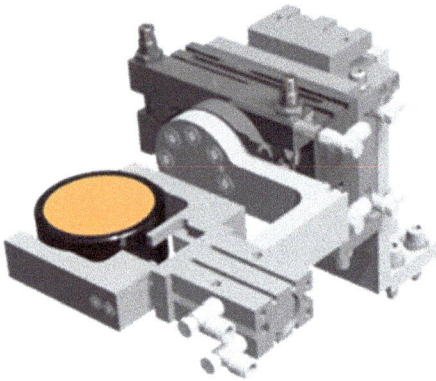

Figure 431 - Upper Flip Station

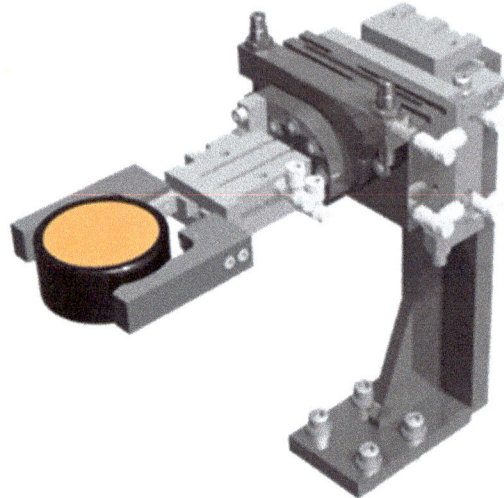

Figure 432 - Lower Flip Station

The actuators also have proximity switches to indicate cylinder positions as listed.

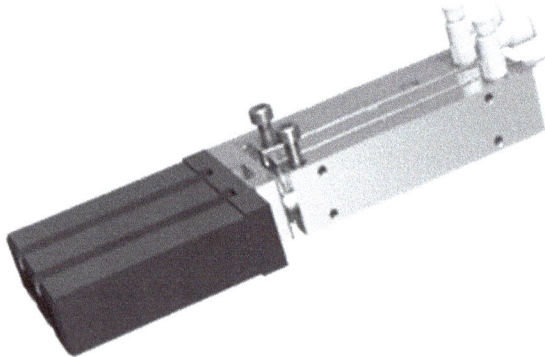

Figure 433 - Candy Dispense Slides

Troubleshooting Exercise

There is extensive fault detection code written that is displayed on the HMI, but for these exercises assume that the technician has no HMI indication of what is wrong. Using the information given in the Schematic Reading chapter and the previous functional descriptions, answer the following questions.

1. What is the problem indicated in Figure 434?

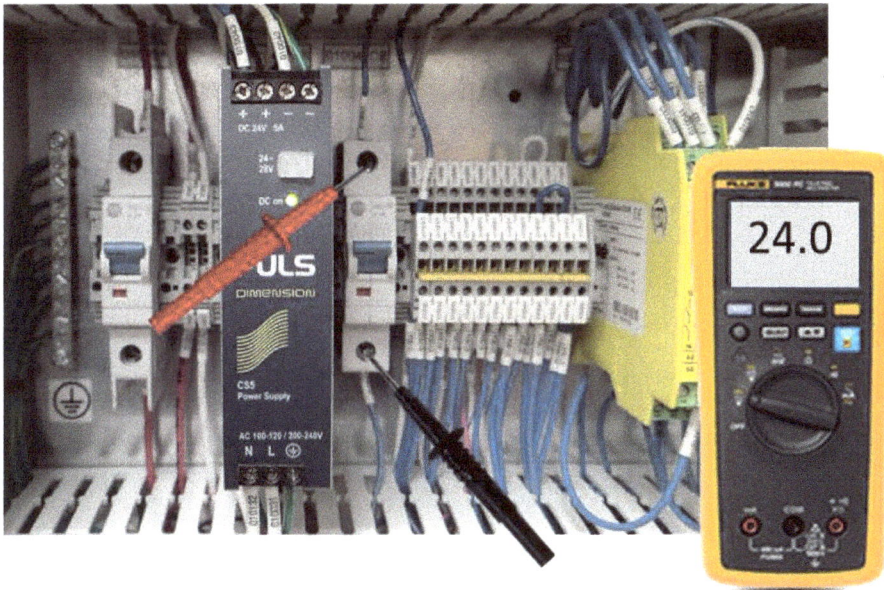

Figure 434 - Troubleshooting Exercise Problem 1

What else could you do to verify your findings? What is the remedy for this problem?

2. The Power On Reset Pushbutton is pressed, but the machine does not energize. The guards all appear to be closed and E-stops pulled out, the safety input cards are as shown in Figure 435. What seems to be the problem?

How would you verify that the problem is what you think it is?

Figure 435 - Troubleshooting Exercise Problem 2

3. What is the problem in Figure 436?

Figure 436 - Troubleshooting Exercise Problem 3

What is the remedy for this problem?

4. The photoelectric sensors on the Sweet Machine are all physically set to be ON when light is detected. There are two lights on each sensor, the green one indicating power and the yellow one indicating light is detected.

There are sensors with reflectors used to detect pucks on the pallet, and diffuse sensors that reflect light off the base of the pallet itself. These are conditioned in the PLC program using the DI_Handler AOI.

Scenario: The Sweet machine has been running in Normal Mode, passing pallets from one station to another continuously. Suddenly the machine stops... conditions are as follows:

All stations have pucks except for the lower conveyor's flip station. There is a pallet at the pre-stop before the station.

Figure 437 - Troubleshooting Exercise Problem 4 Scenario

Figure 437 shows the station with the pallet stopped at the Pre-stop.

Figure 438 - Base Program Routines

The Pre-stop and Stop stations are controlled using the routines in the Base program in Figure 438. Since the LowerFlip_AutoSeq routine appears to be the most likely, you open the logic to find the following in Figure 439:

▲ ᗾ Base
  ◇ Parameters and Local Tags
  ▣ Main
  ᘓ R00_Initialization
  ᗜ R11_Inputs
  ᗜ R17_C1Pick_AutoSeq
  ᗜ R18_C1Place_AutoSeq
  ᗜ R20_LowerFlip_AutoSeq
  ᗜ R22_LeftElevatorAutoSeq
  ᗜ R23_RightElevatorAutoSeq
  ᗜ R25_SweetMode_AutoSeq
  ᗜ R27_UpperFlip_AutoSeq
  ᗜ R30_HomeSeq
  ᗜ R35_RobbyTheSweetRobot
  ᗜ R37_HMISupport
  ᗜ R41_Outputs
  ᗜ R51_Events

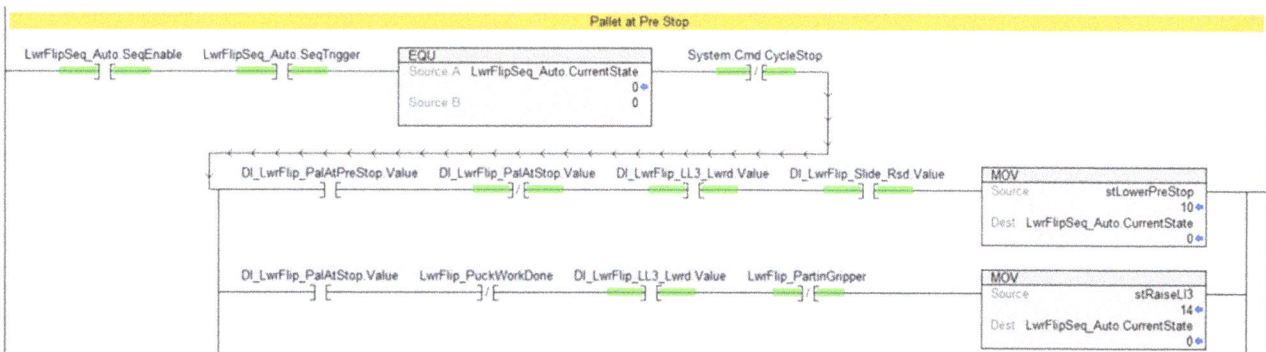

Figure 439 - Troubleshooting Exercise Problem 4A

4A. What step is the sequence in?

Why won't the sequence proceed to the next step? What is the next step number?

What is the address of the component preventing the sequence from proceeding?

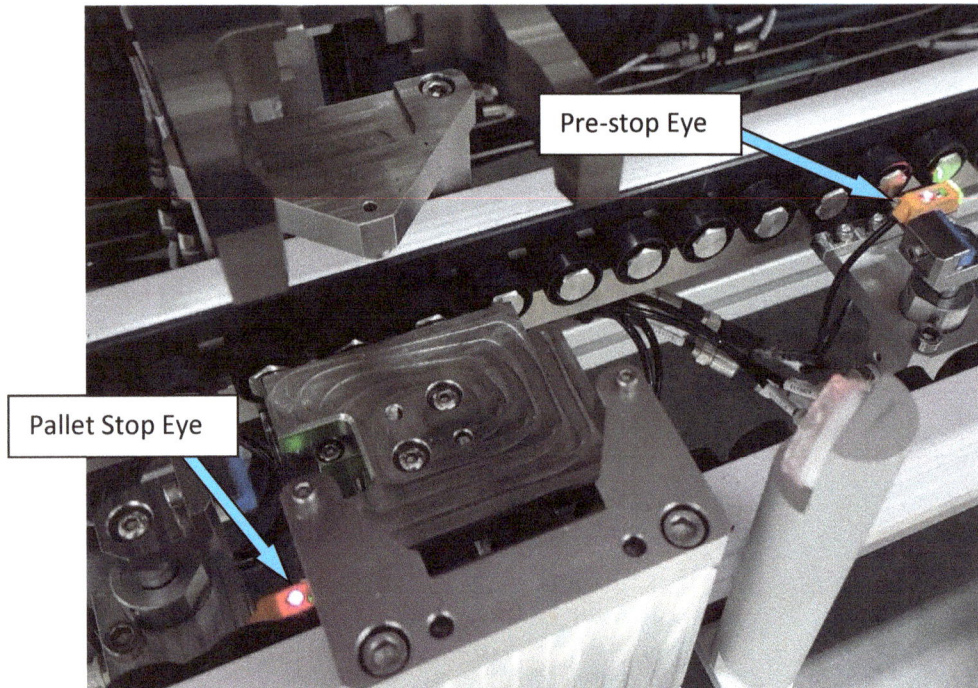

Figure 440 - Pallet Stop Photoeyes

Figure 440 shows the station with no pallets. The sensors point straight up, and the light reflects off the pallet.

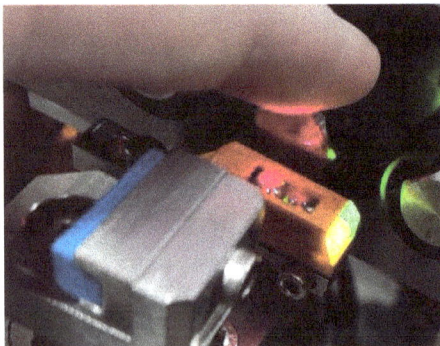

The diffuse photoeye can be triggered by waving a hand in front of it. The power light for the sensor is ON and placing a hand in front of the eye makes the yellow light flash.

*Note: Sensors can be tested safely with guard doors open. Nothing should be able to move.

Figure 441 - Checking Sensor with Finger

4B. Where is the first location that should be checked for a sensor signal?

4C. The light on the sensor works, but the signal is not getting to the PLC. What else can be checked other than electrical sensor wiring? What is an easy way to make this check?

4D. Figure 385 in the Reading Schematics section shows a photoelectric sensor's wiring to the IO Block. List 4 possible points of failure for the electrical signal.

Is there another possible point of failure (besides communications) not included in the wiring?

Disassembling devices and connections is very time consuming. Can you think of a way to make it easier/faster to diagnose wiring problems since most of these type connections are the same?

*There are screens on the HMI that light an indicator if the sensor is on. This is the best way to check the signal all the way to the PLC.

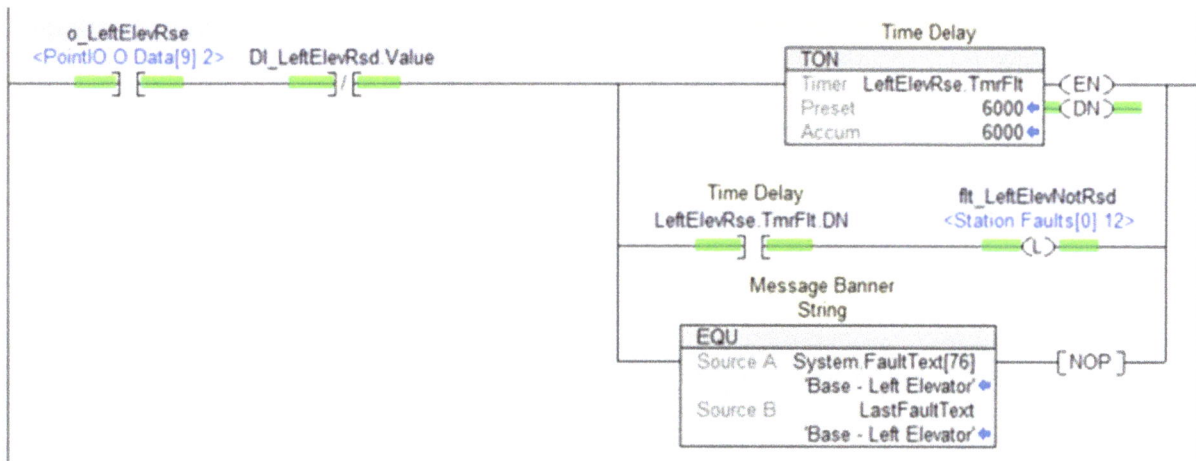

Figure 442 - Troubleshooting Exercise Problem 5

5. What does the logic in Figure 442 indicate? How long is the timer set for?

When the fault is reset and the elevator manually activated using the HMI, it moves very slowly. How can the technician change the speed of the elevator?

6. Look at figure 313 in the Reading Schematics section of the book.

The belt conveyor is not running. The circuit breaker 300081CB indicates that it is not tripped. The control relay 630251CR is energized, and the program shows the conveyor as running. Using a meter, the voltage is as shown in Figure 443. What seems to be the problem?

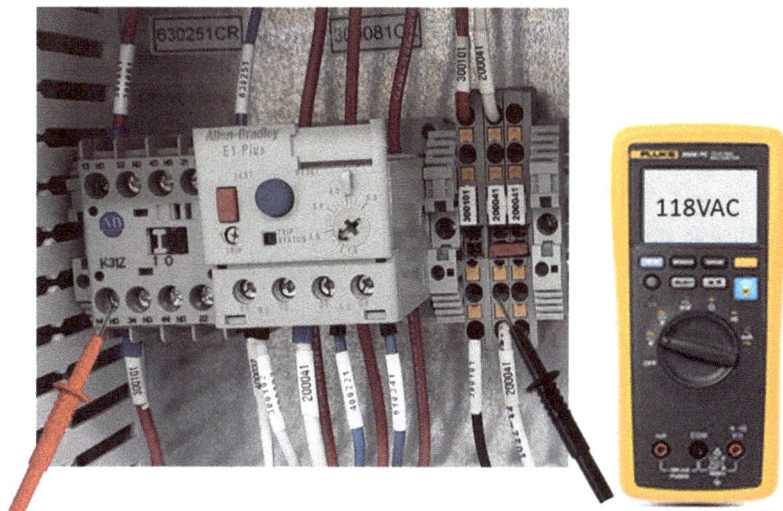

Figure 443 - Troubleshooting Exercise Problem 6

What are some of the ways you could verify your answer?

What are some of the causes and possible remedies for this problem?

7. Hovering the mouse cursor over the devices in the I/O Tree in the PLC software will show the IP address of the robot, which is 10.2.4.95. If the PLC can't see the robot a red "X" will appear over it. Check Figure 414 in the Schematics section for a view of the I/O Tree.

7A. The robot is not responding to signals from the PLC. What is the first step in diagnosing this problem?

The software used to interface with the robot is Epson RC+ Robot Manager. It saves projects in a set of folders that must be imported from the EpsonRC70>Backup folder. You do so and open the project, which is named "Sweet_Machine".

Upon opening the file, you find that the files all seem to be present, and a list of points are as shown in Figure 444.

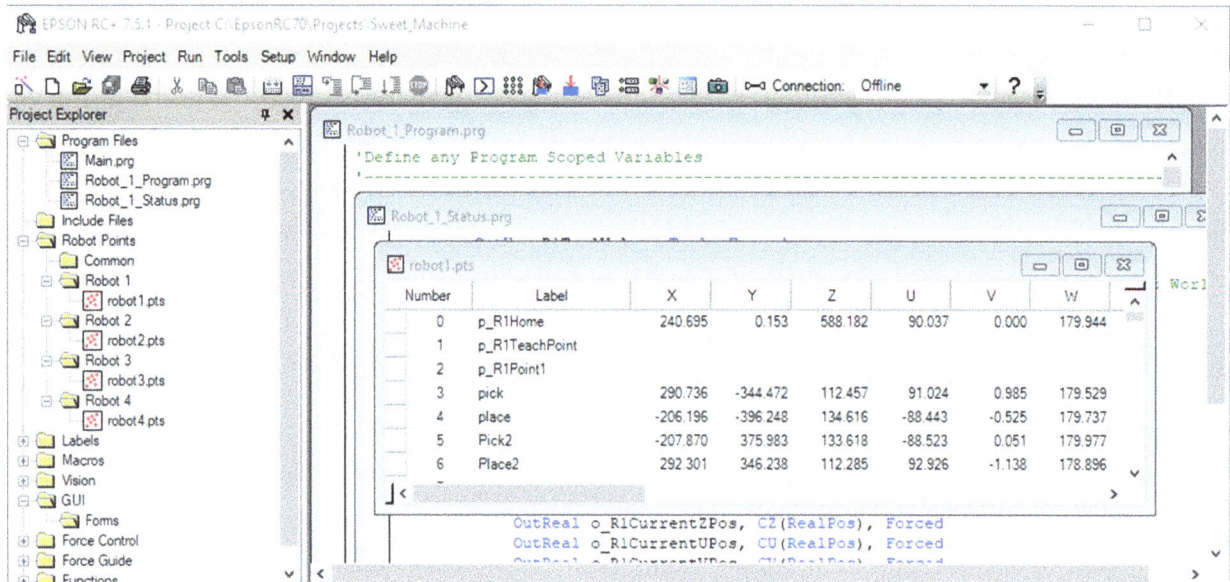

Figure 444 - Epson Robot Software

There is a selection for PC to Controller Communications under the Setup tab in the software that allows a user to connect to the Robot using USB. You are able to download the program to the robot.

There is a screen on the HMI that allows the robot to be jogged as shown in Figure 445.

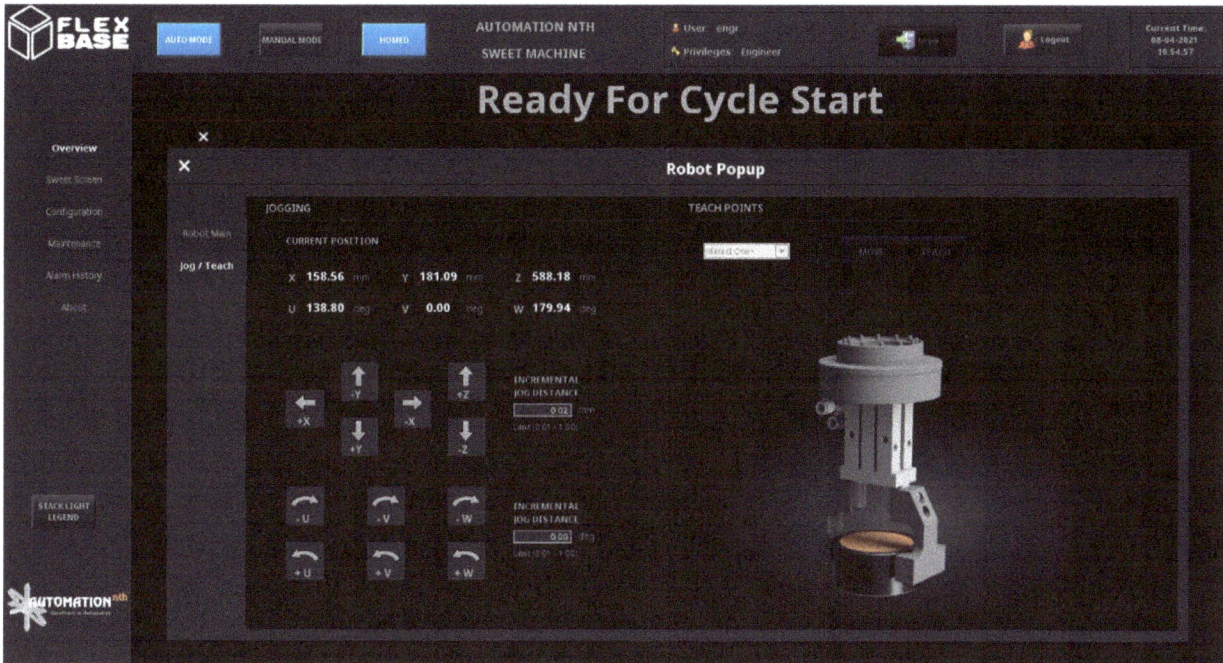

Figure 445 - Robot Jog/Teach Screen

It is clear as you carefully drive the robot close to the pallet points that the positions will need to be re-taught.

7B. What are some physical precautions that should be taken while teaching the robot?

After teaching the new points to the robot, the system is homed. Before placing the machine in AutoCycle, you decide to check the PLC program to see where the robot is about to move.

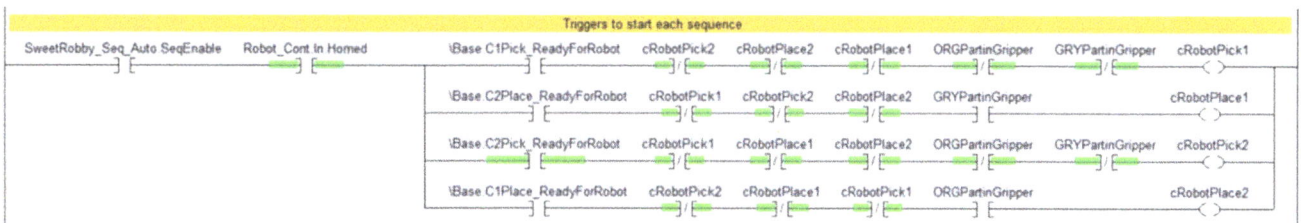

7C. Where will the robot move to if the Auto Sequence is enabled?

Appendix A: Ampacity

Though these tables are taken from reputable sources, designers should consult the appropriate reference when sizing wire for applications. Ampacity ratings depend on temperature, applied voltage, wire length and composition (solid or stranded, number of strands)

Extracted from Model Railroad wire ampacity table.

Copper wire inside enclosure, Low Voltage DC (sumidacrossing.org)

Extracted from Square D Motor Data Calculator based on the National Electrical Code (NEC).

Small Copper Wire Ampacity

AWG	Ampacity
30	0.52
28	0.83
26	1.3
24	2.1
22	5
20	7.5
18	10
16	13
14	17

Copper Wire Ampacity

AWG	Ampacity
14	20
12	25
10	35
8	50
6	65
4	85
3	100
2	115
1	130
1/0	150
2/0	175
3/0	200
4/0	230
250	255
300	285
350	310
400	335
500	380
600	420
700	460
750	475
800	490
900	520
1000	545

Appendix B: Motor Sizing

Three Phase Motors:

Motor data sourced from Square D Motor Data Calculator based on the National Electrical Code (NEC)

	Three Phase Motor Data							
For 60 Hz 1800RPM Standard Squirrel Cage Motors (Non Design E)								
	200(208) Volts				230(240) Volts			
HP	FLA	Min. Copper Wire Size	Circuit Breaker	Fusible Switch	FLA	Min. Copper Wire Size	Circuit Breaker	Fusible Switch
0.50	2.2	14	15	4	2.2	14	15	4
0.75	3.7	14	15	6.25	3.2	14	15	5.6
1.00	4.8	14	15	8	4.2	14	15	8
1.50	6.9	14	15	10	6.0	14	15	10
2.00	7.8	14	15	10	6.8	14	15	10
3.00	11.0	14	20	17.5	9.6	14	20	15
5.00	17.5	12	35	25	15.2	14	30	25
7.50	25.3	10	50	40	22.0	10	45	30
10.00	32.2	8	60	50	28.0	10	60	40
15.00	48.3	6	90	60	42.0	6	80	60
20.00	62.1	4	100	90	54.0	4	90	80
25.00	78.2	3	110	100	68.0	4	100	100
30.00	92.0	2	125	125	80.0	3	110	100
40.00	120.0	1/0	175	175	104.0	1	150	150
50.00	150.0	3/0	200	200	130.0	2/0	200	200
60.00	177.0	4/0	250	250	154.0	3/0	225	200
75.00	221.0	300	300	300	192.0	250	250	300
100.00	285.0	500	400	400	248.0	350	350	350
125.00	359.0	2-4/0	600	500	312.0	2-3/0	450	400
150.00	414.0	2-300	600	600	360.0	2-4/0	600	500
200.00	552.0	2-500	800	N/A	480.0	2-500	800	600

Three Phase Motor Data For 60 Hz 1800RPM Standard Squirrel Cage Motors (Non Design E)								
	460(480) Volts				575(600) Volts			
HP	FLA	Min. Copper Wire Size	Circuit Breaker	Fusible Switch	FLA	Min. Copper Wire Size	Circuit Breaker	Fusible Switch
0.50	1.1	14	15	2	0.9	14	15	1.8
0.75	1.6	14	15	3.2	1.3	14	15	2.5
1.00	2.1	14	15	4	1.7	14	15	3.2
1.50	3.0	14	15	5.6	2.4	14	15	4
2.00	3.4	14	15	6.25	2.7	14	15	5
3.00	4.8	14	15	8	3.9	14	15	6.25
5.00	7.6	14	15	15	6.1	14	15	10
7.50	11.0	14	20	20	9.0	14	15	15
10.00	14.0	14	25	20	11.0	14	20	20
15.00	21.0	10	40	30	17.0	12	35	25
20.00	27.0	10	60	40	22.0	10	45	30
25.00	34.0	8	70	50	27.0	10	60	40
30.00	40.0	8	80	60	32.0	8	60	50
40.00	52.0	6	90	80	41.0	6	80	60
50.00	65.0	4	100	100	52.0	6	90	80
60.00	77.0	3	110	100	62.0	4	100	90
75.00	96.0	1	125	150	77.0	3	110	100
100.00	124.0	2/0	200	175	99.0	1	150	150
125.00	156.0	3/0	225	200	125.0	2/0	200	175
150.00	180.0	4/0	250	250	144.0	3/0	200	200
200.00	240.0	350	350	350	192.0	250	250	300

Single Phase Motors:

Motor data sourced from Square D Motor Data Calculator based on the National Electrical Code (NEC)

	115(120) Volts				230(240) Volts			
Single Phase Motor Data								
For 60 Hz 1800RPM Standard Squirrel Cage Motors (Non Design E)								
HP	**FLA**	**Min. Copper Wire Size**	**Circuit Breaker**	**Fusible Switch**	**FLA**	**Min. Copper Wire Size**	**Circuit Breaker**	**Fusible Switch**
1/6	4.4	14	15	6.25	2.2	14	15	3.2
1/4	5.8	14	15	9	2.9	14	15	4.5
1/3	7.2	14	15	10	3.6	14	15	5.6
1/2	9.8	14	20	15	4.9	14	15	7
3/4	13.8	14	25	20	6.9	14	15	10
1.00	16.0	14	30	25	8.0	14	15	12
1.50	20.0	12	40	30	10.0	14	20	15
2.00	24.0	10	50	30	12.0	14	25	20
3.00	34.0	8	70	50	17.0	12	35	25
5.00	56.0	4	90	80	28.0	10	60	40
7.50	80.0	3	110	100	40.0	8	80	60
10.00	*	*	*	*	50.0	6	90	60

Appendix C: Resistor Color Codes
(Courtesy of Arrow Electronics, www.arrow.com)

How to Read Resistor Color Codes

Appendix D: Field of View (FOV) Chart

Focal Length vs. Distance to Object for lens selection

Object to Camera Face Distance (in)	Lens Focal Length (mm)					
	3.5 mm	6 mm	12 mm	16 mm	25 Mm	55 mm
	ASI2000 Camera Image FOV					
5	10.6 in	6.2 in	3.1 in	2.3 in	1.5 In	0.7 in
6	12.7 in	7.4 in	3.7 in	2.8 in	1.8 In	0.8 in
7	14.8 in	8.6 in	4.3 in	3.2 in	2.1 In	0.9 in
8	16.9 in	9.9 in	4.9 in	3.7 in	2.4 In	1.1 in
12	25.4 in	14.8 in	7.4 in	5.6 in	3.6 In	1.6 in
18	38.1 in	22.2 in	11.1 in	8.3 in	5.3 in	2.4 in
24	50.7 in	29.6 in	14.8 in	11.1 in	7.1 in	3.2 in
30	63.4 in	37.0 in	18.5 in	13.9 in	8.9 in	4.0 in
36	76.1 in	44.4 in	22.2 in	16.7 in	10.7 in	4.8 in
48	101.5 in	59.2 in	29.6 in	22.2 in	14.2 in	6.5 in
60	126.9 in	74.0 in	37.0 in	27.8 in	17.8 in	8.1 in
72	152.2 in	88.8 in	44.4 in	33.3 in	21.3 in	9.7 in
84	177.6 in	103.6 in	51.8 in	38.9 in	24.9 in	11.3 in
96	203.0 in	118.4 in	59.2 in	44.4 in	28.4 in	12.9 in
108	228.3 in	133.2 in	66.6 in	50.0 in	32.0 in	14.5 in
120	253.7 in	148.0 in	74.0 in	55.5 in	35.5 in	16.1 in
132	279.1 in	162.8 in	81.4 in	61.1 in	39.1 in	17.8 in
144	304.5 in	177.6 in	88.8 in	66.6 in	42.6 in	19.4 in

Appendix E: ASCII Tables

Dec	Hx	Oct	Char		Dec	Hx	Oct	Html	Chr	Dec	Hx	Oct	Html	Chr	Dec	Hx	Oct	Html	Chr
0	0	000	NUL	(null)	32	20	040	&#32;	Space	64	40	100	&#64;	@	96	60	140	&#96;	`
1	1	001	SOH	(start of heading)	33	21	041	&#33;	!	65	41	101	&#65;	A	97	61	141	&#97;	a
2	2	002	STX	(start of text)	34	22	042	"	"	66	42	102	&#66;	B	98	62	142	&#98;	b
3	3	003	ETX	(end of text)	35	23	043	&#35;	#	67	43	103	&#67;	C	99	63	143	&#99;	c
4	4	004	EOT	(end of transmission)	36	24	044	&#36;	$	68	44	104	&#68;	D	100	64	144	&#100;	d
5	5	005	ENQ	(enquiry)	37	25	045	&#37;	%	69	45	105	&#69;	E	101	65	145	&#101;	e
6	6	006	ACK	(acknowledge)	38	26	046	&	&	70	46	106	&#70;	F	102	66	146	&#102;	f
7	7	007	BEL	(bell)	39	27	047	'	'	71	47	107	&#71;	G	103	67	147	&#103;	g
8	8	010	BS	(backspace)	40	28	050	&#40;	(	72	48	110	&#72;	H	104	68	150	&#104;	h
9	9	011	TAB	(horizontal tab)	41	29	051	&#41;	)	73	49	111	&#73;	I	105	69	151	&#105;	i
10	A	012	LF	(NL line feed, new line)	42	2A	052	&#42;	*	74	4A	112	&#74;	J	106	6A	152	&#106;	j
11	B	013	VT	(vertical tab)	43	2B	053	&#43;	+	75	4B	113	&#75;	K	107	6B	153	&#107;	k
12	C	014	FF	(NP form feed, new page)	44	2C	054	&#44;	,	76	4C	114	&#76;	L	108	6C	154	&#108;	l
13	D	015	CR	(carriage return)	45	2D	055	&#45;	-	77	4D	115	&#77;	M	109	6D	155	&#109;	m
14	E	016	SO	(shift out)	46	2E	056	&#46;	.	78	4E	116	&#78;	N	110	6E	156	&#110;	n
15	F	017	SI	(shift in)	47	2F	057	&#47;	/	79	4F	117	&#79;	O	111	6F	157	&#111;	o
16	10	020	DLE	(data link escape)	48	30	060	&#48;	0	80	50	120	&#80;	P	112	70	160	&#112;	p
17	11	021	DC1	(device control 1)	49	31	061	&#49;	1	81	51	121	&#81;	Q	113	71	161	&#113;	q
18	12	022	DC2	(device control 2)	50	32	062	&#50;	2	82	52	122	&#82;	R	114	72	162	&#114;	r
19	13	023	DC3	(device control 3)	51	33	063	&#51;	3	83	53	123	&#83;	S	115	73	163	&#115;	s
20	14	024	DC4	(device control 4)	52	34	064	&#52;	4	84	54	124	&#84;	T	116	74	164	&#116;	t
21	15	025	NAK	(negative acknowledge)	53	35	065	&#53;	5	85	55	125	&#85;	U	117	75	165	&#117;	u
22	16	026	SYN	(synchronous idle)	54	36	066	&#54;	6	86	56	126	&#86;	V	118	76	166	&#118;	v
23	17	027	ETB	(end of trans. block)	55	37	067	&#55;	7	87	57	127	&#87;	W	119	77	167	&#119;	w
24	18	030	CAN	(cancel)	56	38	070	&#56;	8	88	58	130	&#88;	X	120	78	170	&#120;	x
25	19	031	EM	(end of medium)	57	39	071	&#57;	9	89	59	131	&#89;	Y	121	79	171	&#121;	y
26	1A	032	SUB	(substitute)	58	3A	072	&#58;	:	90	5A	132	&#90;	Z	122	7A	172	&#122;	z
27	1B	033	ESC	(escape)	59	3B	073	&#59;	;	91	5B	133	&#91;	[	123	7B	173	&#123;	{
28	1C	034	FS	(file separator)	60	3C	074	&#60;	<	92	5C	134	&#92;	\	124	7C	174	&#124;	|
29	1D	035	GS	(group separator)	61	3D	075	&#61;	=	93	5D	135	&#93;	]	125	7D	175	&#125;	}
30	1E	036	RS	(record separator)	62	3E	076	&#62;	>	94	5E	136	&#94;	^	126	7E	176	&#126;	~
31	1F	037	US	(unit separator)	63	3F	077	&#63;	?	95	5F	137	&#95;	_	127	7F	177	&#127;	DEL

Source: www.LookupTables.com

Table 1 - Extended ASCII

128	Ç	144	É	160	á	176	░	192	└	208	╨	224	α	240	≡
129	ü	145	æ	161	í	177	▒	193	┴	209	╤	225	ß	241	±
130	é	146	Æ	162	ó	178	▓	194	┬	210	╥	226	Γ	242	≥
131	â	147	ô	163	ú	179	│	195	├	211	╙	227	π	243	≤
132	ä	148	ö	164	ñ	180	┤	196	─	212	╘	228	Σ	244	⌠
133	à	149	ò	165	Ñ	181	╡	197	┼	213	╒	229	σ	245	⌡
134	å	150	û	166	ª	182	╢	198	╞	214	╓	230	µ	246	÷
135	ç	151	ù	167	º	183	╖	199	╟	215	╫	231	τ	247	≈
136	ê	152	ÿ	168	¿	184	╕	200	╚	216	╪	232	Φ	248	°
137	ë	153	Ö	169	⌐	185	╣	201	╔	217	┘	233	Θ	249	∙
138	è	154	Ü	170	¬	186	║	202	╩	218	┌	234	Ω	250	·
139	ï	155	¢	171	½	187	╗	203	╦	219	█	235	δ	251	√
140	î	156	£	172	¼	188	╝	204	╠	220	▄	236	∞	252	ⁿ
141	ì	157	¥	173	¡	189	╜	205	═	221	▌	237	φ	253	²
142	Ä	158	₧	174	«	190	╛	206	╬	222	▐	238	ε	254	■
143	Å	159	ƒ	175	»	191	┐	207	╧	223	▀	239	∩	255	

Source: www.LookupTables.com

Appendix F: Exercise Answers

Exercise 1:

1. Preventive, Corrective, Predictive.

2. Corrective.

3. CMMS.

Exercise 5:

1. Belt, pulley, motor, electrical connection, filter cleanliness, object in fan.

2. Air cylinder, valve, chain, conveyor belt, motor, electrical connection.

3. Sight, hearing, smell, touch.

Exercise 6:

1. The cylinder will cycle back and forth repetitively.

2. Exhaust, meter out.

3. No, center blocked, 5/3.

4. No, lubricator is not required.

5. 5/3, 4/2, 2 Position Spring Return.

6. A. Photoeye or other sensor.

   B. Not safe, needs guarding.

7. Two hand control for the press valve.

8. Area = 3.14159 x (1 x 1) = 3.14159. 75 psi x 3.14159 = c. 235.6 lbf.

Exercise 7:

1. Contamination of the hydraulic fluid.

2. 1000-5000psi or more.

3. 110-150 degrees Fahrenheit.

4. A pump.

Exercise 8:

1. Resistance, wattage, tolerance.

2. Impedance, reactance.

3. Voltage, current.

4. Convert one voltage to another, or isolate an AC circuit.

5. A circuit breaker can be reset, a fuse is destroyed.

6. A relay or contactor.

7. Switch a signal, amplify or change a waveform, emit or detect light.

8. NPN, PNP.

9. Amplification and Switching

10. Heat

11. Prevent arcing across a coil when power is disconnected.

12. 2250 Ohms, 5%

Exercise 9:

1. I = 1.5/10 = 150 mA.

2. 1000 Ohms.

3. I = 24/1000 = 24 mA.

4. P = (0.024 x 0.024) x 400 = 0.2304 W.

5. 60 Hz.

6. The physical attributes and states of the system or environment.

Exercise 10:

1. The normal operating or "working" area.

2. Signed Integer.

Exercise 11:

1. To transmit and receive a series of bits over a single wire.

2. RS422 is from a single device to another, RS485 can be "multi-drop" to multiple devices.

3. RS = Recommended Standard.

4. It connects a software package and a device, or two different software packages together. Acts as an interpreter.

5. Seven layers.

6. CIP is deterministic

7. A router connects ethernet devices on different networks and can include a firewall.

8. 508

Exercise 12:

1. Decimal: 27,831   Hexadecimal: 6CB7   Octal: 66,267
    This number cannot be converted to BCD, two of the binary groups are higher than 1001.

2. Signed Integer: -27,255

3. BCD: 0100_0001_0111 = 417   Binary: 1 1010 0001   Hexadecimal: 1A1

4. Binary: 10 1010 1001 1110   Decimal: 10,910

5. There are 4 Bytes in a Double Integer.

Exercise 13:

1. A pushbutton or switch.

2. A limit switch.

3. Unshielded proxes have a longer range.

4. Through Beam.

5. Brown.

Exercise 14:

1. Sourcing

2. No. They require an amplifier.

3. A High-Speed Counter card.

4. No.

5. No.

6. Electrical noise, physical linkages and couplings.

7. Data Matrix and QR codes.

8. Active and Passive.

Exercise 15:

1. A system that uses feedback from a sensor to maintain the setpoint of a physical variable.

2. Analog and Pulse.

3. Proportional, Integral and Derivative.

4. Kp, Ki and Kd.

5. Stop and restart the control system.

Exercise 16:

1. Manufacturers, integrators and users of machinery.

2. A risk assessment.

3. Contacts are mechanically linked together and can't switch independently.

4. Category 2.

5. OSSD channel signals are clocked pulses of different frequencies to prevent cross-connecting.

6. To ensure motion control systems come to a safe stop before removing control power.

7. Switched outputs are outputs to actuators where power is removed from the card by the safety circuit

Exercise 17:

1. An overload.

2. 1750 RPM.

3. Switch any two phases.

4. No.

5. Electrical wiring signals, communication commands from a PLC, or the controls on the front of the drive.

6. Preset speeds, fault reset, auto/manual control.

7. R, S and T or L1, L2 and L3; U, V and W or T1, T2 and T3

8. No.

Exercise 18:

1. 200 steps per revolution.

2. Current and temperature.

3. An integrated servo.

4. Movement of one axis depends on the changing position of another, or on a virtual axis.

5. A servo motor has holding torque at zero speed.

6. 2, motor power and feedback

Exercise 19:

1. 6 axes.

2. Yes.

3. Use the teach pendant.

4. Actuator Sensor Interface

5. Robot, PLC, HMI, teach pendant, end effector, safety devices, workpiece fixture, physical guarding, machine vision

Exercise 20:

1. Pixels.

2. 256, from 0-255.

3. Lens, Lighting, Processor, I/O, software, camera, frame grabber, etc.

4. Smart cameras, dedicated processors and computer-based systems.

5. A 16mm lens provides a 5.6" field of view in the longest camera axis, but the entire image may not be visible. A 12mm lens may be needed.

6. Edge tools, measurement tools, blob tools, pattern finders, optical character recognition (OCR), color tools, bar code tools, etc.

7. Look for a change in the image, examine lighting, lenses, position of target.

Exercise 21:

1. Some brands of PLC call a Tag a Symbol, but symbols usually are a shortcut to a numerical data address register.

2. The Main routine.

3. Hardware.

Exercise 22:

1. Ladder Logic (LAD), Instruction List (IL), Function Block Diagram (FBD), Structured Text (ST) and Sequential Function Chart (SFC).

2. Read Inputs, Process Logic, Write Outputs, Housekeeping

3. Yes (Online Edits).

4. The processor will exceed the watchdog timer and fault.

Exercise 23:

1. No.

2. Series (AND), Parallel (OR).

3. One <u>Scan</u>.

4. Either one <u>or</u> two addresses, depending on the type.

5. Latch/Unlatch or Set/Reset

Programming exercises continued on next page >

6.

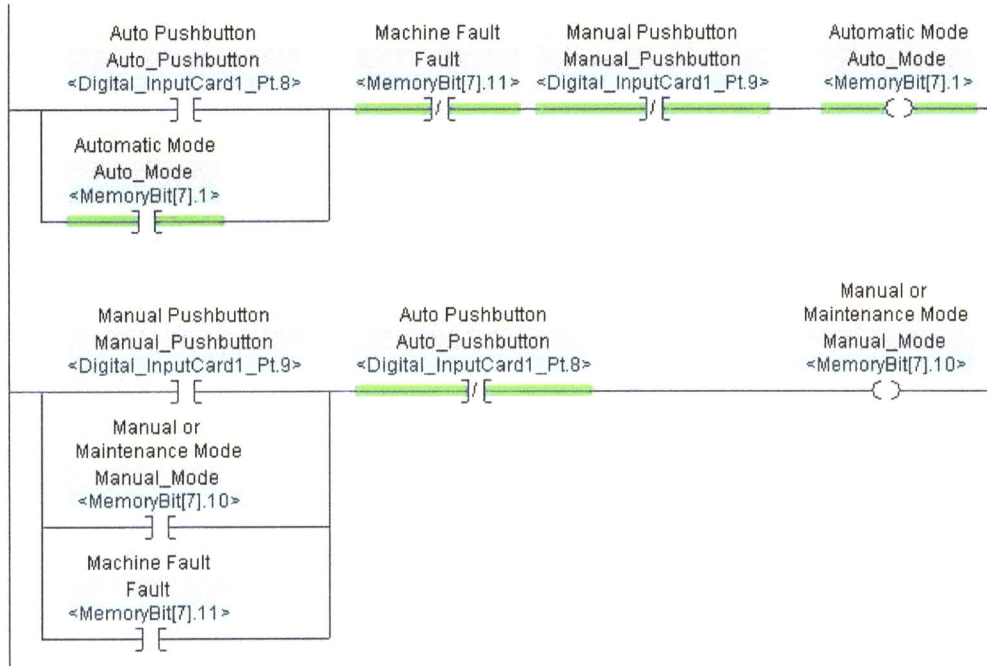

7.

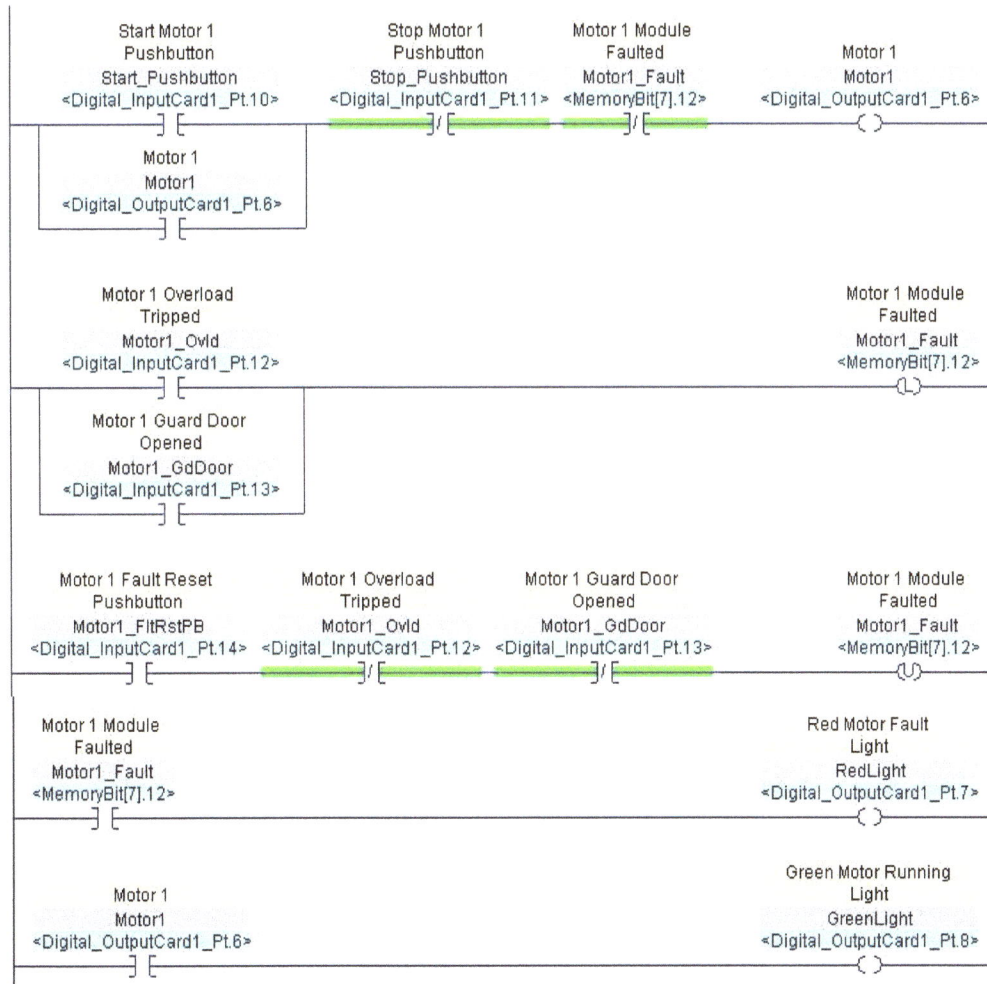

Exercise 24:

1.

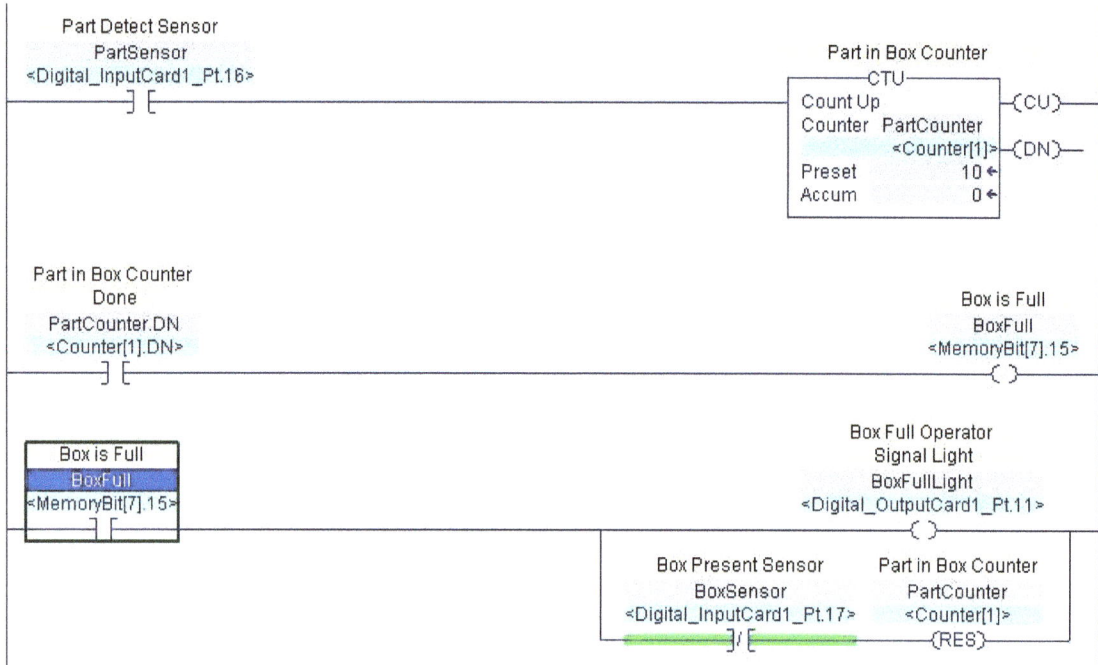

2.

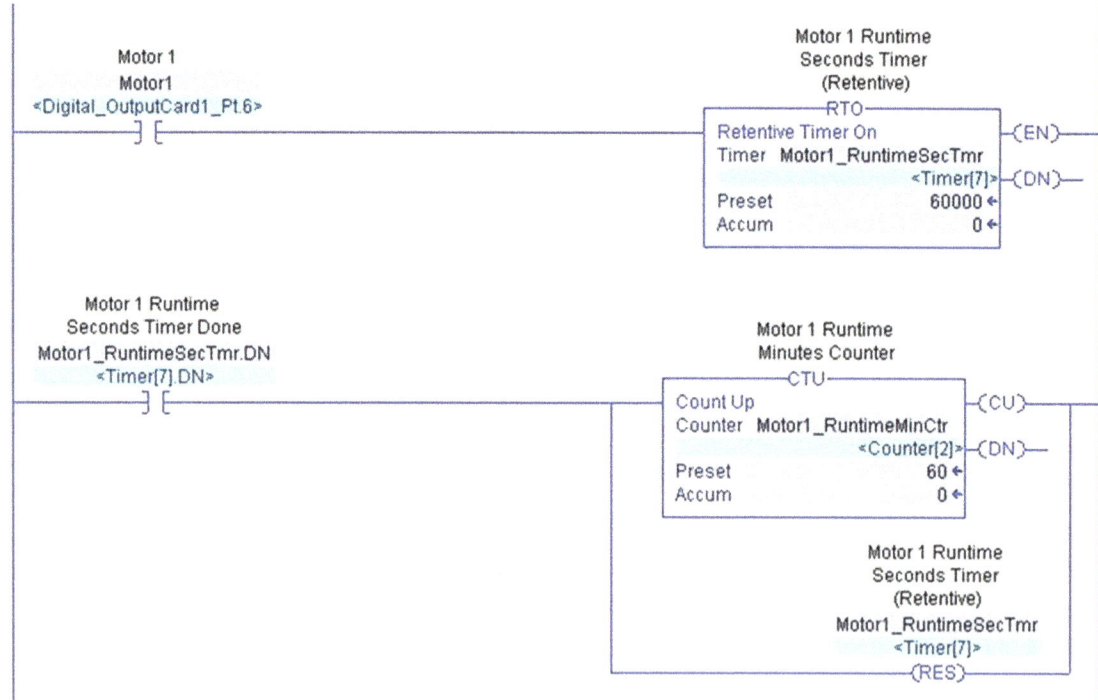

Continued next page:

Exercise 24 # 2 continued:

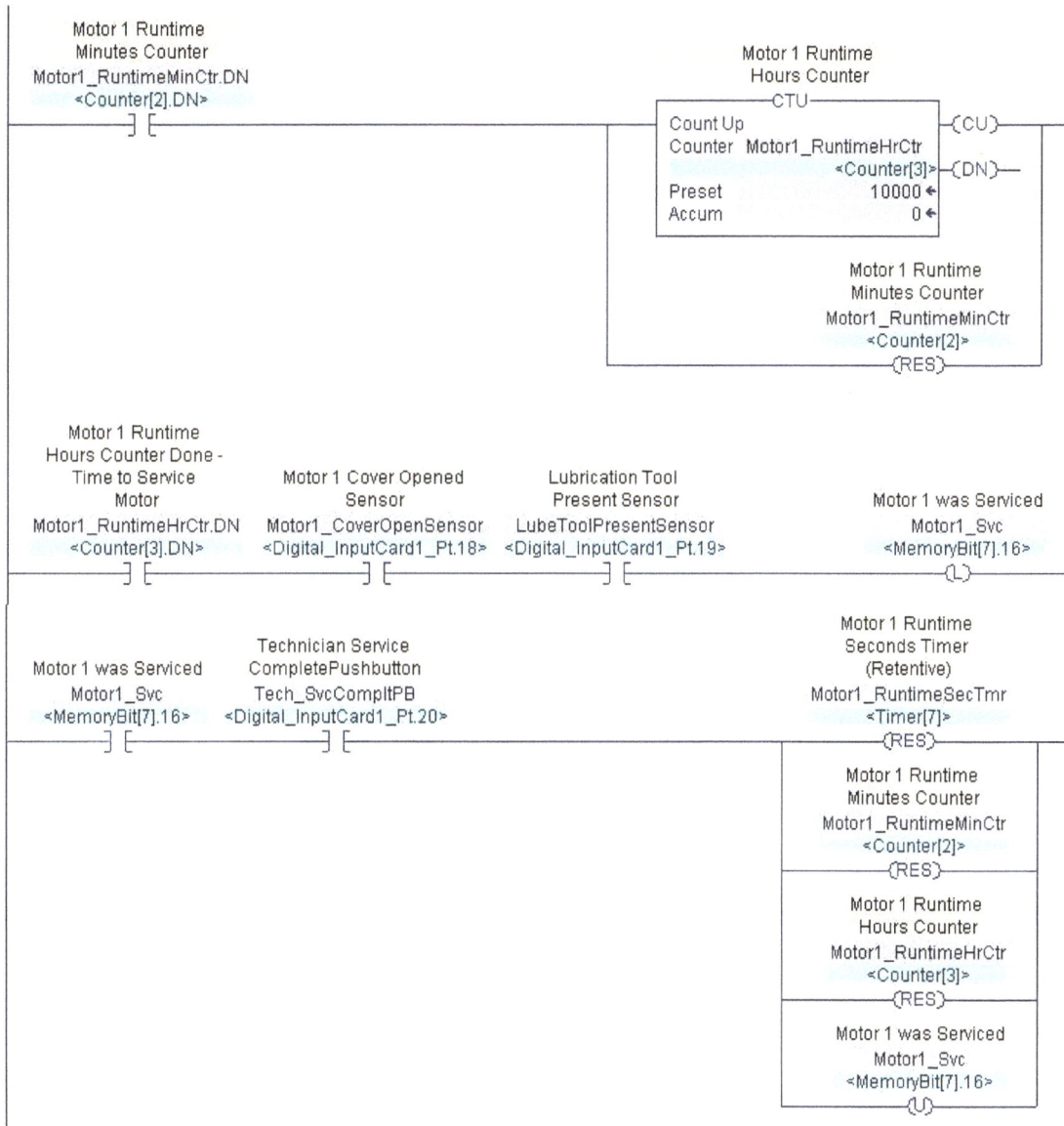

Motor 1 Runtime
Minutes Counter
Motor1_RuntimeMinCtr.DN
<Counter[2].DN>
┤ ├

Motor 1 Runtime
Hours Counter
─────CTU─────
Count Up   ─(CU)─
Counter Motor1_RuntimeHrCtr
                 <Counter[3]>─(DN)─
Preset   10000 ←
Accum   0 ←

Motor 1 Runtime
Minutes Counter
Motor1_RuntimeMinCtr
<Counter[2]>
─(RES)─

Motor 1 Runtime
Hours Counter Done -
Time to Service
Motor   Motor 1 Cover Opened   Lubrication Tool
Motor1_RuntimeHrCtr.DN   Sensor   Present Sensor   Motor 1 was Serviced
<Counter[3].DN>   Motor1_CoverOpenSensor   LubeToolPresentSensor   Motor1_Svc
                         <Digital_InputCard1_Pt.18> <Digital_InputCard1_Pt.19> <MemoryBit[7].16>
┤ ├   ┤ ├   ┤ ├   ─(L)─

                                                                             Motor 1 Runtime
                                                                             Seconds Timer
                         Technician Service   (Retentive)
Motor 1 was Serviced   CompletePushbutton   Motor1_RuntimeSecTmr
Motor1_Svc   Tech_SvcCompltPB   <Timer[7]>
<MemoryBit[7].16>   <Digital_InputCard1_Pt.20>   ─(RES)─
┤ ├   ┤ ├

                                                                             Motor 1 Runtime
                                                                             Minutes Counter
                                                                             Motor1_RuntimeMinCtr
                                                                             <Counter[2]>
                                                                             ─(RES)─

                                                                             Motor 1 Runtime
                                                                             Hours Counter
                                                                             Motor1_RuntimeHrCtr
                                                                             <Counter[3]>
                                                                             ─(RES)─

                                                                             Motor 1 was Serviced
                                                                             Motor1_Svc
                                                                             <MemoryBit[7].16>
                                                                             ─(U)─

Exercise 25:

1.

A:	0	1	1	1	_	1	0	1	0	_	0	0	1	1	_	1	0	1	0	31290
B:	0	0	0	1	_	0	0	0	1	_	0	0	1	1	_	1	0	1	0	4410
Mask:	0	0	0	0	_	0	0	0	0	_	1	1	1	1	_	1	1	1	1	00FF

Yes, the numbers are equal through the mask.

2.

Start Sequence, Turn On Conveyor

```
Machine or System is                              Start Sequence
   Auto Cycling                                       Trigger
   Auto_Cycle          Auto Sequence Step          Start_Seq_Trg            Auto Sequence Step
 <MemoryBit[7].0>            -EQU-               <MemoryBit[7].17>                -MOV-
      ] [             Equal                           ] [                    Move
                      Source A  AutoSequence                                 Source              10
                              <Memory_DWord[2]>
                                       0 ←                                   Dest   AutoSequence
                      Source B           0                                       <Memory_DWord[2]>
                                                                                          0 ←
```

Wait for Part Present, Close Clamp

```
Machine or System is                            Part Detect Sensor
   Auto Cycling                                      PartSensor
   Auto_Cycle          Auto Sequence Step       <Digital_InputCard1_Pt.16>    Auto Sequence Step
 <MemoryBit[7].0>            -EQU-                      ] [                         -MOV-
      ] [             Equal                                                   Move
                      Source A  AutoSequence                                  Source              20
                              <Memory_DWord[2]>
                                       0 ←                                    Dest   AutoSequence
                      Source B          10                                        <Memory_DWord[2]>
                                                                                          0 ←
```

Clamp Closed, Start Drill

```
Machine or System is                             Clamp Closed
   Auto Cycling                                 Proximity Switch
   Auto_Cycle          Auto Sequence Step         ClampClosedPX             Auto Sequence Step
 <MemoryBit[7].0>            -EQU-            <Digital_InputCard1_Pt.21>         -MOV-
      ] [             Equal                             ] [                   Move
                      Source A  AutoSequence                                  Source              30
                              <Memory_DWord[2]>
                                       0 ←                                    Dest   AutoSequence
                      Source B          20                                        <Memory_DWord[2]>
                                                                                          0 ←
```

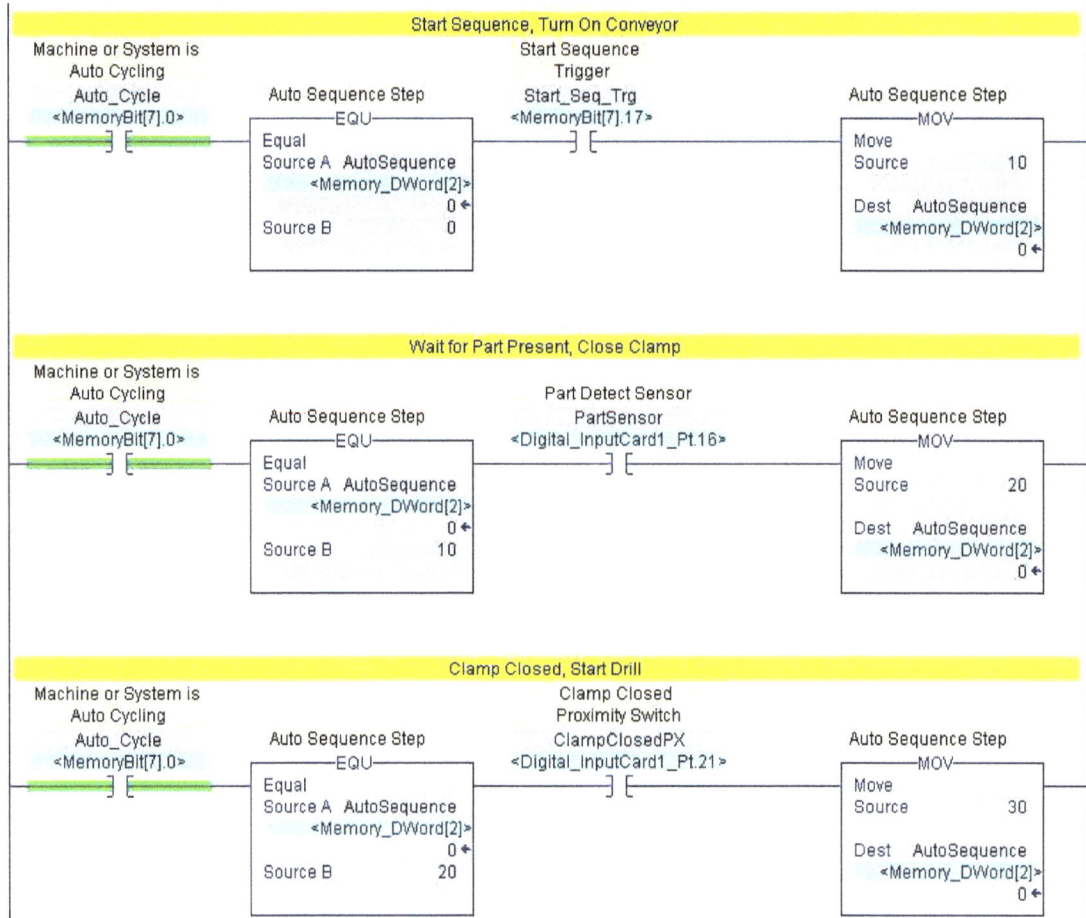

3. Auto sequences often increment by 10 so that an extra step (i.e. Step 15) can be inserted if necessary.

4. A numerical operation affects a single value, while a file operation affects a range of values or a file/structure.

Exercise 26:

1. VFD Calculations:

   1750/31,760 = 0.0551 (Scaling Factor)
   Test: 31,760 * 0.0551 = 1749.976. Close enough! So, <Sensor Value> * 0.0551 = RPM.
   Percent = RPM/Max RPM * 100%

Motor Speed in RPM	Motor Percent Factor	Motor Speed Percent of Max RPM
MUL	DIV	MUL
Multiply	Divide	Multiply
Source A Motor_RawSpeed	Source A Motor_RPM	Source A MotorFactor
<Analog_InputCard3_Ch[1]>	<Memory_REAL[2]>	<Memory_REAL[3]>
0	0.0	0.0
Source B 0.0551	Source B 1750.0	Source B 100
Dest Motor_RPM	Dest MotorFactor	Dest Motor_SpdPct
<Memory_REAL[2]>	<Memory_REAL[3]>	<Memory_REAL[4]>
0.0	0.0	0.0

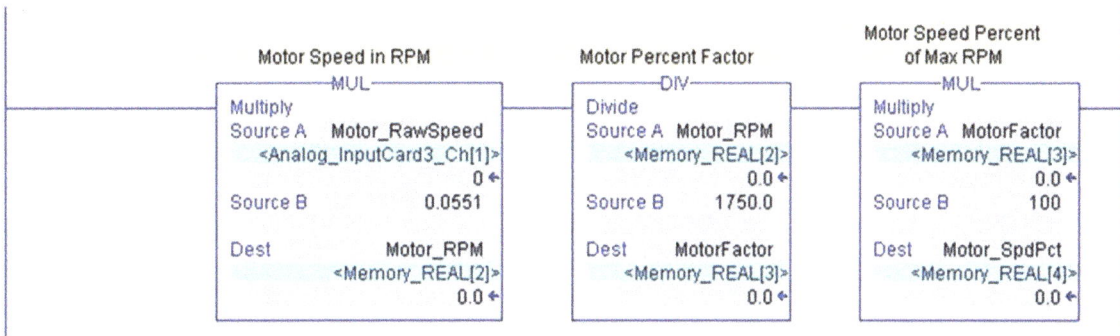

2. Production Calculations:

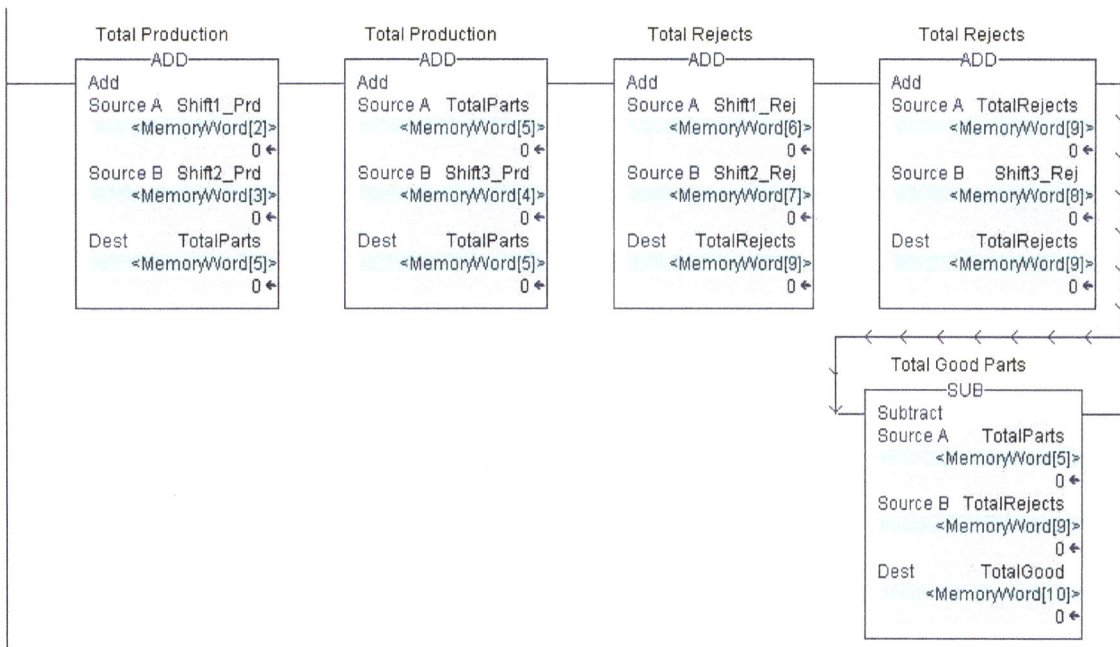

Total Production	Total Production	Total Rejects	Total Rejects
ADD	ADD	ADD	ADD
Add	Add	Add	Add
Source A Shift1_Prd	Source A TotalParts	Source A Shift1_Rej	Source A TotalRejects
<MemoryWord[2]>	<MemoryWord[5]>	<MemoryWord[6]>	<MemoryWord[9]>
0	0	0	0
Source B Shift2_Prd	Source B Shift3_Prd	Source B Shift2_Rej	Source B Shift3_Rej
<MemoryWord[3]>	<MemoryWord[4]>	<MemoryWord[7]>	<MemoryWord[8]>
0	0	0	0
Dest TotalParts	Dest TotalParts	Dest TotalRejects	Dest TotalRejects
<MemoryWord[5]>	<MemoryWord[5]>	<MemoryWord[9]>	<MemoryWord[9]>
0	0	0	0

Total Good Parts
SUB
Subtract
Source A TotalParts
<MemoryWord[5]>
0
Source B TotalRejects
<MemoryWord[9]>
0
Dest TotalGood
<MemoryWord[10]>
0

Exercise 27:

1. Tank Scaling:

   M = (Y2-Y1)/(X2-X1) = (6000-0)/(24780-96) = 0.24307
   B = Y1 – (M * X1) = 0 – (0.24307 * 96) = -23.33472
   Liters = Gallons * 3.78541

```
                Sensor Input x
                   Scalar                    Gallons in Tank              Liters in Tank
                 ┌──MUL──┐                  ┌──ADD──┐                    ┌──MUL──┐
                 Multiply                    Add                          Multiply
                 Source A  Analog_InputCard3_Ch[1]   Source A  Input_x_Scalar     Source A     Gallons
                              15506 ←                    <Memory_REAL[6]>             <Memory_REAL[8]>
                 Source B        Scalar_M                  3769.0435 ←                  3745.7087 ←
                          <Memory_REAL[5]>       Source B       Offset_B       Source B    3.78541
                               0.24307 ←               <Memory_REAL[7]>
                 Dest         Input_x_Scalar              -23.33472 ←         Dest            Liters
                          <Memory_REAL[6]>      Dest           Gallons               <Memory_REAL[9]>
                             3769.0435 ←               <Memory_REAL[8]>                 14179.043 ←
                                                        3745.7087 ←
```

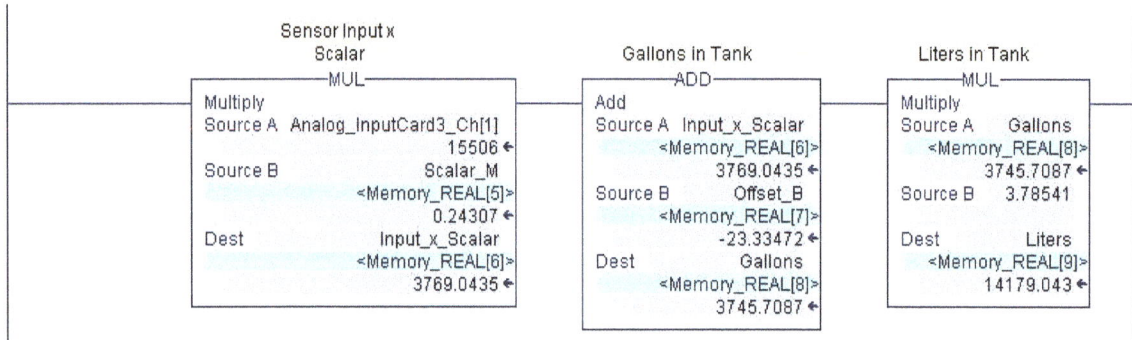

2. Trigonometric functions are usually used in motion control and positioning applications; they calculate geometrical coordinates.

3. 47 _G_ 6F _o_ 6F _o_ 64 _d_ 20 ___ 4A _J_ 6F _o_ 62 _b_ 21 _!_
   Good Job!

4. Yes, Jump instructions can jump backward. If so, a method to exit the loop is necessary, such as incrementing a counter.

5. FIFO: First-In First -Out   LIFO: Last-In First-Out

6. A Sequencer monitors and controls repeatable operations

Exercise 28:

1. False, it affects only the Input Image Table

2. True

3. Q98, "PP04" will be off

4. Q98.5 will be on.

5. Look for coils and operations affecting digital and analog outputs (destructive instructions)

6. Physical inputs and contacts with no coil.

Exercise 29:

1. The Supervisory and Production Level

2. Belt, Roller, Tabletop Chain, Slat, Screw

3. Slitting, Sheeting, Coating, Laminating

4. Bagging, Wrapping, Labeling, Palletizing

5. Clean In Place.

6. ISO 14644-1 and US FED DTD 209E

7. Demonstrate that software produces results that are consistent, quality of product hasn't changed since last validation.

8. Electrical and water utilities both use SCADA for monitoring and control.

Exercise 30:

1. An HMI is a dedicated device for interfacing with a controller, whereas SCADA is software that is deployed on a computer. HMIs can't run other computer applications and generally don't have much extra memory.

2. Changing the visibility, size, position or color of an object on a screen based on a tag.

3. RTU – Remote Terminal Unit, DCS – Distributed Control System

4. Number of Tags used or number of servers where software is installed.

Exercise 31:

1. Availability, Performance and Quality

2. 960 Minutes/850 Minutes =0.8854, **88.54% Availability**
   850 Minutes/742 Parts = 0.8729, **87.29% Performance**
   721 Good Parts/742 Total Parts = 0.9717, **97.17% Quality**

   0.8854 x 0.8729 x 0.9717 = 0.7510, **75.1% OEE**

   16Hrs/24Hrs = 0.6667, **66.67% Loading**
   0.6667 x 0.7510 = 0.5006, **50.06% TEEP**

3. Availability, Unplanned Stops

4. Availability, Planned Stops

5. Incorrect equipment settings, operator errors, bad materials, faulty changeover, lot expiration.

Exercise 32:

1. Because one event occurred after another event does not mean that the first event caused the second. "Correlation does not imply causation".

2. A. Identify and document what the problem is.
    B. Determine why it happened.
    C. Determine a solution to prevent it from happening again.

3. To discover the origin of problems in a system.

4. Start with the simple, begin with a known good state, substitute components, use checklists or flowcharts, reproduce symptoms, use the "half-split" method, use the "five whys", use Ishikawa or "fishbone" diagrams.

Exercise 33:

1. Vernier, Dial and Digital.

2. Micrometers are considered more accurate.

3. To limit the amount of torque applied to the bolt or nut.

Exercise 34:

1. Tap wrench and T-Handle

2. List may include wrenches, pliers, cutters, drills, drivers, saws and more.

3. An "Easy Out".

4. Electrical Outlet, Battery or Pneumatic

5. Heat, caused by excessive speed or lack of lubrication

6. Ferrous and non-ferrous

7. The "kerf"

Exercise 35:

1. Resistance, Voltage and Current.

2. Dial position and leads into correct jack location.

3. Ensure power is disconnected from the circuit, isolate the resistance by disconnecting one end, select correct resistance range.

4. Data logging, thermocouple, diode and transistor measurements, frequency measurements, capture peak voltage.

5. Wear protective gear (PPE), don't hold meter in hand, don't hold leads or probes in two hands, work on de-energized circuits when possible, use rubber mats, lock out equipment.

6. A solenoid operated continuity tester.

7. A Megohmmeter tests resistance of insulators.

Exercise 36:

1. Display analog waveforms and do signal comparisons.

2. Create analog waveforms and signals.

3. Analyze signals across a spectrum of frequencies.

4. High contact resistance, load imbalances, improperly sized conductors, failed components.

Exercise 37:

1. Multimeter, wire strippers, crimpers, screwdrivers, diagonal cutters, wrenches, knockouts, scissors, utility knife, pliers.

2. A jigsaw.

3. Self-laminating, heat shrink, snap on, noodle, adhesive.

4. To reinforce stripped stranded wire.

5. NFPA 70, the National Electrical Code.

6. #10 AWG, Black/Red/Blue

Troubleshooting Exercise:

1. The circuit breaker is on, but there is a voltage drop across its terminals. This seems to be a bad circuit breaker.

Flip the circuit breaker off and back on again to reset it. Read the voltage at the bottom of the circuit breaker relative to a -DC terminal. If it is zero volts with the circuit breaker on, it is not conducting current. Lock out the disconnect, removing power from the machine. Disconnect the wires from the breaker and remove it from the din-rail. Set your meter to Ohms and read the resistance across the breaker with the switch in the up or ON position. If it shows a high resistance or "OL", the breaker is bad. Replace the breaker.

2. Emergency Stop 2 Channels 1 and 2 show as OFF. To be sure, push the E-Stop down and pull it up again to see if the light on Slot 1, indicators 2 and 3 come on. If they do, your problem may be fixed... try pressing the Power On button again. If the lights don't come on, check the contacts on the back of the button with your meter. Other places that can be checked: the terminals in the enclosure (top and bottom), The terminals on the cards.

It is <u>unlikely</u> that both channels/contacts would go bad at the same time.

3. Both proximity switches are on at the same time, this should not be possible. With air on the machine, set each proximity switch to where it is centered on the open and closed positions of the grippers. The grippers should be able to be actuated from the HMI in Manual Mode.

4A. Sequence is at Step 0. It will not proceed because DI_LwrFlip_PalAtPreStop.Value is not ON. The next step of the sequence would be 10. The address of the sensor is: IOBlock3.X10_12_14_16_18_DI2.0

4B. Check the light on IOBlock 3, That is the quickest thing to check.

4C. Check communications between the IO Block and the PLC. An easy way to do this is to see if any other sensor flashes a light on the module or changes the state of its contact in the program.

4D.   1. Male M8 molded connector attached to sensor.

2. Female M8 field wireable that mates with the male.

3. Male field wireable connector that connects to the Y-splitter.

4. The Y-splitter module itself.

Yes, the input point on the IO Block itself could be bad.

To make troubleshooting of connections easier on these circuits, the following might be helpful:

A. Male M8 connector with flying leads – can be used to apply a signal to the black wire, or to check voltage on the blue and brown wires.

B. A battery or 24vdc powered "sensor checker" that sensors can be plugged into to check the sensor itself. This would have a 24vdc powered light that illuminates if there is a signal.

5. The output that raises the elevator was on for a period without the corresponding sensor being made. The time setpoint is six seconds.

To change the elevator speed, adjust the flow controls. For the "raise" motion, adjust the flow control at the top of the cylinder (meter out).

6. The overload is tripped. Voltage to the actual coil of the relay could be checked, the "tripped" indicator should also show red.

Reset the overload and wait a few minutes to allow the overload and motor to cool. Try running the conveyor again. Possible causes could be increased load on the motor (check the belt) or a short in the motor windings. The overload setting could be adjusted slightly higher to check this, and the resistance of the motor windings could be checked.

7A. Try connecting to the ethernet network with a computer and "pinging" the robot's IP address.

7B. While teaching, ensure all guards are closed and that the robot is set to a very slow speed. Make sure you have good visibility of the grippers.

7C. The robot will pick from Conveyor 2.

About the Author

Frank Lamb is an industrial automation consultant and trainer with more than 30 years of experience in controls and machine automation.

From 1996 to 2006, Frank owned and operated Automation Consulting Services, Inc. (ACS), a panel building and system integration company in Knoxville, TN. From 2006 to 2011, he worked as a senior-level project engineer for Wright Industries in Nashville, TN, where he led the design and implementation of large, complex systems and custom machines for multinational corporations and government agencies. In December 2011, Frank re-established Automation Consulting, LLC with a new vision: to use his experience in the field of industrial automation – from electrical, mechanical and controls engineering to project management, training, and machine documentation – to provide expert consulting and training services to manufacturers.

Frank is the president and owner of Automation Consulting, LLC in Lebanon, TN and serves as Lead Trainer for Automation NTH, a control systems integrator in LaVergne, TN. He is a published author of several books including *Industrial Automation: Hands On*, published by McGraw-Hill Professional in 2013 and *Advanced PLC Hardware and Programming*, Automation Consulting, LLC, 2019. He is a United States Air Force veteran, received his BSEE in Electrical and Computer Engineering from the University of Tennessee, and has a Green Belt in Lean Manufacturing/Six Sigma from Purdue University.

www.ingramcontent.com/pod-product-compliance
Lightning Source LLC
Chambersburg PA
CBHW052340210326
41597CB00037B/6201